Introduction to
Compiler Construction
in a Java World

Introduction to
Compiler Construction
in a Java World

Bill Campbell
Swami Iyer
Bahar Akbal-Delibaş

CRC Press
Taylor & Francis Group
Boca Raton London New York

CRC Press is an imprint of the
Taylor & Francis Group, an **informa** business

A CHAPMAN & HALL BOOK

CRC Press
Taylor & Francis Group
6000 Broken Sound Parkway NW, Suite 300
Boca Raton, FL 33487-2742

Version Date: 20121001

International Standard Book Number: 978-1-4398-6088-5 (Hardback)

Library of Congress Cataloging-in-Publication Data

Campbell, Bill, 1950-
  Introduction to compiler construction in a Java world / Bill Campbell, Swami Iyer, Bahar Akbal-Delibas.
    p. cm.
  Includes bibliographical references and index.
  ISBN 978-1-4398-6088-5 (hardcover : alk. paper)
  1. Compilers (Computer programs) 2. Java (Computer program language) I. Iyer, Swami. II.
Akbal-Delibas, Bahar. III. Title.

QA76.73.J38C363 2013
005.4'53--dc23   2012030751

Visit the Taylor & Francis Web site at
http://www.taylorandfrancis.com

and the CRC Press Web site at
http://www.crcpress.com

Dedication

To Nora, Fiona, and Amy for their loving support. — Bill

To my aunts Subbalakshmi and Vasantha for their unfaltering care and affection, and my parents Inthumathi and Raghunathan, and brother Shiva for all their support. — Swami

To my parents Gülseren and Salih for encouraging me to pursue knowledge, and to my beloved husband Adem for always being there when I need him. — Bahar

Contents

List of Figures

Preface

Why Another Compiler Text?

There are lots of compiler texts out there. Some of them are very good. Some of them use Java as the programming language in which the compiler is written. But we have yet to find a compiler text that uses Java everywhere.

Our text is based on examples that make full use of Java:

- Like some other texts, the implementation language is Java. And, our implementation uses Java's object orientation. For example, polymorphism is used in implementing the `analyze()` and `codegen()` methods for different types of nodes in the abstract syntax tree (AST). The lexical analyzer (the token scanner), the parser, and a back-end code emitter are objects.

- Unlike other texts, the example compiler and examples in the chapters are all about compiling Java. Java is the source language. The student gets a compiler for a non-trivial subset of Java, called *j--*; *j--* includes classes, objects, methods, a few simple types, a few control constructs, and a few operators. The examples in the text are taken from this compiler. The exercises in the text generally involve implementing Java language constructs that are not already in *j--*. And, because Java is an object-oriented language, students see how modern object-oriented constructs are compiled.

- The example compiler and exercises done by the student target the Java Virtual Machine (JVM).

- There is a separate back end (discussed in Chapters 6 and 7), which translates a small but useful subset of JVM code to SPIM (Larus, 2000–2010), a simulator for the MIPS RISC architecture. Again, there are exercises for the student so that he or she may become acquainted with a register machine and register allocation.

The student is immersed in Java and the JVM, and gets a deeper understanding of the Java programming language and its implementation.

Why Java?

It is true that most industrial compilers (and many compilers for textbooks) are written in either C or C++, but students have probably been taught to program using Java. And few students will go on to write compilers professionally. So, compiler projects steeped in Java give students experience working with larger, non-trivial Java programs, making them better Java programmers.

A colleague, Bruce Knobe, says that the compilers course is really a software engineering course because the compiler is the first non-trivial program the student sees. In addition, it is a program built up from a sequence of components, where the later components depend on the earlier ones. One learns good software engineering skills in writing a compiler.

Our example compiler and the exercises that have the student extend it follow this model:

- The example compiler for *j--* is a non-trivial program comprising 240 classes and nearly 30,000 lines of code (including comments). The text takes its examples from this compiler and encourages the student to read the code. We have always thought that reading good code makes for better programmers.

- The code tree includes an Ant file for automatically building the compiler.

- The code tree makes use of JUnit for automatically running each build against a set of tests. The exercises encourage the student to write additional tests before implementing new language features in their compilers. Thus, students get a taste of extreme programming; implementing a new programming language construct in the compiler involves

  - Writing tests
  - Refactoring (re-organizing) the code for making the addition cleaner
  - Writing the new code to implement the new construct

The code tree may be used either

- In a simple command-line environment using any text editor, Java compiler, and Java run-time environment (for example, Oracle's Java SE). Ant will build a code tree under either Unix (including Apple's Mac OS X) or a Windows system; likewise, JUnit will work with either system; or

- It can be imported into an integrated development environment such as IBM's freely available Eclipse.

So, this experience makes the student a better programmer. Instead of having to learn a new programming language, the student can concentrate on the more important things: design, organization, and testing. Students get more excited about compiling Java than compiling some toy language.

Why Start with a *j--* Compiler?

In teaching compiler classes, we have assigned programming exercises both

1. Where the student writes the compiler components from scratch, and

2. Where the student starts with the compiler for a base language such as *j--* and implements language extensions.

We have settled on the second approach for the following reasons:

- The example compiler illustrates, in a concrete manner, the implementation techniques discussed in the text and presented in the lectures.

- Students get hands-on experience implementing extensions to *j--* (for example, interfaces, additional control statements, exception handling, doubles, floats and longs, and nested classes) without having to build the infrastructure from scratch.

- Our own work experiences make it clear that this is the way work is done in commercial projects; programmers rarely write code from scratch but work from existing code bases. Following the approach adopted here, the student learns how to fit code into existing projects and still do valuable work.

Students have the satisfaction of doing interesting programming, experiencing what coding is like in the commercial world, and learning about compilers.

Why Target the JVM?

In the first instance, our example compiler and student exercises target the Java Virtual Machine (JVM); we have chosen the JVM as a target for several reasons:

- The original Oracle Java compiler that is used by most students today targets the JVM. Students understand this regimen.

- This is the way many compiler frameworks are implemented today. For example, Microsoft's .NET framework targets the Common Language Runtime (CLR). The byte code of both the JVM and the CLR is (in various instances) then translated to native machine code, which is real register-based computer code.

- Targeting the JVM exposes students to some code generation issues (instruction selection) but not all, for example, not register allocation.

- We think we cannot ask for too much more from students in a one-semester course (but more on this below). Rather than have the students compile toy languages to real hardware, we have them compile a hefty subset of Java (roughly Java version 4) to JVM byte code.

- That students produce real JVM .class files, which can link to any other .class files (no matter how they are produced), gives the students great satisfaction. The class emitter (`CLEmitter`) component of our compiler hides the complexity of .class files.

This having been said, many students (and their professors) will want to deal with register-based machines. For this reason, we also demonstrate how JVM code can be translated to a register machine, specifically the MIPS architecture.

After the JVM – A Register Target

Beginning in Chapter 6, our text discusses translating the stack-based (and so, register-free) JVM code to a MIPS, register-based architecture. Our example translator does only a

limited subset of the JVM, dealing with static classes and methods and sufficient for translating a computation of factorial. But our translation fully illustrates linear-scan register
allocation—appropriate to modern just-in-time compilation. The translation of additional
portions of the JVM and other register allocation schemes, for example, that are based on
graph coloring, are left to the student as exercises. Our JVM-to-MIPS translator framework
also supports several common code optimizations.

Otherwise, a Traditional Compiler Text

Otherwise, this is a pretty traditional compiler text. It covers all of the issues one expects in
any compiler text: lexical analysis, parsing, abstract syntax trees, semantic analysis, code
generation, limited optimization, register allocation, as well as a discussion of some recent
strategies such as just-in-time compiling and hotspot compiling and an overview of some
well-known compilers (Oracle's Java compiler, GCC, the IBM Eclipse compiler for Java
and Microsoft's C# compiler). A seasoned compiler instructor will be comfortable with all
of the topics covered in the text. On the other hand, one need not cover everything in the
class; for example, the instructor may choose to leave out certain parsing strategies, leave
out the JavaCC tool (for automatically generating a scanner and parser), or use JavaCC
alone.

Who Is This Book for?

This text is aimed at upper-division undergraduates or first-year graduate students in a
compiler course. For two-semester compiler courses, where the first semester covers front-
end issues and the second covers back-end issues such as optimization, our book would
be best for the first semester. For the second semester, one would be better off using a
specialized text such as Robert Morgan's *Building an Optimizing Compiler* [Morgan, 1998];
Allen and Kennedy's *Optimizing Compilers for Modern Architectures* [Allen and Kennedy,
2002]; or Muchnick's *Advanced Compiler Design and Implementation* [Muchnick, 1997]. A
general compilers text that addresses many back-end issues is Appel's *Modern Compiler
Implementation in Java* [Appel, 2002]. We choose to consult only published papers in the
second-semester course.

Structure of the Text

Briefly, *An Introduction to Compiler Construction in a Java World* is organized as follows.
In Chapter 1 we describe what compilers are and how they are organized, and we give
an overview of the example j-- compiler, which is written in Java and supplied with the
text. We discuss (lexical) scanners in Chapter 2, parsing in Chapter 3, semantic analysis
in Chapter 4, and JVM code generation in Chapter 5. In Chapter 6 we describe a JVM
code-to-MIPS code translator, with some optimization techniques; specifically, we target

James Larus's SPIM, an interpreter for MIPS assembly language. We introduce register allocation in Chapter 7. In Chapter 8 we discuss several celebrity (that is, well-known) compilers. Most chapters close with a set of exercises; these are generally a mix of written exercises and programming projects.

There are five appendices. Appendix A explains how to set up an environment, either a simple command-line environment or an Eclipse environment, for working with the example *j--* compiler. Appendix B outlines the *j--* language syntax, and Appendix C outlines (the fuller) Java language syntax. Appendix D describes the JVM, its instruction set, and `CLEmitter`, a class that can be used for emitting JVM code. Appendix E describes SPIM, a simulator for MIPS assembly code, which was implemented by James Larus.

How to Use This Text in a Class

Depending on the time available, there are many paths one may follow through this text. Here are two:

- We have taught compilers, concentrating on front-end issues, and simply targeting the JVM interpreter:

    - Introduction. (Chapter 1)
    - Both a hand-written and JavaCC generated lexical analyzer. The theory of generating lexical analyzers from regular expressions; Finite State Automata (FSA). (Chapter 2)
    - Context-free languages and context-free grammars. Top-down parsing using recursive descent and LL(1) parsers. Bottom-up parsing with LR(1) and LALR(1) parser. Using JavaCC to generate a parser. (Chapter 3)
    - Type checking. (Chapter 4)
    - JVM code generation. (Chapter 5)
    - A brief introduction to translating JVM code to SPIM code and optimization. (Chapter 6)

- We have also taught compilers, spending less time on the front end, and generating code both for the JVM and for SPIM, a simulator for a register-based RISC machine:

    - Introduction. (Chapter 1)
    - A hand-written lexical analyzer. (Students have often seen regular expressions and FSA in earlier courses.) (Sections 2.1 and 2.2)
    - Parsing by recursive descent. (Sections 3.1 3.3.1)
    - Type checking. (Chapter 4)
    - JVM code generation. (Chapter 5)
    - Translating JVM code to SPIM code and optimization. (Chapter 6)
    - Register allocation. (Chapter 7)

In either case, the student should do the appropriate programming exercises. Those exercises that are not otherwise marked are relatively straightforward; we assign several of these in each programming set.

Where to Get the Code?

We supply a code tree, containing

- Java code for the example *j--* compiler and the JVM to SPIM translator,

- Tests (both conformance tests and deviance tests that cause error messages to be produced) for the *j--* compiler and a framework for adding additional tests,

- The JavaCC and JUnit libraries, and

- An Ant file for building and testing the compiler.

We maintain a website at `http://www.cs.umb.edu/j--` for up-to-date distributions.

What Does the Student Need?

The code tree may be obtained at `http://www.cs.umb.edu/j--/j--.zip`. Everything else the student needs is freely obtainable on the WWW: the latest version of Java SE is obtainable from Oracle at `http://www.oracle.com/technetwork/java/javase/downloads/index.html`. Ant is available at `http://ant.apache.org/`; Eclipse can be obtained from `http://www.eclipse.org/`; and SPIM, a simulator of the MIPS machine, can be obtained from `http://sourceforge.net/projects/spimsimulator/files/`. All of this may be installed on Windows, Mac OS X, or any Linux platform.

What Does the Student Come Away with?

The student gets hands-on experience working with and extending (in the exercises) a real, working compiler. From this, the student gets an appreciation of how compilers work, how to write compilers, and how the Java language behaves. More importantly, the student gets practice working with a non-trivial Java program of more than 30,000 lines of code.

About the Authors

Bill Campbell is an associate professor in the Department of Computer Science at the University of Massachusetts, Boston. His professional areas of expertise are software engineering, object-oriented analysis, design and programming, and programming language implementation. He likes to write programs and has both academic and commercial experience. He has been teaching compilers for more than twenty years and has written an introductory Java programming text with Ethan Bolker, *Java Outside In* (Cambridge University Press, 2003).

Professor Campbell has worked for (what is now) AT&T and Intermetrics Inc., and has consulted to Apple Computer and Entitlenet. He has implemented a public domain version of the Scheme programming language called UMB Scheme, which is distributed with Linux. Recently, he founded an undergraduate program in information technology.

Dr. Campbell has a bachelor's degree in mathematics and computer science from New York University, 1972; an M.Sc. in computer science from McGill University, 1975; and a PhD in computer science from St. Andrews University (UK), 1978.

Swami Iyer is a PhD candidate in the Department of Computer Science at the University of Massachusetts, Boston. His research interests are in the fields of dynamical systems, complex networks, and evolutionary game theory. He also has a casual interest in theoretical physics. His fondness for programming is what got him interested in compilers and has been working on the *j--* compiler for several years.

He enjoys teaching and has taught classes in introductory programming and data structures at the University of Massachusetts, Boston. After graduation, he plans on pursuing an academic career with both teaching and research responsibilities.

Iyer has a bachelor's degree in electronics and telecommunication from the University of Bombay (India), 1996, and a master's degree in computer science from the University of Massachusetts, Boston, 2001.

Bahar Akbal-Delibaş is a PhD student in the Department of Computer Science at the University of Massachusetts, Boston. Her research interest is in structural bioinformatics, aimed at better understanding the sequence–structure–function relationship in proteins, modeling conformational changes in proteins and predicting protein-protein interactions. She also performed research on software modeling, specifically modeling wireless sensor networks.

Her first encounter with compilers was a frightening experience as it can be for many students. However, soon she discovered how to play with the pieces of the puzzle and saw the fun in programming compilers. She hopes this book will help students who read it the same way. She has been the teaching assistant for the compilers course at the University of Massachusetts, Boston and has been working with the *j--* compiler for several years

Akbal-Delibaş has a bachelor's degree in computer engineering from Fatih University (Turkey), 2004, and a master's degree in computer science from University of Massachusetts, Boston, 2007.

Acknowledgments

We wish to thank students in CS451 and CS651, the compilers course at the University of Massachusetts, Boston, for their feedback on, and corrections to, the text, the example compiler, and the exercises. We would like to thank Kelechi Dike, Ricardo Menard, and Mini Nair for writing a compiler for a subset of C# that was similar to *j--*. We would particularly like to thank Alex Valtchev for his work on both liveness intervals and linear scan register allocation.

We wish to acknowledge and thank both Christian Wimmer for our extensive use of his algorithms in his masters thesis on linear scan [Wimmer, 2004] and James Larus for our use of SPIM, his MIPS simulator [Larus, 2010].

We wish to thank the people at Taylor & Francis, including Randi Cohen, Jessica Vakili, the editors, and reviewers for their help in preparing this text.

Finally, we wish to thank our families and close friends for putting up with us as we wrote the compiler and the text.

Chapter 1

Compilation

1.1 Compilers

A *compiler* is a program that translates a *source program* written in a high-level programming language such as Java, C#, or C, into an equivalent *target program* in a lower, level language such as machine code, which can be executed directly by a computer. This translation is illustrated in Figure 1.1.

Source Language Program → Compiler → Machine Code Program

FIGURE 1.1 Compilation.

By equivalent, we mean *semantics preserving*: the translation should have the same behavior as the original. This process of translation is called *compilation*.

1.1.1 Programming Languages

A *programming language* is an artificial language in which a programmer (usually a person) writes a program to control the behavior of a machine, particularly a computer. Of course, a program has an audience other than the computer whose behavior it means to control; other programmers may read a program to understand how it works, or why it causes unexpected behavior. So, it must be designed so as to allow the programmer to precisely specify what the computer is to do in a way that both the computer and other programmers can understand.

Examples of programming languages are Java, C, C++, C#, and Ruby. There are hundreds, if not thousands, of different programming languages. But at any one time, a much smaller number are in popular use.

Like a natural language, a programming language is specified in three steps:

1. The tokens, or lexemes, are described. Examples are the keyword if, the operator +, constants such as 4 and 'c', and the identifier foo. Tokens in a programming language are like words in a natural language.

2. One describes the syntax of programs and language constructs such as classes, methods, statements, and expressions. This is very much like the syntax of a natural language but much less flexible.

3. One specifies the meaning, or semantics, of the various constructs. The semantics of various constructs is usually described in English.

Some programming languages, like Java, also specify various static type rules, that a program and its constructs must obey. These additional rules are usually specified as part of the semantics.

Programming language designers go to great lengths to precisely specify the structure of tokens, the syntax, and the semantics. The tokens and the syntax are often described using formal notations, for example, regular expressions and context-free grammars. The semantics are usually described in a natural language such as English[1]. A good example of a programming language specification is the Java Language Specification [Gosling et al., 2005].

1.1.2   Machine Languages

A computer's *machine language* or, equivalently, its *instruction set* is designed so as to be easily interpreted by the computer itself. A machine language program consists of a sequence of instructions and operands, usually organized so that each instruction and each operand occupies one or more bytes and so is easily accessed and interpreted. On the other hand, people are not expected to read a machine code program[2]. A machine's instruction set and its behavior are often referred to as its *architecture*.

Examples of machine languages are the instruction sets for both the Intel i386 family of architectures and the MIPS computer. The Intel i386 is known as a complex instruction set computer (CISC) because many of its instructions are both powerful and complex. The MIPS is known as a reduced instruction set computer (RISC) because its instructions are relatively simple; it may often require several RISC instructions to carry out the same operation as a single CISC instruction. RISC machines often have at least thirty-two registers, while CISC machines often have as few as eight registers. Fetching data from, and storing data in, registers are much faster than accessing memory locations because registers are part of the computer processing unit (CPU) that does the actual computation. For this reason, a compiler tries to keep as many variables and partial results in registers as possible.

Another example is the machine language for Oracle's Java Virtual Machine (JVM) architecture. The JVM is said to be *virtual* not because it does not exist, but because it is not necessarily implemented in hardware[3]; rather, it is implemented as a software program. We discuss the implementation of the JVM in greater detail in Chapter 7. But as compiler writers, we are interested in its instruction set rather than its implementation.

Hence the compiler: the compiler transforms a program written in the high-level programming language into a semantically equivalent machine code program.

Traditionally, a compiler analyzes the input program to produce (or synthesize) the output program,

- Mapping names to memory addresses, stack frame offsets, and registers;

- Generating a linear sequence of machine code instructions; and

- Detecting any errors in the program that can be detected in compilation.

Compilation is often contrasted with *interpretation*, where the high-level language program is executed directly. That is, the high-level program is first loaded into the interpreter

---

[1] Although formal notations have been proposed for describing both the type rules and semantics of programming languages, these are not popularly used.

[2] But one can. Tools often exist for displaying the machine code in mnemonic form, which is more readable than a sequence of binary byte values. The Java toolset provides `javap` for displaying the contents of class files.

[3] Although Oracle has experimented with designing a JVM implemented in hardware, it never took off. Computers designed for implementing particular programming languages rarely succeed.

and then executed (Figure 1.2). Examples of programming languages whose programs may be interpreted directly are the UNIX shell languages, such as bash and csh, Forth, and many versions of LISP.

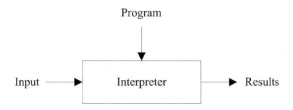

FIGURE 1.2 Interpretation

One might ask, "Why not interpret all programs directly?" There are two answers.

First is performance. Native machine code programs run faster than interpreted high-level language programs. To see why this is so, consider what an interpreter must do with each statement it executes: it must parse and analyze the statement to decode its meaning *every time* it executes that statement; a limited form of compilation is taking place for every execution of every statement. It is much better to translate all statements in a program to *native code* just once, and execute that[4].

Second is secrecy. Companies often want to protect their investment in the programs that they have paid programmers to write. It is more difficult (albeit not impossible) to discern the meaning of machine code programs than of high-level language programs.

But, compilation is not always suitable. The overhead of interpretation does not always justify writing (or, buying) a compiler. An example is the Unix Shell (or Windows shell) programming language. Programs written in shell script have a simple syntax and so are easy to interpret; moreover, they are not executed often enough to warrant compilation. And, as we have stated, compilation maps names to addresses; some dynamic programming languages (LISP is a classic example, but there are a myriad of newer dynamic languages) depend on keeping names around at run-time.

1.2   Why Should We Study Compilers?

So why study compilers? Haven't all the compilers been written? There are several reasons for studying compilers.

1. Compilers are larger programs than the ones you have written in your programming courses. It is good to work with a program that is like the size of the programs you will be working on when you graduate.

2. Compilers make use of all those things you have learned about earlier: arrays, lists, queues, stacks, trees, graphs, maps, regular expressions and finite state automata,

---

[4]Not necessarily always; studies have shown that just-in-time compilation, where a method is translated the first time it is invoked, and then cached, or hotspot compilation, where only code that is interpreted several times is compiled and cached, can provide better performance. Even so, in both of these techniques, programs are partially compiled to some intermediate form such as Oracle's Java Virtual Machine (JVM), or Microsoft's Common Language Runtime (CLR). The intermediate forms are smaller, and space can play a role in run-time performance. We discuss just-in-time compilation and hotspot compilation in Chapter 8.

context-free grammars and parsers, recursion, and patterns. It is fun to use all of these in a real program.

3. You learn about the language you are compiling (in our case, Java).

4. You learn a lot about the target machine (in our case, both the Java Virtual Machine and the MIPS computer).

5. Compilers are still being written for new languages and targeted to new computer architectures. Yes, there are still compiler-writing jobs out there.

6. Compilers are finding their way into all sorts of applications, including games, phones, and entertainment devices.

7. XML. Programs that process XML use compiler technology.

8. There is a mix of theory and practice, and each is relevant to the other.

9. The organization of a compiler is such that it can be written in stages, and each stage makes use of earlier stages. So, compiler writing is a case study in software engineering.

10. Compilers are programs. And writing programs is fun.

1.3   How Does a Compiler Work? The Phases of Compilation

A compiler is usually broken down into several phases—components, each of which performs a specific sub-task of compilation. At the very least, a compiler can be broken into a *front end* and a *back end* (Figure 1.3).

FIGURE 1.3 A compiler: Analysis and synthesis.

The front end takes as input, a high-level language program, and produces as output a representation (another translation) of that program in some intermediate language that lies somewhere between the source language and the target language. We call this the intermediate representation (IR). The back end then takes this intermediate representation of the program as input, and produces the target machine language program.

1.3.1   Front End

A compiler's front end

- Is that part of the compiler that analyzes the input program for determining its meaning, and so

- Is source language dependent (and target machine, or target language independent); moreover, it

- Can be further decomposed into a sequence of analysis phases such as that illustrated in Figure 1.4.

FIGURE 1.4 The front end: Analysis.

The *scanner* is responsible for breaking the input stream of characters into a stream of tokens: identifiers, literals, reserved words, (one-, two-, three-, and four-character) operators, and separators.

The *parser* is responsible for taking this sequence of lexical tokens and parsing against a grammar to produce an abstract syntax tree (AST), which makes the syntax that is implicit in the source program, explicit.

The *semantics phase* is responsible for *semantic analysis*: declaring names in a symbol table, looking up names as they are referenced for determining their types, assigning types to expressions, and checking the validity of types. Sometimes, a certain amount of storage analysis is also done, for example, assigning addresses or offsets to variables (as we do in our j-- compiler). When a programming language allows one to refer to a name that is declared later on in the program, the semantics phase must really involve at least two phases (or two passes over the program).

1.3.2 Back End

A compiler's back end

- Is that part of the compiler that takes the IR and produces (synthesizes) a target machine program having the same meaning, and so

- Is target language dependent (and source language independent); moreover, it

- May be further decomposed into a sequence of synthesis phases such as that illustrated in Figure 1.5.

FIGURE 1.5 The back end: Synthesis.

The *code generation* phase is responsible for choosing what target machine instructions to generate. It makes use of information collected in earlier phases.

The *peephole phase* implements a *peephole optimizer*, which scans through the generated instructions looking locally for wasteful instruction sequences such as branches to branches and unnecessary load/store pairs (where a value is loaded onto a stack or into a register and then immediately stored back at the original location).

Finally, the *object phase* links together any modules produced in code generation and constructs a single machine code executable program.

1.3.3 "Middle End"

Sometimes, a compiler will have an *optimizer*, which sits between the front end and the back end. Because of its location in the compiler architecture, we often call it the "middle end," with a little tongue-in-cheek.

FIGURE 1.6 The "middle end": Optimization.

The purpose of the optimizer (Figure 1.6) is both to improve the IR program and to collect information that the back end may use for producing better code. The optimizer might do any number of the following:

- It might organize the program into what are called basic blocks: blocks of code from which there are no branches out and into which there are no branches.

- From the basic block structure, one may then compute next-use information for determining the lifetimes of variables (how long a variable retains its value before it is redefined by assignment), and loop identification.

- Next-use information is useful for eliminating common sub-expressions and constant folding (for example, replacing x + 5 by 9 when we know x has the value 4). It may also be used for register allocation (deciding what variables or temporaries should be kept in registers and what values to "spill" from a register for making room for another).

- Loop information is useful for pulling loop invariants out of loops and for strength reduction, for example, replacing multiplication operations by (equivalent but less expensive) addition operations.

An optimizer might consist of just one phase or several phases, depending on the optimizations performed. These and other possible optimizations are discussed more fully in Chapters 6 and 7.

1.3.4 Advantages to Decomposition

There are several advantages to separating the front end from the back end:

1. Decomposition reduces complexity. It is easier to understand (and implement) the smaller programs.

2. Decomposition makes it possible for several individuals or teams to work concurrently on separate parts, thus reducing the overall implementation time.

3. Decomposition permits a certain amount of re-use[5] For example, once one has written a front end for Java and a back end for the Intel Core Duo, one need only write a new C front end to get a C compiler. And one need only write a single SPARC back end to re-target both compilers to the Oracle SPARC architecture. Figure 1.7 illustrates how this re-use gives us four compilers for the price of two.

---

[5]This depends on a carefully designed IR. We cannot count the number of times we have written front ends with the intention of re-using them, only to have to rewrite them for new customers (with that same intention!). Realistically, one ends up re-using designs more often than code.

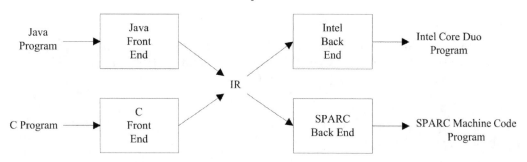

FIGURE 1.7 Re-use through decomposition.

Decomposition was certainly helpful to us, the authors, in writing the *j--* compiler as it allowed us better organize the program and to work concurrently on distinct parts of it.

1.3.5   Compiling to a Virtual Machine: New Boundaries

The Java compiler, the program invoked when one types, for example,

```
> javac MyProgram.java
```

produces a .class file called `MyProgram.class`, that is, a byte code[6] program suitable for execution on a Java Virtual Machine (JVM). The source language is Java; the target machine is the JVM. To execute this .class file, one types

```
> java MyProgram
```

which effectively interprets the JVM program. The JVM is an interpreter, which is implemented based on the observation that almost all programs spend most of their time in a small part of their code. The JVM monitors itself to identify these "hotspots" in the program it is interpreting, and it compiles these critical methods to native code; this compilation is accompanied by a certain amount of in-lining: the replacement of method invocations by the method bodies. The native code is then executed, or interpreted, on the native computer. Thus, the JVM byte code might be considered an IR, with the Java "compiler" acting as the front end and the JVM acting as the back end, targeted to the native machine on which it is running.

The IR analogy to the byte code makes even more sense in Microsoft's Common Language Runtime (CLR) architecture used in implementing its .Net tools. Microsoft has written compilers (or front ends) for Visual Basic, C++, C#, and J++ (a variant of Java), all of which produce byte code targeted for a common architecture (the CLR). Using a technique called *just-in-time (JIT) compilation*, the CLR compiles each method to native code and caches that native code when that method is first invoked. Third parties have implemented other front-end compilers for other programming languages, taking advantage of the existing JIT compilers.

In this textbook, we compile a (non-trivial) subset of Java, which we call *j--*. In the first instance, we target the Oracle JVM. So in a sense, this compiler is a front end. Nevertheless, our compiler implements many of those phases that are traditional to compilers and so it serves as a reasonable example for an introductory compilers course.

The experience in writing a compiler targeting the JVM is deficient in one respect: one does not learn about register allocation because the JVM is a stack-based architecture and has no registers.

---

[6] "byte code" because the program is represented as a sequence of byte instructions and operands (and operands occupy several bytes).

1.3.6   Compiling JVM Code to a Register Architecture

To remedy this deficiency, we (beginning in Chapter 6) discuss the compilation of JVM code to code for the MIPS machine, which is a register-based architecture. In doing this, we face the challenge of mapping possibly many variables to a limited number of fast registers.

One might ask, "Why don't we simply translate *j--* programs to MIPS programs?" After all, C language programs are always translated to native machine code.

The strategy of providing an intermediate virtual machine code representation for one's programs has several advantages:

1. Byte code, such as JVM code or Microsoft's CLR code is quite compact. It takes up less space to store (and less memory to execute) and it is more amenable to transport over the Internet. This latter aspect of JVM code made Java applets possible and accounts for much of Java's initial success.

2. Much effort has been invested in making interpreters like the JVM and the CLR run quickly; their just-in-time compilers are highly optimized. One wanting a compiler for any source language need only write a front-end compiler that targets the virtual machine to take advantage of this optimization.

3. Implementers claim, and performance tests support, that hotspot interpreters, which compile to native code only those portions of a program that execute frequently, actually run faster than programs that have been fully translated to native code. Caching behavior might account for this improved performance.

Indeed, the two most popular platforms (Oracle's Java platform and Microsoft's .NET architecture) follow the strategy of targeting a virtual, stack-based, byte-code architecture in the first instance, and employing either just-in-time compilation or HotSpot compilation for implementing these "interpreters".

1.4   An Overview of the *j--* to JVM Compiler

Our source language, *j--*, is a proper subset of the Java programming language. It has about half the syntax of Java; that is, its grammar that describes the syntax is about half the size of that describing Java's syntax. But *j--* is a non-trivial, object-oriented programming language, supporting classes, methods, fields, message expressions, and a variety of statements, expressions, and primitive types. *j--* is more fully described in Appendix B.

Our *j--* compiler is organized in an object-oriented fashion. To be honest, most compilers are not organized in this way. Nor are they written in languages like Java, but in lower-level languages such as C and C++ (principally for better performance). As the previous section suggests, most compilers are written in a procedural style. Compiler writers have generally bucked the object-oriented organizational style and have relied on the more functional organization described in Section 1.3.

Even so, we decided to structure our *j--* compiler on object-oriented principles. We chose Java as the implementation language because that is the language our students know best and the one (or one like it) in which you will program when you graduate. Also, you are likely to be programming in an object-oriented style.

It has many of the components of a traditional compiler, and its structure is not necessarily novel. Nevertheless, it serves our purposes:

- We learn about compilers.

- We learn about Java. *j--* is a non-trivial subset of Java. The *j--* compiler is written in Java.

- We work with a non-trivial object-oriented program.

1.4.1 *j--* Compiler Organization

Our compiler's structure is illustrated in Figure 1.8.

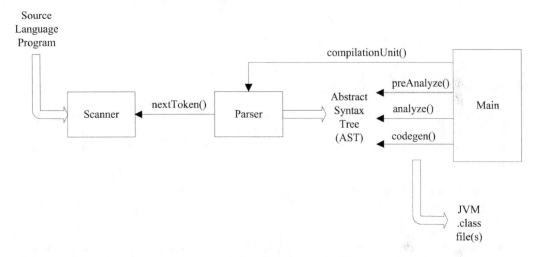

FIGURE 1.8 The *j--* compiler.

The entry point to the *j--* compiler is `Main`[7]. It reads in a sequence of arguments, and then goes about creating a `Scanner` object, for scanning tokens, and a `Parser` object for parsing the input source language program and constructing an abstract syntax tree (AST).

Each node in the abstract syntax tree is an object of a specific type, reflecting the underlying linguistic component or operation. For example, an object of type `JCompilationUnit` sits at the root (the top) of the tree for representing the program being compiled. It has sub-trees representing the package name, list of imported types, and list of type (that is, class) declarations. An object of type `JMultiplyOp` in the AST, for example, represents a multiplication operation. Its two sub-trees represent the two operands. At the leaves of the tree, one finds `JVariable` objects and objects representing constant literals.

Each type of node in the AST defines three methods, each of which performs a specific task on the node, and recursively on its sub-trees:

1. `preAnalyze(Context context)` is defined only for the types of nodes that appear near the top of the AST because *j--* does not implement nested classes. Pre-analysis deals with declaring imported types, defined class names, and class member headers (method headers and fields). This is required because method bodies may make forward references to names declared later on in the input. The context argument is

---

[7]This and other classes related to the compiler are part of the `jminusminus` package under `$j/j--/ src`, where `$j` is the directory that contains the *j--* root directory.

a string of `Context` (or subtypes of `Context`) objects representing the compile-time symbol table of declared names and their definitions.

2. `analyze(Context context)` is defined over all types of AST nodes. When invoked on a node, this method declares names in the symbol table (context), checks types (looking up the types of names in the symbol table), and converts local variables to offsets in a method's run-time local stack frame where the local variables reside.

3. `codegen(CLEmitter output)` is invoked for generating the Java Virtual Machine (JVM) code for that node, and is applied recursively for generating code for any sub-trees. The output argument is a `CLEmitter` object, an abstraction of the output .class file.

Once `Main` has created the scanner and parser,

1. `Main` sends a `compilationUnit()` message to the parser, causing it to parse the program by a technique known as recursive descent, and to produce an AST.

2. `Main` then sends the `preAnalyze()` message to the root node (an object of type `JCompilationUnit`) of the AST. `preAnalyze()` recursively descends the tree down to the class member headers for declaring the types and the class members in the symbol table context.

3. `Main` then sends the `analyze()` message to the root `JCompilationUnit` node, and `analyze()` recursively descends the tree all the way down to its leaves, declaring names and checking types.

4. `Main` then sends the `codegen()` message to the root `JCompilationUnit` node, and `codegen()` recursively descends the tree all the way down to its leaves, generating JVM code. At the start of each class declaration, `codegen()` creates a new `CLEmitter` object for representing a target .class file for that class; at the end of each class declaration, `codegen()` writes out the code to a .class file on the file system.

5. The compiler is then done with its work. If errors occur in any phase of the compilation process, the phase attempts to run to completion (finding any additional errors) and then the compilation process halts.

In the next sections, we briefly discuss how each phase does its work. As this is just an overview and a preview of what is to come in subsequent chapters, it is not important that one understand everything at this point. Indeed, if you understand just 15%, that is fine. The point to this overview is to let you know where stuff is. We have the rest of the text to understand how it all works!

1.4.2   Scanner

The scanner supports the parser. Its purpose is to scan tokens from the input stream of characters comprising the source language program. For example, consider the following source language `HelloWorld` program.

```
import java.lang.System;

public class HelloWorld {
    // The only method.

    public static void main(String[] args) {
```

```
      System.out.println("Hello, World!");
   }
}
```

The scanner breaks the program text into atomic tokens. For example, it recognizes each of `import`, `java`, `.`, `lang`, `.`, `System`, and `;` as being distinct tokens.

Some tokens, such as `java`, `HelloWorld`, and `main`, are identifiers. The scanner categorizes theses tokens as `IDENTIFIER` tokens. The parser uses these category names to identify the kinds of incoming tokens. `IDENTIFIER` tokens carry along their images as attributes; for example, the first `IDENTIFIER` in the above program has `java` as its image. Such attributes are used in semantic analysis.

Some tokens are reserved words, each having its unique name in the code. For example, `import`, `public`, and `class` are reserved word tokens having the names `IMPORT`, `PUBLIC`, and `CLASS`. Operators and separators also have distinct names. For example, the separators `.`, `;`, `{`, `}`, `[`and `]` have the token names `DOT`, `SEMI`, `LCURLY`, `RCURLY`, `LBRACK`, and `RBRACK`, respectively.

Others are literals; for example, the string literal `Hello, World!` comprises a single token. The scanner calls this a `STRING_LITERAL`.

Comments are scanned and ignored altogether. As important as some comments are to a person who is trying to understand a program[8], they are irrelevant to the compiler.

The scanner does not first break down the input program text into a sequence of tokens. Rather, it scans each token on demand; each time the parser needs a subsequent token, it sends the `nextToken()` message to the scanner, which then returns the token id and any image information.

The scanner is discussed in greater detail in Chapter 2.

1.4.3 Parser

The parsing of a *j--* program and the construction of its abstract syntax tree (AST) is driven by the language's syntax, and so is said to be *syntax directed*. In the first instance, our parser is hand-crafted from the *j--* grammar, to parse *j--* programs by a technique known as *recursive descent*.

For example, consider the following grammatical rule describing the syntax for a compilation unit:

compilationUnit ::= [`package` qualifiedIdentifier `;`]
                    {`import` qualifiedIdentifier `;`}
                    {typeDeclaration} EOF

This rule says that a compilation unit consists of

- An optional package clause (the brackets `[]` bracket optional clauses),

- Followed by zero or more import statements (the curly brackets `{}` bracket clauses that may appear zero or more times),

- Followed by zero or more type declarations (in *j--*, these are only class declarations),

- Followed by an end of file (EOF).

---

[8]But we know some who swear by the habit of stripping out all comments before reading a program for fear that those comments might be misleading. When programmers modify code, they often forget to update the accompanying comments.

The tokens PACKAGE, SEMI, IMPORT, and EOF are returned by the scanner.

To parse a compilation unit using the recursive descent technique, one would write a method, call it compilationUnit(), which does the following:

1. If the next (the first, in this case) incoming token were PACKAGE, would scan it (advancing to the next token), invoke a separate method called qualifiedIdentifier() for parsing a qualified identifier, and then we must scan a SEMI (and announce a syntax error if the next token were not a SEMI).

2. While the next incoming token is an IMPORT, scan it and invoke qualifiedIdentifier () for parsing the qualified identifier, and then we must again scan a SEMI. We save the (imported) qualified identifiers in a list.

3. While the next incoming token is not an EOF, invoke a method called typeDeclaration () for parsing the type declaration (in *j--* this is only a class declaration), and we must scan a SEMI. We save all of the ASTs for the type declararations in a list.

4. We must scan the EOF.

Here is the Java code for compilationUnit(), taken directly from Parser.

```java
public JCompilationUnit compilationUnit() {
    int line = scanner.token().line();
    TypeName packageName = null; // Default
    if (have(PACKAGE)) {
        packageName = qualifiedIdentifier();
        mustBe(SEMI);
    }
    ArrayList<TypeName> imports = new ArrayList<TypeName>();
    while (have(IMPORT)) {
        imports.add(qualifiedIdentifier());
        mustBe(SEMI);
    }
    ArrayList<JAST> typeDeclarations = new ArrayList<JAST>();
    while (!see(EOF)) {
        JAST typeDeclaration = typeDeclaration();
        if (typeDeclaration != null) {
            typeDeclarations.add(typeDeclaration);
        }
    }
    mustBe(EOF);
    return new JCompilationUnit(scanner.fileName(), line,
        packageName, imports, typeDeclarations);
}
```

In Parser, see() is a Boolean method that looks to see whether or not its argument matches the next incoming token. Method have() is the same, but has the side-effect of scanning past the incoming token when it does match its argument. Method mustBe() requires that its argument match the next incoming token, and raises an error if it does not.

Of course, the method typeDeclaration() recursively invokes additional methods for parsing the HelloWorld class declaration; hence the technique's name: recursive descent. Each of these parsing methods produces an AST constructed from some particular type of node. For example, at the end of compilationUnit(), a JCompilationUnit node is created for encapsulating any package name (none here), the single import (having its own AST), and a single class declaration (an AST rooted at a JClassDeclaration node).

Parsing in general, and recursive descent in particular, are discussed more fully in Chapter 3.

1.4.4 AST

An abstract syntax tree (AST) is just another representation of the source program. But it is a representation that is much more amenable to analysis. And the AST makes explicit that syntactic structure which is implicit in the original source language program. The AST produced for our `HelloWorld` program from Section 1.4.2 is illustrated in Figure 1.9. The boxes in the figure represent `ArrayLists`.

All classes in the *j--* compiler that are used to represent nodes in the AST extend the abstract class `JAST` and have names beginning with the letter J. Each of these classes implements the three methods required for compilation:

1. `preAnalyze()` for declaring types and class members in the symbol table;

2. `analyze()` for declaring local variables and typing all expressions; and

3. `codegen()` for generating code for each sub-tree.

We discuss these methods briefly below; and in greater detail later on in this book. But before doing that, we must first briefly discuss how we build a symbol table and use it for declaring (and looking up) names and their types.

1.4.5 Types

As in Java, *j--* names and values have types. A type indicates how something can behave. A `boolean` behaves differently from an `int`; a `Queue` behaves differently from a `Hashtable`. Because *j--* (like Java) is *statically typed*, its compiler must determine the types of all names and expressions. So we need a representation for types.

Java already has a representation for its types: objects of type `java.lang.Class` from the Java API. Because *j--* is a subset of Java, why not use class `Class`? The argument is more compelling because *j--*'s semantics dictate that it may make use of classes from the Java API, so its type representation must be compatible with Java's.

But, because we want to define our own functionality for types, we encapsulate the `Class` objects within our own class called `Type`. Likewise, we encapsulate `java.lang.reflect.Method`, `java.lang.reflect.Constructor`, `java.lang.reflect.Field`, and `java.lang.reflect.Member` within our own classes, `Method`, `Constructor`, `Field`, and `Member`, respectively[9]. And we define a sufficiently rich set of operations on these representational classes.

There are places, for example in the parser, where we want to denote a type by its name before that types is known or defined. For this we introduce `TypeName` and (because we need array types) `ArrayTypeName`. During the analysis phase of compilation, these type denotations are resolved: they are looked up in the symbol table and replaced by the actual `Types` they denote.

1.4.6 Symbol Table

During semantic analysis, the compiler must construct and maintain a symbol table in which it declares names. Because *j--* (like Java) has a nested scope for declared names, this symbol table must behave like a pushdown stack.

---

[9]These private classes are defined in the `Type.java` file, together with the public class `Type`. In the code tree, we have chosen to put many private classes in the same file in which their associated public class is defined.

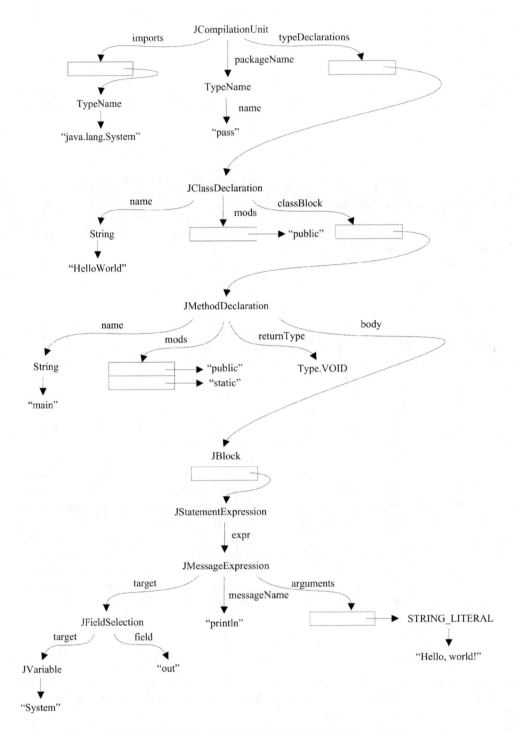

FIGURE 1.9 An AST for the `HelloWorld` program.

In the *j--* compiler, this symbol table is represented as a singly-linked list of `Context` objects, that is, objects whose types extend the `Context` class. Each object in this list represents some area of scope and contains a mapping from names to definitions. Every context object maintains three pointers: one to the object representing the surrounding context, one to the object representing the compilation unit context (at the root), and one to the enclosing class context.

For example, there is a `CompilationUnitContext` object for representing the scope comprising the program, that is, the entire compilation unit. There is a `ClassContext` object for representing the scope of a class declaration. The `ClassContext` has a reference to the defining class type; this is used to determine where we are (that is, in which class declaration the compiler is in) for settling issues such as accessibility.

There is a `MethodContext` (a subclass of `LocalContext`) for representing the scopes of methods and, by extension, constructors. Finally, a `LocalContext` represents the scope of a block, including those blocks that enclose method bodies. Here, local variable names are declared and mapped to `LocalVariableDefns`.

1.4.7 `preAnalyze()` and `analyze()`

`preAnalyze()` is a first pass at type checking. Its purpose is to build that part of the symbol table that is at the top of the AST, to declare both imported types and types introduced by class declarations, and to declare the members declared in those classes. This first pass is necessary for declaring names that may be referenced before they are defined. Because *j--* does not support nested classes, this pass need not descend into the method bodies.

`analyze()` picks up where `preAnalyze()` left off. It continues to build the symbol table, decorating the AST with type information and enforcing the *j--* type rules. The `analyze()` phase performs other important tasks:

- Type checking: `analyze()` computes the type for every expression, and it checks its type when a particular type is required.

- Accessibility: `analyze()` enforces the accessibility rules (expressed by the modifiers `public`, `protected`, and `private`) for both types and members.

- Member finding: `analyze()` finds members (messages in message expressions, based on signature, and fields in field selections) in types. Of course, only the compile-time member name is located; polymorphic messages are determined at run-time.

- Tree rewriting: `analyze()` does a certain amount of AST (sub) tree rewriting. Implicit field selections (denoted by identifiers that are fields in the current class) are made explicit, and field and variable initializations are rewritten as assignment statements after the names have been declared.

1.4.8 Stack Frames

`analyze()` also does a little storage allocation. It allocates positions in the method's current stack frame for formal parameters and (other) local variables.

The JVM is a stack machine: all computations are carried out atop the run-time stack. Each time a method is invoked, the JVM allocates a *stack frame*, a contiguous block of memory locations on top of the run-time stack. The actual arguments substituted for formal parameters, the values of local variables, and temporary results are all given positions within this stack frame. Stack frames for both a static method and for an instance method are illustrated in Figure 1.10.

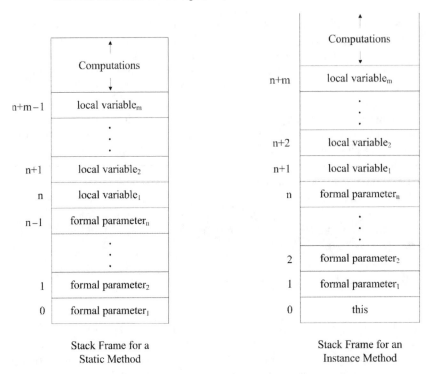

FIGURE 1.10 Run-time stack frames in the JVM.

In both frames, locations are set aside for n formal parameters and m local variables; n, m, or both may be 0. In the stack frame for a static method, these locations are allocated at offsets beginning at 0. But in the invocation of an instance method, the instance itself, that is, `this`, must be passed as an argument, so in an instance method's stack frame, location 0 is set aside for `this`, and parameters and local variables are allocated offset positions starting at 1. The areas marked "computations" in the frames are memory locations set aside for run-time stack computations within the method invocation.

While the compiler cannot predict how many stack frames will be pushed onto the stack (that would be akin to solving the halting problem), it can compute the offsets of all formal parameters and local variables, and compute how much space the method will need for its computations, in each invocation.

1.4.9 `codegen()`

The purpose of `codegen()` is to generate JVM byte code from the AST, based on information computed by `preAnalyze()` and `analyze()`. `codegen()` is invoked by `Main`'s sending the `codegen()` message to the root of the AST, and `codegen()` recursively descends the AST, generating byte code.

The format of a JVM class file is rather arcane. For this reason, we have implemented a tool, `CLEmitter` (and its associated classes), to ease the generation of types (for example, classes), members, and code. `CLEmitter` may be considered an abstraction of the JVM class file; it hides many of the gory details. `CLEmitter` is described further in Appendix D.

`Main` creates a new `CLEmitter` object. `JClassDeclaration` adds a new class, using `addClass()`. `JFieldDeclaration` writes out the fields using `addField()`.

JMethodDeclarations and JConstructorDeclarations add themselves, using `addMethod`(), and then delegate their code generation to their bodies. It is not rocket science.

The code for `JMethodDeclaration.codegen()` illustrates what these `codegen()` methods look like:

```
public void codegen(CLEmitter output) {
    output.addMethod(mods, name, descriptor, null, false);
    if (body != null) {
        body.codegen(output);
    }

    // Add implicit RETURN
    if (returnType == Type.VOID) {
        output.addNoArgInstruction(RETURN);
    }
}
```

In general, we generate only the class headers, members, and their instructions and operands. `CLEmitter` takes care of the rest. For example, here is the result of executing

```
> javap HelloWorld
```

where `javap` is a Java tool that disassembles `HelloWorld.class`:

```
public class HelloWorld extends java.lang.Object
{
  public HelloWorld();
  Code:
   Stack=1, Locals=1, Args_size=1
   0:  aload_0
   1:  invokespecial    #8;  //Method java/lang/Object."<init>":()V
   4:  return

  public static void main(java.lang.String[]);
  Code:
   Stack=2, Locals=1, Args_size=1
   0:  getstatic        #17; //Field java/lang/System.out:
                             //Ljava/io/PrintStream;
   3:  ldc              #19; //String Hello, World!
   5:  invokevirtual    #25; //Method java/io/PrintStream.println:
                             //(Ljava/lang/String;)V
   8:  return
}
```

We have shown only the instructions; tables such as the constant table have been left out of our illustration.

In general, `CLEmitter` does the following for us:

- Builds the constant table and generates references that the JVM can use to reference names and constants; one need only generate the instructions and their operands, using names and literals.

- Computes branch offsets and addresses; the user can use mnemonic labels.

- Computes the argument and local variable counts and the stack space a method requires to do computation.

- Constructs the complete class file.

The `CLEmitter` is discussed in more detail in Appendix D. JVM code generation is discussed more fully in Chapter 5.

1.5   *j--* Compiler Source Tree

The zip file *j--*.zip containing the *j--* distribution can be downloaded from http://www.cs.umb.edu/j--. The zip file may be unzipped into any directory of your choosing. Throughout this book, we refer to this directory as $j.

For a detailed description of what is in the software bundle; how to set up the compiler for command-line execution; how to set up, run, and debug the software in Eclipse[10]; and how to add *j--* test programs to the test framework, see Appendix A.

$j/j--/src/jminusminus contains the source files for the compiler, where jminusminus is a package. These include

- Main.java, the driver program;

- a hand-written scanner (Scanner.java) and parser (Parser.java);

- J*.java files defining classes representing the AST nodes;

- CL*.java files supplying the back-end code that is used by *j--* for creating JVM byte code; the most important file among these is CLEmitter.java, which provides the interface between the front end and back end of the compiler;

- S*.java files that translate JVM code to SPIM files (SPIM is an interpreter for the MIPS machine's symbolic assembly language);

- j--.jj, the input file to JavaCC[11] containing the specification for generating (as opposed to hand-writing) a scanner and parser for the *j--* language; JavaCCMain, the driver program that uses the scanner and parser produced by JavaCC; and

- Other Java files providing representation for types and the symbol table.

$j/j--/bin/j-- is a script to run the compiler. It has the following command-line syntax:

```
Usage: j-- <options> <source file>
where possible options include:
  -t Only tokenize input and print tokens to STDOUT
  -p Only parse input and print AST to STDOUT
  -pa Only parse and pre-analyze input and print AST to STDOUT
  -a Only parse, pre-analyze, and analyze input and print AST to STDOUT
  -s <naive|linear|graph> Generate SPIM code
  -r <num> Max. physical registers (1-18) available for allocation; default=8
  -d <dir> Specify where to place output files; default=.
```

For example, the j– program $j/j--/tests/pass/HelloWorld.java can be compiled using *j--* as follows:

```
> $j/j--/bin/j-- $j/j--/tests/pass/HelloWorld.java
```

to produce a HelloWorld.class file under pass folder within the current directory, which can then be run as

```
> java pass.HelloWorld
```

to produce as output,

```
> Hello, World!
```

---

[10]An open-source IDE; http://www.eclipse.org.
[11]A scanner and parser generator for Java; http://javacc.dev.java.net/.

Enhancing *j--*

Although *j--* is a subset of Java, it provides an elaborate framework with which one may add new Java constructs to *j--*. This will be the objective of many of the exercises in this book. In fact, with what we know so far about *j--*, we are already in a position to start enhancing the language by adding new albeit simple constructs to it.

As an illustrative example, we will add the division[12] operator to *j--*. This involves modifying the scanner to recognize / as a token, modifying the parser to be able to parse division expressions, implementing semantic analysis, and finally, code generation for the division operation.

In adding new language features to *j--*, we advocate the use of the Extreme Programming[13] (XP) paradigm, which emphasizes writing tests before writing code. We will do exactly this with the implementation of the division operator.

Writing Tests

Writing tests for new language constructs using the *j--* test framework involves

- Writing *pass* tests, which are *j--* programs that can successfully be compiled using the *j--* compiler;

- Writing JUnit test cases that would run these pass tests;

- Adding the JUnit test cases to the *j--* test suite; and finally

- Writing *fail* tests, which are erroneous *j--* programs. Compiling a fail test using *j--* should result in the compiler's reporting the errors and gracefully terminating without producing any .class files for the erroneous program.

We first write a pass test `Division.java` for the division operator, which simply has a method `divide()` that accepts two arguments x and y, and returns the result of dividing x by y. We place this file under the `$j/j--/tests/pass` folder; `pass` is a package.

```
package pass;

public class Division {
    public int divide(int x, int y) {
        return x / y;
    }
}
```

Next, we write a JUnit test case `DivisionTest.java`, with a method `testDivide()` that tests the `divide()` method in `Division.java` with various arguments. We place this file under `$j/j--/tests/junit` folder; `junit` is a package.

```
public class DivisionTest extends TestCase {
    private Division division;

    protected void setUp() throws Exception {
        super.setUp();
        division = new Division();
    }

    protected void tearDown() throws Exception {
        super.tearDown();
```

---

[12]We only handle integer division since *j--* supports only `int`s as numeric types.
[13]`http://www.extremeprogramming.org/`.

```
    }

    public void testDivide() {
        this.assertEquals(division.divide(0, 42), 0);
        this.assertEquals(division.divide(42, 1), 42);
        this.assertEquals(division.divide(127, 3), 42);
    }
}
```

Now that we have a test case for the division operator, we must register it with the *j--* test suite by making the following entry in the suite() method of junit. JMinusMinusTestRunner.

```
TestSuite suite = new TestSuite();
...
suite.addTestSuite(DivisionTest.class);
return suite;
```

j-- supports only int as a numeric type, so the division operator can operate only on ints. The compiler should thus report an error if the operands have incorrect types; to test this, we add the following fail test Division.java and place it under the $j/j--/tests/ fail folder; fail is a package.

```
package fail;

import java.lang.System;

public class Division {
    public static void main(String[] args) {
        System.out.println('a' / 42);
    }
}
```

Changes to Lexical and Syntactic Grammars

Appendix B specifies both the lexical and the syntactic grammars for the *j--* language; the former describes how individual tokens are composed and the latter describes how these tokens are put together to form language constructs. Chapters 2 and 3 describe such grammars in great detail.

The lexical and syntactic grammars for *j--* are also available in the files $j/j--/ lexicalgrammar and $j/j--/grammar, respectively. For every language construct that is newly added to *j--*, we strongly recommend that these files be modified accordingly so that they accurately describe the modified syntax of the language. Though these files are for human consumption alone, it is a good practice to keep them up-to-date.

For the division operator, we add a line describing the operator to $j/j--/ lexicalgrammar under the *operators* section.

```
DIV ::= "/"
```

where DIV is the kind of the token and "/" is its image (string representation).

Because the division operator is a multiplicative operator, we add it to the grammar rule describing multiplicative expressions in the $j/j--/grammar file.

```
multiplicativeExpression ::= unaryExpression // level 2
                               {(STAR | DIV) unaryExpression}
```

The level number in the above indicates operator precedence. Next, we discuss the changes in the *j--* codebase to get the compiler to support the division operation.

Changes to Scanner

Here we only discuss the changes to the hand-written scanner. Scanners can also be generated; this is discussed in Chapter 2. Before changing `Scanner.java`, we must register `DIV` as a new token, so we add the following to the `TokenKind` enumeration in the `TokenInfo.java` file.

```
enum TokenKind {
    EOF("<EOF>"),
    ...,
    STAR("*"),
    DIV("/"),
    ...
}
```

The method that actually recognizes and returns tokens in the input is `getNextToken()`. Currently, `getNextToken()` does not recognize / as an operator, and reports an error when it encounters a single / in the source program. In order to recognize the operator, we replace the `getNextToken()` code in `Scanner`.

```
if (ch == '/') {
    nextCh();
    if (ch == '/') {
        // CharReader maps all new lines to '\n'
        while (ch != '\n' && ch != EOFCH) {
            nextCh();
        }
    }
    else {
        reportScannerError(
            "Operator / is not supported in j--.");
    }
}
```

with the following.

```
if (ch == '/') {
    nextCh();
    if (ch == '/') {
        // CharReader maps all new lines to '\n'
        while (ch != '\n' && ch != EOFCH) {
            nextCh();
        }
    }
    else {
        return new TokenInfo(DIV, line);
    }
}
```

Changes to Parser

Here we only discuss the changes to the hand-written parser. Parsers can also be generated; this is discussed in Chapter 3. We first need to define a new AST node to represent the division expression. Because the operator is a multiplicative operator like *, we can model the AST for the division expression based on the one (`JMultiplyOp`) for *. We call the new AST node `JDivideOp`, and because division expression is a binary expression (one with two operands), we define it in `JBinaryExpression.java` as follows:

```
class JDivideOp extends JBinaryExpression {
    public JDivideOp(int line, JExpression lhs, JExpression rhs) {
```

```
            super(line, "/", lhs, rhs);
    }

    public JExpression analyze(Context context) {
        return this;
    }

    public void codegen(CLEmitter output) {

    }
}
```

To parse expressions involving division operator, we modify the `multiplicativeExpres` - `sion()` method in `Parser.java` as follows:

```
private JExpression multiplicativeExpression() {
    int line = scanner.token().line();
    boolean more = true;
    JExpression lhs = unaryExpression();
    while (more) {
        if (have(STAR)) {
            lhs = new JMultiplyOp(line, lhs,
                unaryExpression());
        }
        else if (have(DIV)) {
            lhs = new JDivideOp(line, lhs,
                unaryExpression());
        }
        else {
            more = false;
        }
    }
    return lhs;
}
```

Semantic Analysis and Code Generation

Since `int` is the only numeric type supported in *j--*, analyzing the division operator is trivial. It involves analyzing its two operands, making sure each type is `int`, and setting the resulting expression's type to `int`. We thus implement `analyze()` in the `JDivideOp` AST as follows:

```
public JExpression analyze(Context context) {
    lhs = (JExpression) lhs.analyze(context);
    rhs = (JExpression) rhs.analyze(context);
    lhs.type().mustMatchExpected(line(), Type.INT);
    rhs.type().mustMatchExpected(line(), Type.INT);
    type = Type.INT;
    return this;
}
```

Generating code for the division operator is also trivial. It involves generating (through delegation) code for its operands and emitting the JVM (IDIV) instruction for the (integer) division of two numbers. Hence the following implementation for `codegen()` in JDivideOp.

```
public void codegen(CLEmitter output) {
    lhs.codegen(output);
    rhs.codegen(output);
    output.addNoArgInstruction(IDIV);
}
```

The IDIV instruction is a zero-argument instruction. The operands that it operates on must to be loaded on the operand stack prior to executing the instruction.

Testing the Changes

Finally, we need to test the addition of the new (division operator) construct to *j--*. This can be done at the command prompt by running

```
> ant
```

which compiles our tests using the hand-written scanner and parser, and then tests them. The results of compiling and running the tests are written to the console (STDOUT).

Alternatively, one could compile and run the tests using Eclipse; Appendix A describes how.

1.6 Organization of This Book

This book is organized like a compiler. You may think of this first chapter as the main program, the driver if you like. It gives the overall structure of compilers in general, and of our *j--* compiler in particular.

In Chapter 2 we discuss the scanning of tokens, that is, lexical analysis.

In Chapter 3 we discuss context-free grammars and parsing. We first address the recursive descent parsing technique, which is the strategy the parser uses to parse *j--*. We then go on to examine the LL and LR parsing strategies, both of which are used in various compilers today.

In Chapter 4 we discuss type checking, or semantic analysis. There are two passes required for this in the *j--* compiler, and we discuss both of them. We also discuss the use of attribute grammars for declaratively specifying semantic analysis.

In Chapter 5 we discuss JVM code generation. Again we address the peculiarities of code generation in our *j--* compiler, and then some other more general issues in code generation.

In Chapter 6 we discuss translating JVM code to instructions native to a MIPS computer; MIPS is a register-based RISC architecture. We discuss what is generally called optimization, a process by which the compiler produces better (that is, faster and smaller) target programs. Although our compiler has no optimizations other than register allocation, a general introduction to them is important.

In Chapter 7 register allocation is the principal challenge.

In Chapter 8 we discuss several celebrity compilers.

Appendix A gives instructions on setting up a *j--* development environment.

Appendix B contains the lexical and syntactic grammar for *j--*.

Appendix C contains the lexical and syntactic grammar for Java.

Appendix D describes the CLEmitter interface and also provides a group-wise summary of the JVM instruction set.

Appendix E describes James Larus's SPIM simulator for the MIPS family of computers and how to write *j--* programs that target SPIM.

1.7 Further Readings

The Java programming language is fully described in [Gosling et al., 2005]. The Java Virtual Machine is described in [Lindholm and Yellin, 1999].

Other classic compiler texts include [Aho et al., 2007], [Appel, 2002], [Cooper and Torczon, 2011], [Allen and Kennedy, 2002], and [Muchnick, 1997].

A reasonable introduction to testing is [Whittaker, 2003]. Testing using the JUnit framework is nicely described in [Link and Fröhlich, 2003] and [Rainsberger and Stirling, 2005]. A good introduction to extreme programming, where development is driven by tests, is [Beck and Andres, 2004].

1.8 Exercises

Exercise 1.1. We suggest you use either Emacs or Eclipse for working with the *j--* compiler. In any case, you will want to get the *j--* code tree onto your own machine. If you choose to use Eclipse, do the following.

a. Download Eclipse and install it on your own computer. You can get Eclipse from `http ://www.eclipse.org`.

b. Download the *j--* distribution from `http://www.cs.umb.edu/j--/`.

c. Follow the directions in Appendix A for importing the *j--* code tree as a project into Eclipse.

Exercise 1.2. Now is a good time to begin browsing through the code for the *j--* compiler. Locate and browse through each of the following classes.

a. `Main`

b. `Scanner`

c. `Parser`

d. `JCompilationUnit`

e. `JClassDeclaration`

f. `JMethodDeclaration`

g. `JVariableDeclaration`

h. `JBlock`

i. `JMessageExpression`

j. `JVariable`

k. `JLiteralString`

The remaining exercises may be thought of as optional. Some students (and their professors) may choose to go directly to Chapter 2. Exercises 1.3 through 1.9 require studying the compiler in its entirety, if only cursorily, and then making slight modifications to it. Notice that, in these exercises, many of the operators have different levels of precedence, just as * has a different level of precedence in *j--* than does +. These levels of precedence are captured in the Java grammar (in Appendix C); for example, the parser uses one method to parse expressions involving * and /, and another to parse expressions involving + and -.

Exercise 1.3. To start, follow the process outlined in Section 1.5 to implement the Java remainder operator %.

Exercise 1.4. Implement the Java shift operators, <<, >>, and >>>.

Exercise 1.5. Implement the Java bitwise inclusive or operator, |.

Exercise 1.6. Implement the Java bitwise exclusive or operator, ^.

Exercise 1.7. Implement the Java bitwise and operator, &.

Exercise 1.8. Implement the Java unary bitwise complement operator ~, and the Java unary + operator. What code is generated for the latter?

Exercise 1.9. Write tests for all of the exercises (1.3 through 1.8) done above. Put these tests where they belong in the code tree and modify the JUnit framework in the code tree for making sure they are invoked.

Exercise 1.10 through 1.16 are exercises in *j--* programming. *j--* is a subset of Java and is described in Appendix B.

Exercise 1.10. Write a *j--* program `Fibonacci.java` that accepts a number n as input and outputs the nth Fibonacci number.

Exercise 1.11. Write a *j--* program `GCD.java` that accepts two numbers a and b as input, and outputs the Greatest Common Divisor (GCD) of a and b. Hint: Use the Euclidean algorithm[14]

Exercise 1.12. Write a *j--* program `Primes.java` that accepts a number n as input, and outputs the all the prime numbers that are less than or equal to n. Hint: Use the Sieve of Eratosthenes algorithm[15] For example,

```
> java Primes 11
```

should output

```
> 2 3 5 7 11
```

Exercise 1.13. Write a *j--* program `Date.java` that accepts a date in "yyyy-mm-dd" format as input, and outputs the date in "Month Day, Year" format. For example[16],

```
> java Date 1879-03-14
```

should output

```
> March 14, 1879
```

---

[14]See http://en.wikipedia.org/wiki/Euclidean_algorithm.
[15]See http://en.wikipedia.org/wiki/Sieve_of_Eratosthenes.
[16]March 14, 1879, is Albert Einstein's birthday.

Exercise 1.14. Write a *j--* program `Palindrome.java` that accepts a string as input, and outputs the string if it is a palindrome (a string that reads the same in either direction), and outputs nothing if it is not. The program should be case-insensitive to the input. For example[17]:

```
> java Palindrome Malayalam
```

should output:

```
> Malayalam
```

Exercise 1.15. Suggest enhancements to the *j--* language that would simplify the implementation of the programs described in the previous exercises (1.10 through 1.14).

Exercise 1.16. For each of the *j--* programs described in Exercises 1.10 through 1.14, write a JUnit test case and integrate it with the *j--* test framework (Appendix A describes how this can be done).

Exercises 1.17 through 1.25 give the reader practice in reading JVM code and using `CLEmitter` for producing JVM code (in the form of .class files). The JVM and the `CLEmitter` are described in Appendix D.

Exercise 1.17. Disassemble (Appendix A describes how this can be done) a Java class (say `java.util.ArrayList`), study the output, and list the following:

- Major and minor version

- Size of the constant pool table

- Super class

- Interfaces

- Field names, their access modifiers, type descriptors, and their attributes (just names)

- Method names, their access modifiers, descriptors, exceptions thrown, and their method and code attributes (just names)

- Class attributes (just names)

Exercise 1.18. Compile `$j/j--/tests/pass/HelloWorld.java` using the *j--* compiler and Oracle's *javac* compiler. Disassemble the class file produced by each and compare the output. What differences do you see?

Exercise 1.19. Disassemble the class file produced by the *j--* compiler for `$j/j--/tests/pass/Series.java`, save the output in `Series.bytecode`. Add a single-line (`//...`) comment for each JVM instruction in `Series.bytecode` explaining what the instruction does.

Exercise 1.20. Write the following class names in internal form:

- `java.lang.Thread`

- `java.util.ArrayList`

- `java.io.FileNotFoundException`

---

[17]Malayalam is the language spoken in Kerala, a southern Indian state.

- jminusminus.Parser

- Employee

Exercise 1.21. Write the method descriptor for each of the following constructors/method declarations:

- public Employee(String name)...

- public Coordinates(float latitude, float longitude)...

- public Object get(String key)...

- public void put(String key, Object o)...

- public static int[] sort(int[] n, boolean ascending)...

- public int[][] transpose(int[][] matrix)...

Exercise 1.22. Write a program (Appendix A describes how this can be done) GenGCD.java that produces, using CLEmitter, a GCD.class file with the following methods:

```
// Returns the Greatest Common Divisor (GCD) of a and b.
public static int compute(int a, int b) {
    ...
}
```

Running GCD as follows:

```
> java GCD 42 84
```

should output

```
> 42
```

Modify GenGCD.java to handle java.lang.NumberFormatException that Integer.parseInt() raises if the argument is not an integer, and in the handler, print an appropriate error message to STDERR.

Exercise 1.23. Write a program GenPrimality.java that produces, using CLEmitter, Primality.class, Primality1.class, Primality2.class, and Primality3.class files, where Primality.class is an interface with the following method:

```
// Returns true if the specified number is prime, false
// otherwise.
public boolean isPrime(int n);
```

and Primality1.class, Primality2.class, and Primality3.class are three different implementations of the interface. Write a j-- program TestPrimality.java that has a test driver for the three implementations.

Exercise 1.24. Write a program GenWC.java that produces, using CLEmitter, a WC.class which emulates the UNIX command wc that displays the number of lines, words, and bytes contained in each input file.

Exercise 1.25. Write a program GenGravity.java that produces, using CLEmitter, a Gravity.class file which computes the acceleration g due to gravity at a point on the surface of a massive body. The program should accept the mass M of the body, and the distance r of the point from body's center as input. Use the following formula for computing g:

$$g = \frac{GM}{r^2},$$

where $G = 6.67 \times 10^{-11} \frac{Nm^2}{kg^2}$, is the universal gravitational constant.

Chapter 2

Lexical Analysis

2.1 Introduction

The first step in compiling a program is to break it into tokens. For example, given the *j--*
program

```
package pass;

import java.lang.System;

public class Factorial {

    // Two methods and a field

    public static int factorial(int n) {
        if (n <= 0)
            return 1;
        else
            return n * factorial(n - 1);
    }

    public static void main(String[] args) {
        int x = n;
        System.out.println(x + "! = " + factorial(x));
    }

    static int n = 5;
}
```

we want to produce the sequence of tokens package, pass, ;, import, java, ., lang, .,
System, ;, public, class, Factorial, {, public, static, int, factorial, (, int, n,), {,
if, (, n, <=, 0,), }, return, 1, ;, else, return, n, *, factorial, (, n, -, 1,), }, ;, },
public, static, void, main, (, String, [,], args,), }, {, int, x, =, n, ;, System, ., out,
., println, (, x, +, "!=", +, factorial, (, x,),), }, ;, }, static, int, n, =, 5, ;, and }.

Notice how we have broken down the program's text into its component elements. We
call these the *lexical tokens* (or lexical elements) which are described by the language's
lexical syntax. Notice also how the comment has been ignored in the sequence of tokens;
this is because comments have no meaning to the compiler or to the results of executing
the program. We call this process of recognizing tokens *lexical analysis*, or more simply
scanning.

The lexical syntax of a programming language is usually described separately. For example, in *j--* we describe identifiers as "a letter, underscore, or dollar-sign, followed by zero
or more letters, digits, underscores or dollar-signs." We can also describe these tokens more
formally, using what are called *regular expressions*. We visit regular expressions later in
Section 2.3. For now, let us be a little less formal.

29

In describing our lexical tokens, we usually separate them into categories. In our example program, `public`, `class`, `static`, and `void` are categorized as reserved words. `Factorial`, `main`, `String`, `args`, `System`, `out`, and `println` are all identifiers; an identifier is a token in its own right. The string `!=` is a literal, a string literal in this instance. The rest are operators and separators; notice that we distinguish between single-character operators such as `+` and multi-character operators like `>=`. In describing the tokens, we can usually simply list the different reserved words and operators, and describe the structure of identifiers and literals.

2.2　Scanning Tokens

We call the program that breaks the input stream of characters into tokens a *lexical analyzer* or, less formally, a *scanner*.

A scanner may be hand-crafted, that is, a program written by the compiler writer; or it may be generated automatically from a specification consisting of a sequence of regular expressions. The lexical analyzer that we describe in this section will be hand-crafted. We look at generated scanners later.

When we look at the problem of producing a sequence of tokens from a program like that above, we must determine where each token begins and ends. Clearly, *white space* (blanks, tabs, new lines, etc.) plays a role; in the example above it is used to separate the reserved word `public` from `class`, and `class` from the identifier `Factorial`. But not all tokens are necessarily separated by white space. Whether or not we have come to the end of a token in scanning an input stream of characters depends on what sort of token we are currently scanning. For example, in the context of scanning an identifier, if we come across a letter, that letter is clearly part of the identifier. On the other hand, if we are in the process of scanning an integer value (consisting of the digits 0 to 9), then that same letter would indicate that we have come to the end of our integer token. For this reason, we find it useful to describe our scanner using a *state transition diagram*.

Identifiers and Integer Literals

For example, consider the state transition diagram in Figure 2.1. It may be used for recognizing *j*-- identifiers and decimal integers.

In a state transition diagram, the nodes represent states, and directed edges represent moves from one state to another depending on what character has been scanned. If a character scanned does not label any of the edges, then the unlabeled edge is taken (and the input is not advanced). Think of it as a machine that recognizes (scans) identifiers and integer literals.

Consider the case when the next token in the input is the identifier `x1`. Beginning in the *start* state, the machine considers the first character, `x`; because it is a letter, the machine scans it and goes into the *id* state (that is, a state in which we are recognizing an identifier). Seeing the next `1`, it scans that digit and goes back into the *id* state. When the machine comes across a character that is neither a letter nor a digit, nor an underscore, nor a dollar sign, it takes the unmarked edge and goes into the *idEnd* state (a final state indicated by a double circle) without scanning anything.

If the first character were the digit `0` (zero), the machine would scan the zero and go directly into the (final) *intEnd* state. On the other hand, if the first character were a non-zero digit, it would scan it and go into the *integer* state. From there it would scan succeeding

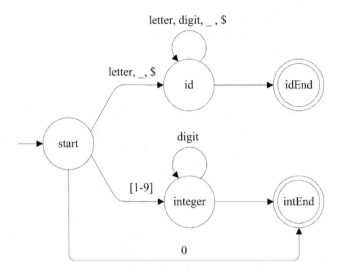

FIGURE 2.1 State transition diagram for identifiers and integers.

digits and repeatedly go back into the *integer* state until a non-digit character is reached; at this point the machine would go into the (final) *intEnd* state without scanning another character.

An advantage of basing our program on such a state transition diagram is that it takes account of *state* in deciding what to do with the next input character. Clearly, what the scanner does with an incoming letter depends on whether it is in the *start* state (it would go into the *id* state), the *id* state (it would remain there), or the *integer* state (where it would go into the *intEnd* state, recognizing that it has come to the end of the integer).

It is relatively simple to implement a state transition diagram in code. For example, the code for our diagram above might look something like

```
if (isLetter(ch) || ch == '_' || ch == '$') {
    buffer = new StringBuffer();
    while (isLetter(ch) || isDigit(ch) || ch == '_' || ch == '$'){
        buffer.append(ch);
        nextCh();
    }
    return new TokenInfo(IDENTIFIER, buffer.toString(), line);
}
else if (ch == '0') {
    nextCh();
    return new TokenInfo(INT_LITERAL, "0", line);
}
else if (isDigit(ch)){
    buffer = new StringBuffer();
    while (isDigit(ch)) {
        buffer.append(ch);
        nextCh();
    }
    return new TokenInfo(INT_LITERAL, buffer.toString(), line);
}
```

Choices translate to if-statements and cycles translate to while-statements. Notice that the `TokenInfo` object encapsulates the value of an integer as the string of digits denoting it. For example, the number 6449 is represented as the `String` "6449". Translating this to binary is done later, during code generation.

In the code above, `TokenInfo` is the type of an object that encapsulates a representation of the token found, its image (if any), and the number of the line on which it was found.

Reserved Words

There are two ways of recognizing reserved words. In the first, we complicate the state transition diagram for recognizing reserved words and distinguishing them from (non-reserved) identifiers. For example, the state transition diagram fragment in Figure 2.2 recognizes reserved words (and identifiers) beginning with the letter 'i' or the letter 'n'.

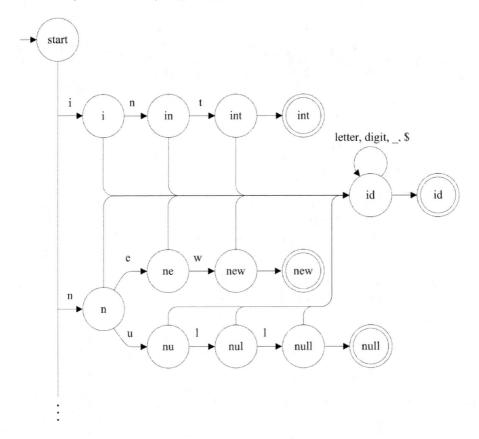

FIGURE 2.2 A state transition diagram that distinguishes reserved words from identifiers.

The code corresponding to this might look something like

```
...
else if (ch == 'n') {
    buffer.append(ch);
    nextCh();
    if (ch == 'e') {
        buffer.append(ch);
        nextCh();
            if (ch == 'w') {
            buffer.append(ch);
            nextCh();
            if (!isLetter(ch) && !isDigit(ch) &&
                    ch != '_' && ch != '$') {
```

```
                    return new TokenInfo(NEW, line);
            }
        }
    }
    else if (ch == 'u') {
        buffer.append(ch);
        nextCh();
        if (ch == 'l') {
            buffer.append(ch);
            nextCh();
            if (ch == 'l') {
                buffer.append(ch);
                nextCh();
                if (!isLetter(ch) && !isDigit(ch) &&
                        ch != '_' && ch != '$') {
                    return new TokenInfo(NULL, line);
                }
            }
        }
    }
    while (isLetter(ch) || isDigit(ch) ||
            ch == '_' || ch == '$') {
        buffer.append(ch);
        nextCh();
    }
    return new TokenInfo(IDENTIFIER, buffer.toString(),
                        line);
}
else ...
```

Unfortunately, such state transition diagrams and the corresponding code are too complex.[1] Imagine the code necessary for recognizing all reserved words.

A second way is much more straightforward for hand-written token scanners. We simply recognize a simple identifier, which may or may not be one of the reserved words. We look up that identifier in a table of reserved words. If it is there in the table, then we return the corresponding reserved word. Otherwise, we return the (non-reserved) identifier. The state transition diagram then looks something like that in Figure 2.3.

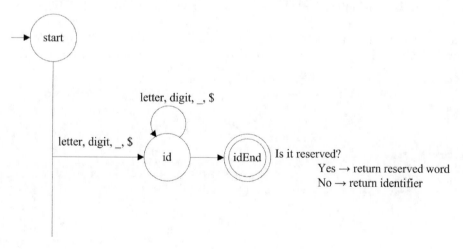

FIGURE 2.3 Recognizing words and looking them up in a table to see if they are reserved.

---

[1]Unless the diagrams (and corresponding code) are automatically generated from a specification; we do this later in the chapter when we visit regular expressions and finite state automata.

The code corresponding to this logic would look something like the following:

```
    if (isLetter(ch) || ch == '_' || ch == '$') {
        buffer = new StringBuffer();
        while (isLetter(ch) || isDigit(ch) ||
                ch == '_' || ch == '$'){
            buffer.append(ch);
            nextCh();
        }
        String identifier = buffer.toString();
        if (reserved.containsKey(identifier)) {
            return new TokenInfo(reserved.get(identifier),
                                    line);
        }
        else {
            return new TokenInfo(IDENTIFIER, identifier,
                                    line);
        }
    }
```

This relies on a map (hash table), `reserved`, mapping reserved identifiers to their representations:

```
reserved = new Hashtable<String, Integer>();
reserved.put("abstract", ABSTRACT);
reserved.put("boolean", BOOLEAN);
reserved.put("char", CHAR);
...
reserved.put("while", WHILE);
```

We follow this latter method, of looking up identifiers in a table of reserved words, in our hand-written lexical analyzer.

Separators and Operators

The state transition diagram deals nicely with operators. We must be careful to watch for certain multi-character operators. For example, the state transition diagram fragment for recognizing tokens beginning with ';', '==', '=', '!', or '*' would look like that in Figure 2.4.

The code corresponding to the state transition diagram in Figure 2.4 would look like the following. Notice the use of the switch-statement for deciding among first characters.

```
switch (ch) {
...
case ';':
    nextCh();
    return new TokenInfo(SEMI, line);
case '=':
    nextCh();
    if (ch == '=') {
        nextCh();
        return new TokenInfo(EQUAL, line);
    }
    else {
        return new TokenInfo(ASSIGN, line);
    }
case '!':
    nextCh();
    return new TokenInfo(LNOT, line);
case '*':
```

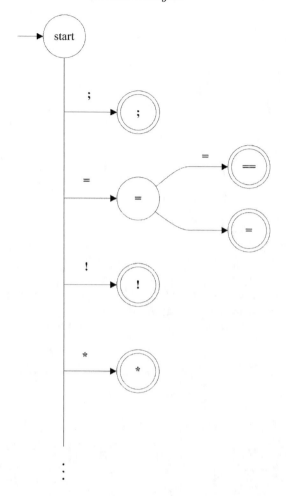

FIGURE 2.4 A state transition diagram for recognizing the separator ; and the operators ==, =, !, and *.

```
    nextCh();
    return new TokenInfo(STAR, line);
...
}
```

White Space

Before attempting to recognize the next incoming token, one wants to skip over all white space. In *j--*, as in Java, white space is defined as the ASCII SP characters (spaces), HT (horizontal tabs), FF (form feeds), and line terminators; in *j--* (as in Java), we can denote these characters as ' ', 't', 'f', 'b', 'r', and 'n', respectively. Skipping over white space is done from the start state, as illustrated in Figure 2.5.

The code for this is simple enough, and comes at the start of a method for reading the next incoming token:

```
while (isWhitespace(ch)) {
    nextCh();
}
```

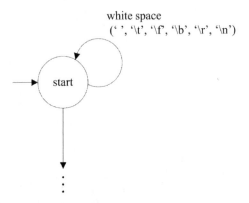

FIGURE 2.5 Dealing with white space.

Comments

Comments can be considered a special form of white space because the compiler ignores them. A *j--* comment extends from a double-slash, //, to the end of the line. This complicates the skipping of white space somewhat, as illustrated in Figure 2.6.

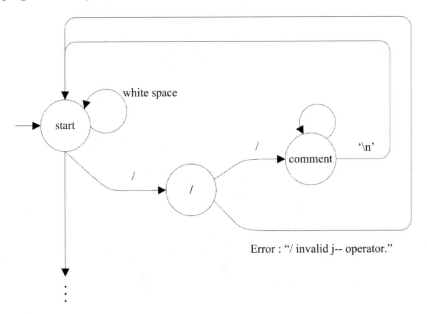

FIGURE 2.6 Treating one-line (// ...) comments as white space.

Notice that a / operator on its own is meaningless in *j--*. Adding it (for denoting division) is left as an exercise. But notice that when coming upon an erroneous single /, the lexical analyzer reports the error and goes back into the start state in order to fetch the next valid token. This is all captured in the code:

```
boolean moreWhiteSpace = true;
while (moreWhiteSpace) {
    while (isWhitespace(ch)) {
        nextCh();
    }
```

```
    if (ch == '/') {
        nextCh();
        if (ch == '/') {
            // CharReader maps all new lines to '\n'
            while (ch != '\n' && ch != EOFCH) {
                nextCh();
            }
        }
        else {
            reportScannerError(''Operator / is not supported in j--.'');
        }
    }
    else {
        moreWhiteSpace = false;
    }
}
```

There are other kinds of tokens we must recognize as well, for example, String literals and character literals. The code for recognizing all tokens appears in file `Scanner.java`; the principal method of interest is `getNextToken()`. This file is part of source code of the *j--* compiler that we discussed in Chapter 1. At the end of this chapter you will find exercises that ask you to modify this code (as well as that of other files) for adding tokens and other functionality to our lexical analyzer.

A pertinent quality of the lexical analyzer described here is that it is hand-crafted. Although writing a lexical analyzer by hand is relatively easy, particularly if it is based on a state transition diagram, it is prone to error. In a later section we shall learn how we may automatically produce a lexical analyzer from a notation based on regular expressions.

2.3 Regular Expressions

Regular expressions comprise a relatively simple notation for describing patterns of characters in text. For this reason, one finds them in text processing tools such as text editors. We are interested in them here because they are also convenient for describing lexical tokens.

Definition 2.1. We say that a *regular expression* defines a *language* of strings over an *alphabet*. Regular expressions may take one of the following forms:

1. If a is in our alphabet, then the regular expression a describes the language consisting of the string a. We call this language $L(a)$.

2. If r and s are regular expressions, then their concatenation rs is also a regular expression describing the language of all possible strings obtained by concatenating a string in the language described by r, to a string in the language described by s. We call this language $L(rs)$.

3. If r and s are regular expressions, then the alternation $r|s$ is also a regular expression describing the language consisting of all strings described by either r or s. We call this language $L(r|s)$.

4. If r is a regular expression, the repetition[2] $r*$ is also a regular expression describing the language consisting of strings obtained by concatenating zero or more instances of strings described by r together. We call this language $L(r*)$.

---

[2]Also known as the *Kleene closure*.

Notice that $r^0 = \epsilon$, the empty string of length 0; $r^1 = r$, $r^2 = rr$, $r^3 = rrr$, and so on; $r*$ denotes an infinite number of finite strings.

5. ϵ is a regular expression describing the language containing only the empty string.

6. Finally, if r is a regular expression, then (r) is also a regular expression denoting the same language. The parentheses serve only for grouping.

Example. So, for example, given an alphabet $\{0, 1\}$,

1. 0 is a regular expression describing the single string 0

2. 1 is a regular expression describing the single string 1

3. 0|1 is a regular expression describing the language of two strings 0 and 1

4. (0|1) is a regular expression describing the (same) language of two strings 0 and 1

5. (0|1)* is a regular expression describing the language of all strings, including the empty string, of 1's and 0's: ϵ, 0, 1, 00, 01, 10, 11, 000, 001, 010, 011, ..., 000111, ...

6. 1(0|1)* is a regular expression describing the language of all strings of 1's and 0's that start with a 1.

7. 0|1(0|1)* is a regular expression describing the language consisting of all binary numbers (excluding those having unnecessary leading zeros).

Notice that there is an order of precedence in the construction of regular expressions: repetition has the highest precedence, then concatenation, and finally alternation. So, 01 * 0|1* is equivalent to $(0(1*)0)|(1*)$. Of course, parentheses may always be used to change the grouping of sub-expressions.

Example. Given an alphabet $\{a, b\}$,

1. $a(a|b)*$ denotes the language of non-empty strings of a's and b's, beginning with an a

2. $aa|ab|ba|bb$ denotes the language of all two-symbol strings over the alphabet

3. $(a|b)*ab$ denotes the language of all strings of a's and b's, ending in ab (this includes the string ab itself)

As in programming, we often find it useful to give names to things. For example, we can define D=1|2|3|4|5|6|7|8|9. Then we can say 0|D(D|0)* denotes the language of natural numbers, which is the same as 0|(1|2|3|4|5|6|7|8|9)(1|2|3|4|5|6|7|8|9|0)*.

There are all sorts of extensions to the notation of regular expressions, all of which are shorthand for standard regular expressions. For example, in the *Java Language Specification* [Gosling et al., 2005], [0-9] is shorthand for (0|1|2|3|4|5|6|7|8|9), and [a-z] is shorthand for (a|b|c|d|e|f|g|h|i|j|k|l|m|n|o|p|q|r|s|t|u|v|w|x|y|z).

Other notations abound. For example, there are the POSIX extensions [IEEE, 2004], which allow the square bracket notation above, ? for optional, +, *, etc. JavaCC uses its own notation. In our appendices, we use the notation used by [Gosling et al., 2005]. The important thing is that all of these extensions are simply shorthand for regular expressions that may be written using the notation described in Definition 2.1.

In describing the lexical tokens of programming languages, one uses some standard input character set as one's alphabet, for example, the 128 character ASCII set, the 256 character extended ASCII set, or the much larger Unicode set. Java works with Unicode, but aside from identifiers, characters, and string literals, all input characters are ASCII, making implementations compatible with legacy operating systems. We do the same for *j--*.

Example. The reserved words may be described simply by listing them. For example,

```
      abstract
 |  boolean
 |  char
 .
 .
 .
 |  while
```

Likewise for operators. For example,

```
      =
 |  ==
 |  >
 .
 .
 .
 |  *
```

Identifiers are easily described; for example,

```
([a-zA-Z] | _ | $)([a-zA-Z0-9] | _ | $)*
```

which is to say, an identifier begins with a letter, an underscore, or a dollar sign, followed by zero or more letters, digits, underscores, and dollar signs.

A full description of the lexical syntax for *j--* may be found in Appendix B. In the next section, we formalize state transition diagrams.

2.4   Finite State Automata

It turns out that for any language described by a regular expression, there is a state transition diagram that can parse strings in this language. These are called *finite state automata*.

Definition 2.2. A *finite state automaton* (FSA) F is a quintuple $F = (\Sigma, S, s_0, M, F)$ where

- Σ (pronounced sigma) is the input alphabet.

- S is a set of states.

- $s_0 \in S$ is a special start state.

- M is a set of moves or state transitions of the form

$$m(r, a) = s \text{ where } r, s \in S, a \in \Sigma$$

  read as, "if one is in state r, and the next input symbol is a, scan the a and move into state s."

- $F \in S$ is a set of final states.

A finite state automaton is just a formalization of the state transition diagrams we saw in Section 2.2. We say that a finite state automaton recognizes a language. A sentence over the alphabet Σ is said to be in the language recognized by the FSA if, starting in the start state, a set of moves based on the input takes us into one of the final states.

Example. Consider the regular expression, $(a|b)a*b$. This describes a language over the alphabet $\{a, b\}$; it is the language consisting of all strings starting with either an a or a b, followed by zero or more a's, and ending with a b.

An FSA F that recognizes this same language is $F = (\Sigma, S, s_0, M, F)$, where $\Sigma = \{a, b\}, S = \{0, 1, 2\}, s_0 = 0, M = \{m(0, a) = 1, m(0, b) = 1, m(1, a) = 1, m(1, b) = 2\}, F = \{2\}$.

The corresponding state transition diagram is shown in Figure 2.7.

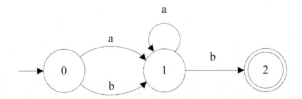

FIGURE 2.7 An FSA recognizing $(a|b)a*b$.

An FSA recognizes strings in the same way that state transition diagrams do. For example, given the input sentence *baaab* and beginning in the start state 0, the following moves are prescribed:

- $m(0, b) = 1 \implies$ in state 0 we scan a b and go into state 1,

- $m(1, a) = 1 \implies$ in state 1 we scan an a and go back into state 1,

- $m(1, a) = 1 \implies$ in state 1 we scan an a and go back into state 1 (again),

- $m(1, a) = 1 \implies$ in state 1 we scan an a and go back into state 1 (again), and

- $m(1, b) = 2 \implies$ finally, in state 1 we scan a b and go into the final state 2.

Each move scans the corresponding character. Because we end up in a final state after scanning the entire input string of characters, the string is accepted by our FSA.

The question arises, given the regular expression, have we a way of automatically generating the FSA? The answer is yes! But first we must discuss two categories of automata: Non-deterministic Finite-State Automata (NFA) and Deterministic Finite-state Automata (DFA).

2.5   Non-Deterministic Finite-State Automata (NFA) versus Deterministic Finite-State Automata (DFA)

The example FSA given above is actually a *deterministic* finite-state automaton.

Definition 2.3. A *deterministic finite-state automaton (DFA)* is an automaton where there are no ϵ-moves (see below), and there is a unique move from any state, given a single input symbol a. That is, there **cannot** be two moves:

$$m(r, a) = s$$
$$m(r, a) = t$$

where $s \neq t$. So, from any state there is at most one state that we can go into, given an incoming symbol.

Definition 2.4. A *non-deterministic finite-state automaton* (NFA) is a finite state automaton that allows either of the following conditions.

- More than one move from the same state, on the same input symbol, that is,

$$m(r, a) = s,$$
$$m(r, a) = t, \text{ for states } r, s \text{ and } t \text{ where } s \neq t.$$

- An ϵ-move defined on the empty string ϵ, that is,

$$m(r, \epsilon) = s,$$

which says we can move from state r to state s without scanning any input symbols.

An example of a deterministic finite-state automaton is $N = (\Sigma, S, s_0, M, F)$, where $\Sigma = \{a, b\}, S = \{0, 1, 2\}, s_0 = 0, M = \{m(0, a) = 1, m(0, b) = 1, m(1, a) = 1, m(1, b) = 1, m(1, \epsilon) = 0, m(1, b) = 2\}, F = \{2\}$ and is illustrated by the diagram in Figure 2.8. This NFA recognizes all strings of a's and b's that begin with an a and end with a b. Like any FSA, an NFA is said to recognize an input string if, starting in the start state, there exists a set of moves based on the input that takes us into one of the final states.

But this automaton is definitely not deterministic. Being in state 1 and seeing b, we can go either back into state 1 or into state 2. Moreover, the automaton has an ϵ-move.

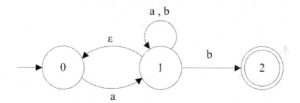

FIGURE 2.8 An NFA.

Needless to say, a lexical analyzer based on a non-deterministic finite state automaton requires backtracking, where one based on a deterministic finite-state automaton does not. One might ask why we are at all interested in NFA. Our only interest in non-deterministic finite-state automata is that they are an intermediate step from regular expressions to deterministic finite-state automata.

2.6 Regular Expressions to NFA

Given any regular expression R, we can construct a non-deterministic finite state automaton N that recognizes the same language; that is, $L(N) = L(R)$. We show that this is true by using what is called *Thompson's construction*:

1. If the regular expression r takes the form of an input symbol, a, then the NFA that recognizes it has two states: a *start* state and a *final* state, and a move on symbol a from the start state to the final state.

FIGURE 2.9 Scanning symbol *a*.

2. If N_r and N_s are NFA recognizing the languages described by the regular expressions r and s, respectively, then we can create a new NFA recognizing the language described by rs as follows. We define an ϵ-move from the final state of N_r to the start state of N_s. We then choose the start state of N_r to be our new start state, and the final state of N_s to be our new final state.

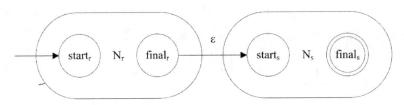

FIGURE 2.10 Concatenation *rs*.

3. If N_r and N_s are NFA recognizing the languages described by the regular expressions r and s, respectively, then we can create a new NFA recognizing the language described by $r|s$ as follows. We define a new start state, having ϵ-moves to each of the start states of N_r and N_s, and we define a new final state and add ϵ-moves from each of N_r and N_s to this state.

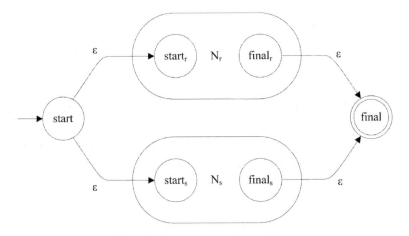

FIGURE 2.11 Alternation *r*|*s*.

4. If N_r is an NFA recognizing that language described by a regular expression r, then we construct a new NFA recognizing $r*$ as follows. We add an ϵ-move from N_r's final state back to its start state. We define a new start state and a new final state, we add ϵ-moves from the new start state to both N_r's start state and the new final state, and we define an ϵ-move from N_r's final state to the new final state.

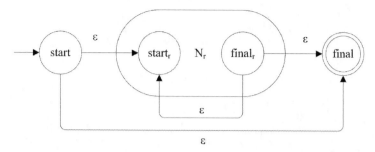

FIGURE 2.12 Repetition $r*$.

5. If r is ϵ, then we just need an ϵ-move from the start state to the final state.

FIGURE 2.13 ϵ-move.

6. If N_r is our NFA recognizing the language described by r, then N_r also recognizes the language described by (r). Parentheses only group expressions.

Example. As an example, reconsider the regular expression $(a|b)a*b$. We decompose this regular expression, and display its syntactic structure in Figure 2.14.

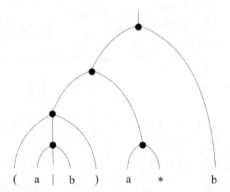

FIGURE 2.14 The syntactic structure for $(a|b)a*b$.

We can construct our NFA based on this structure, beginning with the simplest components, and putting them together according to the six rules above.

- We start with the first a and b; the automata recognizing these are easy enough to construct using rule 1 above.

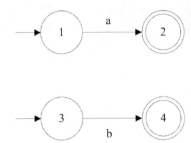

- We then put them together using rule 3 to produce an NFA recognizing $a|b$.

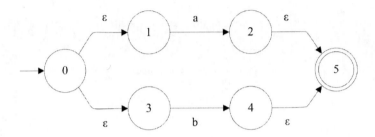

- The NFA recognizing $(a|b)$ is the same as that recognizing $a|b$, by rule 6. An NFA recognizing the second instance of a is simple enough, by rule 1 again.

- The NFA recognizing $a*$ can be constructed from that recognizing a, by applying rule 4.

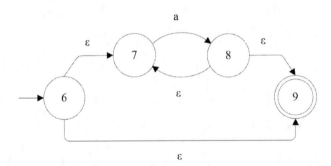

- We then apply rule 2 to construct an NFA recognizing the concatenation $(a|b)a*$.

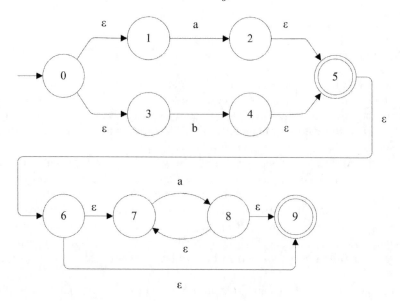

- An NFA recognizing the second instance of *b* is simple enough, by rule 1 again.

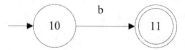

- Finally, we can apply rule 2 again to produce an NFA recognizing the concatenation of (*a*|*b*)*a*∗ and *b*, that is (*a*|*b*)*a*∗*b*. This NFA is illustrated, in Figure 2.15.

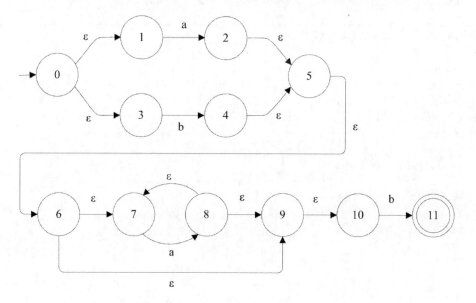

FIGURE 2.15 An NFA recognizing (*a*|*b*)*a*∗*b*.

2.7   NFA to DFA

Of course, any NFA will require backtracking. This requires more time and, because we in practice wish to collect information as we recognize a token, is impractical. Fortunately, for any non-deterministic finite automaton (NFA), there is an equivalent deterministic finite automaton (DFA). By equivalent, we mean a DFA that recognizes the same language. Moreover, we can show how to construct such a DFA.

In general, the DFA that we will construct is always in a state that simulates all the possible states that the NFA could possibly be in having scanned the same portion of the input. For this reason, we call this a powerset construction[3].

For example, consider the NFA constructed for $(a|b)a*b$ illustrated in Figure 2.15. The start state of our DFA, call it s_0, must reflect all the possible states that our NFA can be in before any character is scanned; that is, the NFA's start state 0, and all other states reachable from state 0 on ϵ-moves alone: 1 and 3. Thus, the start state in our new DFA is $s_0 = \{0, 1, 3\}$.

This computation of all states reachable from a given state s based on ϵ-moves alone is called taking the ϵ-*closure* of that state.

Definition 2.5. The ϵ-closure(s) for a state s includes s and all states reachable from s using ϵ-moves alone. That is, for a state $s \in S$, ϵ-closure(s) = $\{s\} \cup \{r \in S|$, there is a path of only ϵ-moves from s to $r\}$.

We will also be interested in the ϵ-closure over a set of states.

Definition 2.6. The ϵ-closure(S) for a set of states S includes s and all states reachable from any state s in S using ϵ-moves alone.

Algorithm 2.1 computes ϵ-closure(S) where S is a set of states.

Algorithm 2.1 ϵ-closure(S) for a Set of States S

Input: a set of states, S
Output: ϵ-closure(S)
  Stack P.addAll(S) // a stack containing all states in S
  Set C.addAll(S) // the closure initially contains the states in S
  while ! P.empty() **do**
    $s \leftarrow P$.pop()
    for r in $m(s, \epsilon)$ **do**
      // $m(s, \epsilon)$ is a set of states
      if $r \notin C$ **then**
        P.push(r)
        C.add(r)
      end if
    end for
  end while
  return C

Given Algorithm 2.1, the algorithm for finding the ϵ-closure for a single state is simple. Algorithm 2.2 does this.

---

[3]The technique is also known as a *subset construction*; the states in the DFA are a subset of the powerset of the set of states in the NFA

Algorithm 2.2 ϵ-closure(s) for a State s

Input: a state, s
Output: ϵ-closure(s)
  Set S.add(s) // $S = \{s\}$
  return ϵ-closure(S)

Returning to our example, from the start state s_0, and scanning the symbol a, we shall want to go into a state that reflects all the states we could be in after scanning an a in the NFA: 2, and then (via ϵ-moves) 5, 6, 7, 9, and 10. Thus,

$$m(s_0, a) = s_1, \text{ where}$$
$$s_1 = \epsilon\text{-closure}(2) = \{2, 5, 6, 7, 9, 10\}.$$

Similarly, scanning a symbol b in state s_0, we get

$$m(s_0, b) = s_2, \text{where}$$
$$s_2 = \epsilon\text{-closure}(4) = \{4, 5, 6, 7, 9, 10\}.$$

From state s_1, scanning an a, we have to consider where we could have gone from the states $\{2, 5, 6, 7, 9, 10\}$ in the NFA. From state 7, scanning an a, we go into state 8, and then (by ϵ-moves) 7, 9, and 10. Thus,

$$m(s_1, a) = s_3, \text{where}$$
$$s_3 = \epsilon\text{-closure}(8) = \{7, 8, 9, 10\}.$$

Now, from state s_1, scanning b, we have

$$m(s_1, b) = s_4, \text{where}$$
$$s_4 = \epsilon\text{-closure}(11) = \{11\}$$

because there are no ϵ-moves out of state 11.

From state s_2, scanning an a takes us into a state reflecting 8, and then (by ϵ-moves) 7, 9, and 10, generating a candidate state, $\{7, 8, 9, 10\}$.

But this is a state we have already seen, namely s_3. Scanning a b, from state s_2, takes us into a state reflecting 11, generating the candidate state, $\{11\}$.

But this is s_4. Thus,

$$m(s_2, a) = s_3, \text{ and}$$
$$m(s_2, b) = s_4.$$

From state s_3 we have a similar situation. Scanning an a takes us back into s_3. Scanning a b takes us into s_4. So,

$$m(s_3, a) = s_3, \text{ and}$$
$$m(s_3, b) = s_4.$$

There are no moves at all out of state s_4. So we have found all of our transitions and all of our states. Of course, the alphabet in our new DFA is the same as that in the original NFA.

But what are the final states? Because the states in our DFA mirror the states in our original NFA, any state reflecting (derived from a state containing) a final state in the NFA is a final state in the DFA. In our example, only s_4 is a final state because it contains (the final) state 11 from the original NFA.

Putting all of this together, a DFA derived from our NFA for $(a|b)a*b$ is illustrated in Figure 2.16.

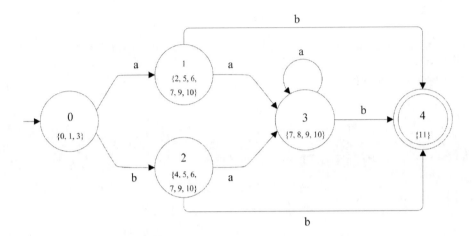

FIGURE 2.16 A DFA recognizing $(a|b)a*b$.

We can now give the algorithm for constructing a DFA that is equivalent to an NFA.

2.8   Minimal DFA

So, how do we come up with a smaller DFA that recognizes the same language? Given an input string in our language, there must be a sequence of moves taking us from the start state to one of the final states. And, given an input string that is not in our language, there cannot be such a sequence; we must get stuck with no move to take or end up in a non-final state.

Clearly, we must combine states if we can. Indeed, we would like to combine as many states together as we can. So the states in our new DFA are partitions of the states in the original (perhaps larger) DFA.

A good strategy is to start with just one or two partitions of the states, and then split states when it is necessary to produce the necessary DFA. An obvious first partition has two sets: the set of final states and the set of non-final states; the latter could be empty, leaving us with a single partition containing all states.

For example, consider the DFA from Figure 2.16, partitioned in this way. The partition into two sets of states is illustrated in Figure 2.17.

The two states in this new DFA consist of the start state, $\{0, 1, 2, 3\}$ and the final state $\{4\}$. Now we must make sure that, in each of these states, the move on a particular symbol reflects a move in the old DFA. That is, from a particular partition, each input symbol must move us to an identical partition.

Algorithm 2.3 NFA to DFA Construction

Input: an NFA, $N = (\Sigma, S, s_0, M, F)$
Output: DFA, $D = (\Sigma, S_D, s_{D0}, M_D, F_D)$
  Set $S_{D0} \leftarrow \epsilon\text{-closure}(s_0)$
  Set $S_D.\text{add}(S_{D0})$
  Moves M_D
  Stack $stk.\text{push}(S_{D0})$
  $i \leftarrow 0$
  while ! $stk.\text{empty}()$ **do**
    $t \leftarrow stk.\text{pop}()$
    for a in Σ **do**
      $S_{Di+1} \leftarrow \epsilon\text{-closure}(m(t, a));$
      if $S_{Di+1} \neq \{\}$ **then**
        if $S_{Di+1} \in S_D$ **then**
          // We have a new state
          $S_D.\text{add}(S_{Di+1})$
          $stk.\text{push}(S_{Di+1})$
          $i \leftarrow i + 1$
          $M_D.\text{add}(M_D(t, a) = i)$
        else if $\exists j, S_j \in S_D \wedge S_{Di+1} = S_j$ **then**
          // In the case that the state already exists
          $M_D.\text{add}(M_D(t, a) = j)$
        end if
      end if
    end for
  end while
  Set F_D
  for s_D in S_D **do**
    for s in s_D **do**
      if $s \in F$ **then**
        $F_D.\text{add}(s_D)$
      end if
    end for
  end for
  return $D = (\Sigma, S_D, s_{D0}, M_D, F_D)$

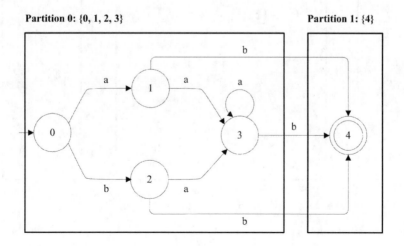

FIGURE 2.17 An initial partition of DFA from Figure 2.16.

For example, beginning in any state in the partition $\{0, 1, 2, 3\}$, an a takes us to one of the states in $\{0, 1, 2, 3\}$;

$$m(0, a) = 1,$$
$$m(1, a) = 3,$$
$$m(2, a) = 3, \text{ and}$$
$$m(3, a) = 3.$$

So, our partition $\{0, 1, 2, 3\}$ is fine so far as moves on the symbol a are concerned. For the symbol b,

$$m(0, b) = 2,$$

but

$$m(1, b) = 4,$$
$$m(2, b) = 4, \text{ and}$$
$$m(3, b) = 4.$$

So we must split the partition $\{0, 1, 2, 3\}$ into two new partitions, $\{0\}$ and $\{1, 2, 3\}$. The question arises: if we are in state s, and for an input symbol a in our alphabet there is no defined move,

$$m(s, a) = t,$$

What do we do? We can invent a special dead state d, so that we can say

$$m(s, a) = d,$$

Thus defining moves from all states on all symbols in the alphabet.

Now we are left with a partition into three sets: $\{0\}$, $\{1, 2, 3\}$, and $\{4\}$, as is illustrated in Figure 2.18.

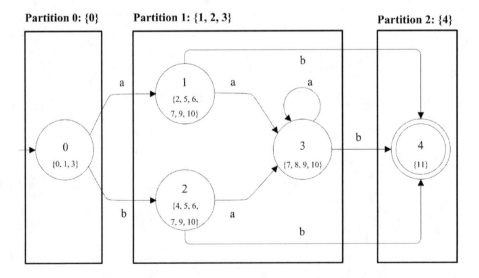

FIGURE 2.18 A second partition of DFA from Figure 2.16.

We need not worry about $\{0\}$ and $\{4\}$ as they contain just one state and so correspond to (those) states in the original machine. So we consider $\{1, 2, 3\}$ to see if it is necessary to split it. But, as we have seen,

$$m(1, a) = 3,$$
$$m(2, a) = 3, \text{ and}$$
$$m(3, a) = 3.$$

Also,

$$m(1, b) = 4,$$
$$m(2, b) = 4, \text{ and}$$
$$m(3, b) = 4.$$

Thus, there is no further state splitting to be done, and we are left with the smaller DFA in Figure 2.19.

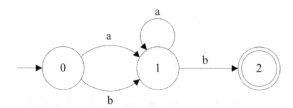

FIGURE 2.19 A minimal DFA recognizing $(a|b)a*b$.

The algorithm for minimizing a DFA is built around this notion of splitting states.

Algorithm 2.4 Minimizing a DFA

Input: a DFA, $D = (\Sigma, S, s_0, M, F)$
Output: a partition of S
   Set *partition* $\leftarrow \{S - F, F\}$ // start with two sets: the non-final states and the final states
   // Splitting the states
   while splitting occurs **do**
      for Set *set* in *partition* **do**
         if *set*.size() > 1 **then**
            for Symbol a in Σ **do**
               // Determine if moves from this 'state' force a split
               State $s \leftarrow$ a state chosen from set S
               targetSet $\leftarrow$ the set in the partition containing $m(s, a)$
               Set *set1* $\leftarrow \{$states s from set S, such that $m(s, a) \in targetSet\}$
               Set *set2* $\leftarrow \{$states s from set S, such that $m(s, a) \notin targetSet\}$
               if *set2* $\neq \{\}$ **then**
                  // Yes, split the states.
                  replace *set* in *partition* by *set1* and *set2* and break out of the for-loop to
                  continue with the next set in the partition
               end if
            end for
         end if
      end for
   end while

Then, renumber the states and re-compute the moves for the new (possibly smaller) set of states, based on the old moves on the original set of states.

Let us quickly run through one additional example, starting from a regular expression, producing an NFA, then a DFA, and finally a minimal DFA.

Example. Consider the regular expression, $(a|b)*baa$. Its syntactic structure is illustrated in Figure 2.20.

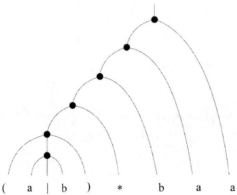

FIGURE 2.20 The syntactic structure for $(a|b)*baa$.

Given this, we apply the Thompson's construction for producing the NFA illustrated in Figure 2.21.

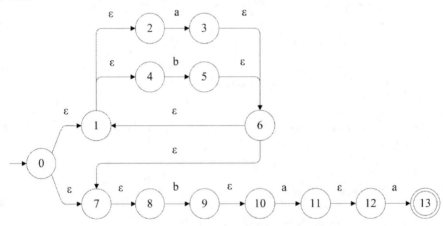

FIGURE 2.21 An NFA recognizing $(a|b)*baa$.

Using the powerset construction method, we derive a DFA having the following states:

$$s_0 : \{0, 1, 2, 4, 7, 8\},$$
$$m(s_0, a) : \{1, 2, 3, 4, 6, 7, 8\} = s_1,$$
$$m(s_0, b) : \{1, 2, 4, 5, 6, 7, 8, 9, 10\} = s_2,$$
$$m(s_1, a) : \{1, 2, 3, 4, 6, 7, 8\} = s_1,$$
$$m(s_1, b) : \{1, 2, 4, 5, 6, 7, 8, 9, 10\} = s_2,$$
$$m(s_2, a) : \{1, 2, 3, 4, 6, 7, 8, 11, 12\} = s_3,$$
$$m(s_2, b) : \{1, 2, 4, 5, 6, 7, 8, 9, 10\} = s_2, \text{ and}$$
$$m(s_3, a) : \{1, 2, 3, 4, 6, 7, 8, 13\} = s_4.$$

The DFA itself is illustrated in Figure 2.22.

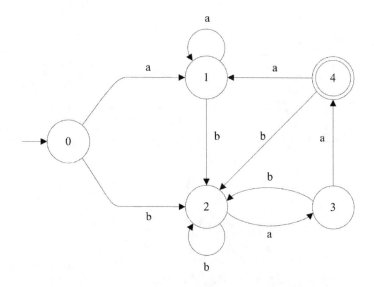

FIGURE 2.22 A DFA recognizing $(a|b)*baa$.

Finally, we use partitioning to produce the minimal DFA illustrated in Figure 2.23.

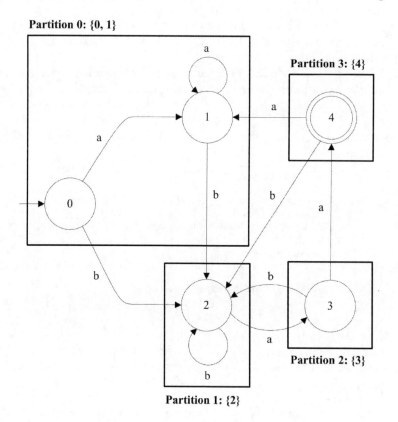

FIGURE 2.23 Partitioned DFA from Figure 2.22.

We re-number the states to produce the equivalent DFA shown in Figure 2.24.

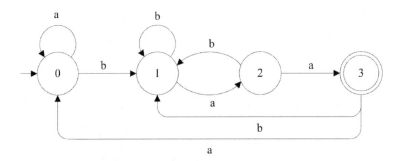

FIGURE 2.24 A minimal DFA recognizing $(a|b)*baa$.

2.9 JavaCC: Tool for Generating Scanners

JavaCC (the CC stands for compiler-compiler) is a tool for generating lexical analyzers from regular expressions, and parsers from context-free grammars. In this section we are interested in the former; we visit the latter in the next chapter.

A lexical grammar specification takes the form of a set of regular expressions and a set of *lexical states*; from any particular state, only certain regular expressions may be matched in scanning the input. There is a standard DEFAULT state, in which scanning generally begins. One may specify additional states as required.

Scanning a token proceeds by considering all regular expressions in the current state and choosing that which consumes the greatest number of input characters. After a match, one can specify a state in which the scanner should go into; otherwise the scanner stays in the current state.

There are four *kinds* of regular expressions, determining what happens when the regular expression has been matched:

1. SKIP: throws away the matched string.

2. MORE: continues to the next state, taking the matched string along.

3. TOKEN: creates a token from the matched string and returns it to the parser (or any caller).

4. SPECIAL_TOKEN: creates a special token that does not participate in the parsing.

For example, a SKIP can be used for ignoring white space:

```
SKIP: {" "|"\t"|"\n"|"\r"|"\f"}
```

This matches one of the white space characters and throws it away; because we do not specify a next state, the scanner remains in the current (DEFAULT) state.

We can deal with single-line comments with the following regular expressions:

```
MORE: { "//": IN_SINGLE_LINE_COMMENT }
<IN_SINGLE_LINE_COMMENT>
SPECIAL_TOKEN: { <SINGLE_LINE_COMMENT: "\n"|"\r"|"\r\n" > : DEFAULT }
<IN_SINGLE_LINE_COMMENT>
MORE: { < ~[] > }
```

Matching the // puts the scanner into the IN_SINGLE_LINE_COMMENT state. The next two regular expressions apply only to this state. The first matches an end of line and returns it as a special token (which is not seen by the parser); it then puts the scanner back into the DEFAULT state. The second matches anything else and throws it away; because no next state is specified, the scanner remains in the IN_SINGLE_LINE_COMMENT state. An alternative regular expression dealing with single-line comments is simpler[4]:

```
SPECIAL_TOKEN: {
  <SINGLE_LINE_COMMENT: "//" (~["\n","\r"])* ("\n"|"\r"|"\r\n")>
}
```

One may easily specify the syntax of reserved words and symbols by spelling them out, for example,

```
TOKEN: {
  < ABSTRACT: "abstract" >
| < BOOLEAN: "boolean" >
...

| < COMMA: "," >
| < DOT: "." >
}
```

The Java identifier preceding the colon, for example, ABSTRACT, BOOLEAN, COMMA, and DOT, represents the token's kind. Each token also has an image that holds onto the actual input string that matches the regular expression following the colon.

A more interesting token is that for scanning identifiers:

```
TOKEN: {
  < IDENTIFIER: (<LETTER>|"_"|"$") (<LETTER>|<DIGIT>|"_"|"$")* >
| < #LETTER: ["a"-"z","A"-"Z"] >
| < #DIGIT: ["0"-"9"] >
}
```

This says that an IDENTIFIER is a letter, underscore, or dollar sign, followed by zero or more letters, digits, underscores and dollar signs. Here, the image records the identifier itself. The # preceding LETTER and DIGIT indicates that these two identifiers are private to the scanner and thus unknown to the parser.

Literals are also relatively straightforward:

```
TOKEN: {
  < INT_LITERAL: ("0" | <NON_ZERO_DIGIT> (<DIGIT>)*) >
| < #NON_ZERO_DIGIT: ["1"-"9"] >
| < CHAR_LITERAL: "'" (<ESC> | ~["'","\\","\n","\r"]) "'" >
| < STRING_LITERAL: "\"" (<ESC> | ~["\"","\\","\n","\r"])* "\"" >
| < #ESC: "\\" ["n","t","b","r","f","\\","'","\""] >
}
```

JavaCC takes a specification of the lexical syntax and produces several Java files. One of these, TokenManager.java, defines a program that implements a state machine; this is our scanner.

To see the entire lexical grammar for *j--*, read the JavaCC input file, j--.jj, in the jminusminus package; the lexical grammar is close to the top of that file.

---

[4]Both implementations of the single-line comment come from the examples and documentation distributed with JavaCC. This simpler one comes from the **TokenManager** mini-tutorial at `https://javacc.dev.java.net/doc/tokenmanager.html`.

JavaCC has many additional features for specifying, scanning, and dealing with lexical tokens [Copeland, 2007] and [Norvell, 2011]. For example, we can make use of a combination of lexical states and lexical actions to deal with nested comments[5]. Say comments were defined as beginning with (* and ending with *); nested comments would allow one to nest them to any depth, for example (* ...(*...*)...(*...*)...*). Nested comments are useful when commenting out large chunks of code, which may contain nested comments.

To do this, we include the following code at the start of our lexical grammar specification; it declares a counter for keeping track of the nesting level.

```
TOKEN_MGR_DECLS: {
    int commentDepth;
}
```

When we encounter a (* in the standard DEFAULT state, we use a lexical action to initialize the counter to one and then we enter an explicit COMMENT state.

```
SKIP:{ "(*" { commentDepth = 1; }: COMMENT }
```

Every time we encounter another (* in this special COMMENT state, we bump up the counter by one.

```
<COMMENT> SKIP : { "(*" { commentDepth +=1; } }
```

Every time we encounter a closing *), we decrement the counter and either switch back to the standard DEFAULT state (upon reaching a depth of zero) or remain in the special COMMENT state.

```
<COMMENT> SKIP : { "*)"
    commentDepth -= 1;
    SwitchTo( commentDepth == 0 ? DEFAULT : COMMENT ); } }
```

Once we have skipped the outermost comment, the scanner will go about finding the first legitimate token. But to skip all other characters while in the COMMENT state, we need another rule:

```
<COMMENT> SKIP: { < ~[] > }
```

2.10   Further Readings

The lexical syntax for Java may be found in [Gosling et al., 2005]; this book is also published online at http://docs.oracle.com/javase/specs/.

For a more rigorous presentation of finite state automata and their proofs, see [Sipser, 2006] or [Linz, 2011]. There is also the classic [Hopcroft and Ullman, 1969].

JavaCC is distributed with both documentation and examples; see https://javacc.dev.java.net. Also see [Copeland, 2007] for a nice guide to using JavaCC.

Lex is a classic lexical analyzer generator for the C programming language. The best description of its use is still [Lesk and Schmidt, 1975]. An open-source implementation called Flex, originally written by Vern Paxton is [Paxton, 2008].

---

[5]This example is from [Norvell, 2011].

2.11 Exercises

Exercise 2.1. Consult Chapter 3 (Lexical Structure) of *The Java Language Specification* [Gosling et al., 2005]. There you will find a complete specification of Java's lexical syntax.

a. Make a list of all the keywords that are in Java but not in *j--*.

b. Make a list of the escape sequences that are in Java but are not in *j--*.

c. How do Java identifiers differ from *j--* identifiers?

d. How do Java integer literals differ from *j--* integer literals?

Exercise 2.2. Draw the state transition diagram that recognizes Java multi-line comments, beginning with a /* and ending with */.

Exercise 2.3. Draw the state transition diagram for recognizing all Java integer literals, including octals and hexadecimals.

Exercise 2.4. Write a regular expression that describes the language of all Java integer literals.

Exercise 2.5. Draw the state transition diagram that recognizes all Java numerical literals (both integers and floating point).

Exercise 2.6. Write a regular expression that describes all Java numeric literals (both integers and floating point).

Exercise 2.7. For each of the following regular expressions, use Thompson's construction to derive a non-deterministic finite automaton (NFA) recognizing the same language.

a. aaa

b. $(ab)*ab$

c. $a*bc*d$

d. $(a|bc*)a*$

e. $(a|b)*$

f. $a*|b*$

g. $(a*|b*)*$

h. $((aa)*(ab)*(ba)*(bb)*)*$

Exercise 2.8. For each of the NFA's in the previous exercise, use powerset construction for deriving an equivalent deterministic finite automaton (DFA).

Exercise 2.9. For each of the DFA's in the previous exercise, use the partitioning method to derive an equivalent minimal DFA.

The following exercises ask you to modify the hand-crafted scanner in the *j--* compiler for recognizing new categories of tokens. For each of these, write a suitable set of tests, then add the necessary code, and run the tests.

Exercise 2.10. Modify Scanner in the *j--* compiler to scan (and ignore) Java multi-line comments.

Exercise 2.11. Modify Scanner in the *j--* compiler to recognize and return all Java operators.

Exercise 2.12. Modify Scanner in the *j--* compiler to recognize and return all Java reserved words.

Exercise 2.13. Modify Scanner in the *j--* compiler to recognize and return Java double precision literal (returned as DOUBLE_LITERAL).

Exercise 2.14. Modify Scanner in the *j--* compiler to recognize and return all other literals in Java, for example, FLOAT_LITERAL, LONG_LITERAL, etc.

Exercise 2.15. Modify Scanner in the *j--* compiler to recognize and return all other representations of integers (hexadecimal, octal, etc.).

The following exercises ask you to modify the j--.jj file in the *j--* compiler for recognizing new categories of tokens. For each of these, write a suitable set of tests, then add the necessary code, and run the tests. Consult Appendix A to learn how tests work.

Exercise 2.16. Modify the j--.jj file in the *j--* compiler to scan (and ignore) Java multi-line comments.

Exercise 2.17. Modify the j--.jj file in the *j--* compiler to deal with nested Java multi-line comments, using lexical states and lexical actions.

Exercise 2.18. Re-do Exercise 2.17, but insuring that any nested parentheses inside the comment are balanced.

Exercise 2.19. Modify the j--.jj file in the *j--* compiler to recognize and return all Java operators.

Exercise 2.20. Modify the j--.jj file in the *j--* compiler to recognize and return all Java reserved words.

Exercise 2.21. Modify the j--.jj file in the *j--* compiler to recognize and return Java double precision literal (returned as DOUBLE_LITERAL).

Exercise 2.22. Modify the j--.jj file in the *j--* compiler to recognize and return all other literals in Java, for example, FLOAT_LITERAL, LONG_LITERAL, etc.

Exercise 2.23. Modify the j--.jj file in the *j--* compiler to recognize and return all other representations of integers (hexadecimal, octal, etc.).

Chapter 3

Parsing

3.1 Introduction

Once we have identified the tokens in our program, we then want to determine its syntactic structure. That is, we want to put the tokens together to make the larger syntactic entities: expressions, statements, methods, and class definitions. This process of determining the syntactic structure of a program is called *parsing*.

First, we wish to make sure the program is syntactically valid—that it conforms to the grammar that describes its syntax. As the parser parses the program it should identify syntax errors and report them and the line numbers they appear on. Moreover, when the parser does find a syntax error, it should not just stop, but it should report the error and gracefully recover so that it may go on looking for additional errors.

Second, the parser should produce some representation of the parsed program that is suitable for semantic analysis. In the *j--* compiler, we produce an abstract syntax tree (AST).

For example, given the *j--* program we saw in Chapter 2,

```
package pass;f

import java.lang.System;

public class Factorial {

    // Two methods and a field

    public static int factorial(int n) {
        if (n <= 0)
            return 1;
        else
            return n * factorial(n - 1);
    }

    public static void main(String[] args) {
        int x = n;
        System.out.println(x + "! = " + factorial(x));
    }

    static int n = 5;
}
```

we would like to produce an abstract syntax tree such as that in Figure 3.1 (the boxes represent `ArrayLists`).

Notice that the nodes in the AST represent our syntactic objects. The tree is rooted at a `JCompilationUnit`, the syntactic object representing the program that we are compiling. The directed edges are labeled by the names of the fields they represent; for example, the `JCompilationUnit` has a package name, a list (an `ArrayList`) of imported types, and a list (an `ArrayList`) of class declarations—in this case just one.

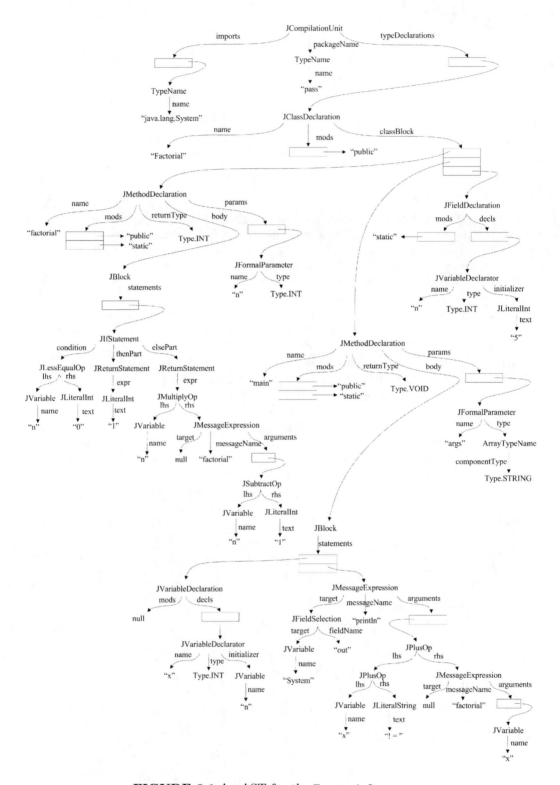

FIGURE 3.1 An AST for the Factorial program

A BNF grammar describes the syntax of programs in a programming language. For example, Appendix B at the end of this book uses BNF for describing the syntax of *j--* and Appendix C describes the syntax of Java. BNF is described in the next section. The AST captures the essential syntax of our program.

Why are we interested in a tree representation for our program? Because it is easier to analyze and decorate (with type information) a tree than it is to do the same with text. The abstract syntax tree makes that syntax, which is implicit in the program text, explicit. This is the purpose of parsing.

Before discussing how we might go about parsing *j--* programs, we must first discuss context-free grammars and the languages they describe.

3.2 Context-Free Grammars and Languages

3.2.1 Backus–Naur Form (BNF) and Its Extensions

The grammars that we use to describe programming languages are inherently recursive and so are best described by what we call context-free grammars. For example, the context-free rule

$$S ::= \text{if } (E) \ S \tag{3.1}$$

says that, if E is an expression and S is a statement,

$$\text{if } (E) \ S$$

is also a statement. Or, we can read the rule as, a statement S can be written as an `if`, followed by a parenthesized expression E and finally a statement S. That is, we can read the "::=" as "can be written as a". There are abbreviations possible in the notation. For example, we can write

$$
\begin{aligned}
S ::= &\ \text{if } (E) \ S \\
      &\ | \ \text{if } (E) \ S \ \text{else } S
\end{aligned}
\tag{3.2}
$$

as shorthand for

$$
\begin{aligned}
S ::= \text{if } (E) \ S \\
S ::= \text{if } (E) \ S \ \text{else } S
\end{aligned}
\tag{3.3}
$$

That is, we can read the "|" as "or". So, a statement S can be written either as an `if` followed by a parenthesized expression E, and a statement S; or as an `if`, followed by a parenthesized expression E, a statement S, an `else`, and another statement S.

This notation for describing programming language is called Backus-Naur Form (BNF) for John Backus and Peter Naur who used it to describe the syntax for Algol 60 in the Algol 60 Report [Backus et al., 1963].

The rules have all sorts of names: BNF-rules, rewriting rules, production rules, productions, or just rules (when the meaning is clear from the context). We use the terms production rule or simply rule.

There exist many extensions to BNF[1], made principally for reasons of expression and

---

[1] BNF with these extensions is called extended BNF or EBNF.

clarity. Our notation for describing *j--* syntax follows that used for describing Java [Gosling et al., 2005].

In our grammar, the square brackets indicate that a phrase is optional, for example, (3.2) and (3.3) could have been written as

$$S ::= \text{ if } (E) \ S \ [\text{else } S] \tag{3.4}$$

which says an S may be written as if (E) S, optionally followed by else S. Curly braces denote the Kleene closure, indicating that the phrase may appear zero or more times. For example,

$$E ::= T \ \{+ \ T\} \tag{3.5}$$

which says that an expression E may be written as a term T, followed by zero or more occurrences of + followed by a term T:

$$T$$
$$T + T$$
$$T + T + T$$
$$\dots$$

Finally, one may use the alternation sign | inside right-hand sides, using parentheses for grouping, for example,

$$E ::= T \ \{(+ \ | \ -) \ T\} \tag{3.6}$$

meaning that the additive operator may be either + or −, allowing for

$$T + T - T + T$$

This extended BNF allows us to describe such syntax as that for a *j--* compilation unit,

$$\text{compilationUnit} ::= [\text{package qualifiedIdentifier ;}] \tag{3.7}$$
$$\{\text{import qualifiedIdentifier ;}\}$$
$$\{\text{typeDeclaration}\} \ \text{EOF}$$

which says that a *j--* compilation unit may optionally have a package clause at the top, followed by zero or more imports, followed by zero or more type (for example, class) declarations[2]. The EOF represents the end-of-file.

These extensions do not increase the expressive power of BNF because they can be expressed in classic BNF. For example, if we allow the empty string ϵ on the right-hand side of a production rule, then the optional

$$[X \ Y \ Z]$$

can be expressed as T, where T is defined by the rules

$$T ::= X \ Y \ Z$$
$$T ::= \epsilon$$

---

[2]Yes, syntactically this means that the input file can be empty. We often impose additional rules, enforced later, saying that there must be at most one public class declared.

and T occurs in no other rule. Likewise,

$$\{X \; Y \; Z\}$$

can be expressed as U, where U is defined by the rules

$$U \; ::= \; X \; Y \; Z \; U$$
$$U \; ::= \; \epsilon$$

and U appears nowhere else in the grammar. Finally,

$$(X \mid Y \mid Z)$$

can be expressed as V, where V is defined by the rules

$$
\begin{aligned}
V \; ::= \; & X \\
\mid \; & Y \\
\mid \; & Z
\end{aligned}
$$

and V appears nowhere else in the grammar.

Even though these abbreviations express the same sorts of languages as classic BNF, we use them for convenience. They make for more compact and more easily understood grammars.

3.2.2   Grammar and the Language It Describes

Definition 3.1. A context-free grammar is a tuple, $G = (N, T, S, P)$, where

- N is a set of non-terminal symbols, sometimes called non-terminals;

- T is a set of terminal symbols, sometimes called terminals;

- $S \in N$ is a designated non-terminal, called the start symbol,; and

- P is a set of production rules, sometimes called productions or rules.

For example, a context-free grammar that describes (very) simple arithmetic expressions is $G = (N, T, S, P)$, where $N = \{E, T, F\}$ is the set of non-terminals, $T = \{\texttt{+}, \texttt{*}, \texttt{(}, \texttt{)}, \texttt{id}\}$ is the set of terminals, $S = E$ is the start symbol, and

$$
\begin{aligned}
P = \{ & E ::= E \texttt{ + } T, \\
& E ::= T, \\
& T ::= T \texttt{ * } F, \\
& T ::= F, \\
& F ::= (E), \\
& F ::= \texttt{id} \}
\end{aligned}
\tag{3.8}
$$

A little less formally, we can denote the same grammar simply as a sequence of production rules. For example, (3.9) denotes the same grammar as does (3.8).

$$E ::= E + T$$
$$E ::= T$$
$$T ::= T * F$$
$$T ::= F$$
$$F ::= (E)$$
$$F ::= \text{id}$$

(3.9)

We may surmise that the symbols (here E, T, and F) that are defined by at least one production rule are non-terminals. Conventionally, the first defined non-terminal (that is, E here) is our start symbol. Those symbols that are not defined (here +, *, (,), and id) are terminals.

The start symbol is important to us because it is from this symbol, using the production rules, that can generate strings in a language. For example, because we designate E to be the start symbol in the grammar above, we can record a sequence of applications of the production rules, starting from E to the sentence id + id * id as follows:

$$E \Rightarrow E + T$$
$$\Rightarrow T + T$$
$$\Rightarrow F + T$$
$$\Rightarrow \text{id} + T$$
$$\Rightarrow \text{id} + T * F$$
$$\Rightarrow \text{id} + F * F$$
$$\Rightarrow \text{id} + \text{id} * F$$
$$\Rightarrow \text{id} + \text{id} * \text{id}$$

We call this a *derivation*. When a string of symbols derives another string of symbols in one step, that is, by a single application of a production rule, we say that first string *directly derives* the second string. For example,

E directly derives $E + T$
$E + T$ directly derives $T + T$
$T + T$ directly derives $F + T$
and so on ...

When one string can be rewritten as another string, using zero or more production rules from the grammar, we say the first string *derives* the second string. For this *derives* relation, we usually use the symbol $\overset{*}{\Rightarrow}$. For example,

$E \overset{*}{\Rightarrow} E$ (in zero steps)
$E \overset{*}{\Rightarrow} \text{id} + F * F$
$T + T \overset{*}{\Rightarrow} \text{id} + \text{id} * \text{id}$

We say the *language* $L(G)$ that is described by a grammar G consists of all the strings (sentences) comprised of only terminal symbols that can be derived from the start symbol. A little more formally, we can express the language $L(G)$ for a grammar G with start symbol S and terminals T as

$$L(G) = \{w \mid S \overset{*}{\Rightarrow} w \ \text{and} \ w \in T^*\}.$$

(3.10)

For example, in our grammar above,

$E \overset{*}{\Rightarrow} \text{id} + \text{id} * \text{id}$
$E \overset{*}{\Rightarrow} \text{id}$
$E \overset{*}{\Rightarrow} (\text{id} + \text{id}) * \text{id}$

so, $L(G)$ includes each of

```
id + id * id
id
(id + id) * id
```

and infinitely more finite sentences.

We are interested in languages, those strings of terminals that can be derived from a grammar's start symbol. There are two kinds of derivation that will be important to us when we go about parsing these languages: left-most derivations and right-most derivations.

A *left-most derivation* is a derivation in which at each step, the next string is derived by applying a production rule for rewriting the *left-most* non-terminal. For example, we have already seen a left-most derivation of id + id * id:

$$
\begin{aligned}
\underline{E} &\Rightarrow \underline{E} + T \\
&\Rightarrow \underline{T} + T \\
&\Rightarrow \underline{F} + T \\
&\Rightarrow \text{id} + \underline{T} \\
&\Rightarrow \text{id} + \underline{T} * F \\
&\Rightarrow \text{id} + \underline{F} * F \\
&\Rightarrow \text{id} + \text{id} * \underline{F} \\
&\Rightarrow \text{id} + \text{id} * \text{id}
\end{aligned}
$$

Here we have underlined the left-most non-terminal in each string of symbols in the derivation to show that it is indeed a left-most derivation.

A *right-most derivation* is a derivation in which at each step, the next string is derived by applying a production rule for rewriting the *right-most* non-terminal. For example, the right-most derivation of id + id * id would go as follows:

$$
\begin{aligned}
\underline{E} &\Rightarrow E + \underline{T} \\
&\Rightarrow E + T * \underline{F} \\
&\Rightarrow E + \underline{T} * \text{id} \\
&\Rightarrow E + \underline{F} * \text{id} \\
&\Rightarrow \underline{E} + \text{id} * \text{id} \\
&\Rightarrow \underline{T} + \text{id} * \text{id} \\
&\Rightarrow \underline{F} + \text{id} * \text{id} \\
&\Rightarrow \text{id} + \text{id} * \text{id}
\end{aligned}
$$

We use the term *sentential form* to refer to any string of (terminal and non-terminal) symbols that can be derived from the start symbol. So, for example in the previous derivation,

$$
\begin{aligned}
&E \\
&E + T \\
&E + T * F \\
&E + T * \text{id} \\
&E + F * \text{id} \\
&E + \text{id} * \text{id} \\
&T + \text{id} * \text{id} \\
&F + \text{id} * \text{id} \\
&\text{id} + \text{id} * \text{id}
\end{aligned}
$$

are all sentential forms. Clearly, any sentential form consisting solely of terminal symbols is a sentence in the language.

An alternative representation of a derivation is the *parse tree*, a tree that illustrates the derivation and the structure of an input string (at the leaves) from a start symbol (at the root). For example, Figure 3.2 shows the parse tree for id + id * id.

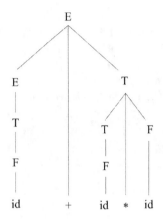

FIGURE 3.2 A parse tree for id + id * id.

Consider the *j*-- grammar in Appendix B. In this grammar, compilationUnit is the start symbol; from it one may derive syntactically legal *j*-- programs. Not every syntactically legal *j*-- program is in fact a bona fide *j*-- program. For example, the grammar permits such fragments as

```
5 * true;
```

The *j*-- grammar in fact describes a superset of the programs in the *j*-- language. To restrict this superset to the set of legal *j*-- programs, we need additional rules, such as type rules requiring, for example, that the operands to * must be numbers. Type rules are enforced during semantic analysis.

3.2.3 Ambiguous Grammars and Unambiguous Grammars

Given a grammar G, if there exists a sentence s in $L(G)$ for which there are more than one left-most derivations in G (or equivalently, either more than one right-most derivations, or more than one parse tree for s in G), we say that the sentence s is *ambiguous*. Moreover, if a grammar G describes at least one ambiguous sentence, the grammar G is *ambiguous (grammar)*. If there is no such sentence, that is, if every sentence is derivable by a unique left-most (or right-most) derivation (or has a unique parse tree), we say the grammar is *unambiguous*.

Example. Consider the grammar

$$E ::= E + E \mid E * E \mid (E) \mid \text{id} \tag{3.11}$$

and, consider the sentence id + id * id. One left-most derivation for this sentence is

$$
\begin{aligned}
\underline{E} &\Rightarrow \underline{E} + E \\
&\Rightarrow \text{id} + \underline{E} \\
&\Rightarrow \text{id} + \underline{E} * E \\
&\Rightarrow \text{id} + \text{id} * \underline{E} \\
&\Rightarrow \text{id} + \text{id} * \text{id}
\end{aligned}
$$

Another left-most derivation of the same sentence is

$$\underline{E} \Rightarrow \underline{E} * E$$
$$\Rightarrow \underline{E} + E * E$$
$$\Rightarrow \texttt{id} + \underline{E} * E$$
$$\Rightarrow \texttt{id} + \texttt{id} * \underline{E}$$
$$\Rightarrow \texttt{id} + \texttt{id} * \texttt{id}$$

Therefore, the grammar is ambiguous. It is also the case that the sentence has two right-most derivations in the grammar:

$$\underline{E} \Rightarrow E + \underline{E}$$
$$\Rightarrow E + \underline{E} * \underline{E}$$
$$\Rightarrow E + \underline{E} * \texttt{id}$$
$$\Rightarrow \underline{E} + \texttt{id} * \texttt{id}$$
$$\Rightarrow \texttt{id} + \texttt{id} * \texttt{id}$$

and

$$\underline{E} \Rightarrow E * \underline{E}$$
$$\Rightarrow \underline{E} * \texttt{id}$$
$$\Rightarrow E + \underline{E} * \texttt{id}$$
$$\Rightarrow \underline{E} + \texttt{id} * \texttt{id}$$
$$\Rightarrow \texttt{id} + \texttt{id} * \texttt{id}$$

These two right-most derivations for the same sentence also show the grammar is ambiguous. Finally, the two parse trees, illustrated in Figure 3.3, for the same sentence also demonstrate ambiguity.

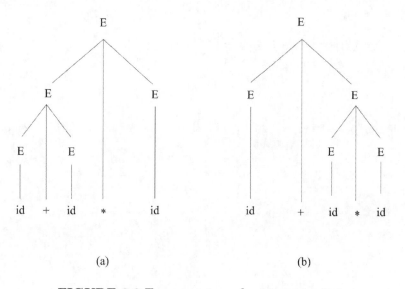

(a)   (b)

FIGURE 3.3 Two parse trees for `id + id * id`.

Clearly, we would rather have unambiguous grammars describe our programming language, because ambiguity in the parsing can lead to ambiguity in assigning semantics (meaning) to programs. For example, in `id + id * id`, which operation is applied first: addition or multiplication? From our math days, we would like the multiplication operator * to be more binding than the addition operator +. Notice that the grammar in (3.11) does not capture this precedence of operators in the way that the grammar in (3.9) does.

Example. As another example, consider the grammar describing conditional statements

$$S ::= \texttt{if } (E) \ S$$
$$\quad | \ \texttt{if } (E) \ S \texttt{ else } S \qquad\qquad (3.12)$$
$$\quad | \ \texttt{s}$$
$$E ::= \ \texttt{e}$$

Here, the token e represents an arbitrary expression and the s represents a (non-conditional) statement.

Consider the sentence

$$\texttt{if (e) if (e) s else s} \qquad\qquad (3.13)$$

If we look at this sentence carefully, we see that it nests one conditional statement within another. One might ask: To which if does the else belong? We cannot know; the sentence is ambiguous. More formally, there exist two left-most derivations for this sentence:

$$\underline{S} \Rightarrow \texttt{if } (\underline{E}) \ S \texttt{ else } S$$
$$\Rightarrow \texttt{if (e) } \underline{S} \texttt{ else } S$$
$$\Rightarrow \texttt{if (e) if } (\underline{E}) \ S \texttt{ else } S$$
$$\Rightarrow \texttt{if (e) if (e) } \underline{S} \texttt{ else } S$$
$$\Rightarrow \texttt{if (e) if (e) s else } \underline{S}$$
$$\Rightarrow \texttt{if (e) if (e) s else s}$$

and

$$\underline{S} \Rightarrow \texttt{if } (\underline{E}) \ S$$
$$\Rightarrow \texttt{if (e) } \underline{S}$$
$$\Rightarrow \texttt{if (e) if } (\underline{E}) \ S \texttt{ else } S$$
$$\Rightarrow \texttt{if (e) if (e) } \underline{S} \texttt{ else } S$$
$$\Rightarrow \texttt{if (e) if (e) s else } \underline{S}$$
$$\Rightarrow \texttt{if (e) if (e) s else s}$$

The two differing parse trees in Figure 3.4 for the same sentence (3.13) also demonstrate ambiguity.

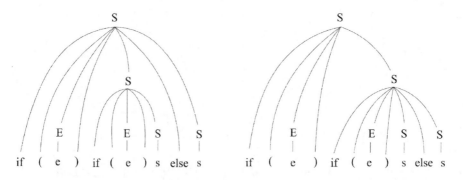

FIGURE 3.4 Two parse trees for if (e) if (e) s else s.

If unambiguous grammars are so important, why do we see so many languages having this ambiguous conditional statement? All of *j--*, Java, C, and C++ include precisely this same ambiguous conditional construct. One could easily modify the syntax of conditionals to remove the ambiguity. For example, the programming languages S-Algol and Algol-S define conditionals so:

$$S ::= \texttt{if } E \texttt{ do } S$$
$$| \texttt{ if } E \texttt{ then } S \texttt{ else } S \tag{3.14}$$
$$| \texttt{ s}$$
$$E ::= \texttt{e}$$

Here, the `do` indicates the simpler conditional (having no `else` clause) while the `then` indicates the presence of an `else` clause. A simple solution but it does change the language.

But programmers have become both accustomed to (and fond of) the ambiguous conditional. So both programmers and compiler writers have learned to live with it. Programming language reference manuals include extra explanations, such as *In the conditional statement, the* `else` *clause goes with the nearest preceding* `if`.

And compiler writers handle the rule as a special case in the compiler's parser, making sure that an `else` is grouped along with the closest preceding `if`.

The *j--* grammar (and the Java grammar) have another ambiguity, which is even more difficult. Consider the problem of parsing the expression

```
x.y.z.w
```

Clearly, `w` is a field; if it were a method expression, then that would be evident in the syntax:

```
x.y.z.w()
```

But, what about `x.y.z`? There are several possibilities depending on the types of the names `x`, `y`, and `z`.

- If `x` is the name of a local variable, it might refer to an object with a field `y`, referring to another object with a field `z`, referring to yet another object having our field `w`. In this case, the expression would be parsed as a cascade of field selection expressions.

- Alternatively, `x.y` might be the name of a package in which the class `z` is defined. In this case, we would parse the expression by first parsing `x.y.z` as a fully qualified class and then parse `w` as a static field selection of that class.

- Other possibilities, parsing various permutations of (possibly qualified) class names and field selection operations, also exist.

The parser cannot determine just how the expression `x.y.z` is parsed because types are not decided until after it has parsed the program and constructed its abstract syntax tree (AST). So the parser represents `x.y.z` in the AST by an `AmbiguousName` node. Later on, after type declarations have been processed, the compiler rewrites this node as the appropriate sub-tree.

These two examples of ambiguity in *j--* (and in Java) make the point that in compiling everyday programming languages, one must deal with messiness. One relies on formal techniques where practical, because formal techniques lead to more correct compilers. But once in awhile, formality fails and one must deal with special cases.

In general, there is no algorithm that can tell us whether or not an arbitrary grammar is ambiguous. That is, ambiguity is not decidable. But there are large classes of grammars (and so programming languages) that both can be shown to be decidable and for which efficient parsers can be automatically constructed. These classes of grammars are large enough to describe the many programming languages in use today. We look at these classes of grammars in the next sections.

Because of their recursive nature, parsing languages described by context-free grammars requires the use of a pushdown stack. In general, we can parse any language that is described by a context-free grammar but more often than not, the algorithms that do so involve backtracking.

But there are classes of grammars whose languages may be parsed *without* backtracking; the parsing is said to be *deterministic*. At each step of the parsing algorithm, the next step may be determined simply by looking at both the state of the pushdown stack and the incoming symbol (as the input sentence is scanned from left to right). These deterministic parsers may be roughly divided into two categories:

- Top-down parsing: where the parser begins at the start symbol and, step-by-step derives the input sentence and producing either a parse tree or (more likely) an abstract syntax tree (AST) from the root down to its leaves.

- Bottom-up parsing: where the parser begins at the input sentence and, scanning it from left to right, applies the production rules, replacing right-hand sides with left-hand-sides, in a step-by-step fashion for reducing the sentence to obtain the start symbol. Here the parse tree or abstract syntax tree is built from the bottom leaves up to the top root.

3.3   Top-Down Deterministic Parsing

There are two popular top-down deterministic parsing strategies: parsing by recursive descent and LL(1) parsing. Both parsers scan the input from left to right, looking at and scanning just one symbol at a time.

In both cases, the parser starts with the grammar's start symbol as an initial goal in the parsing process; that is, the goal is to scan the input sentence and parse the syntactic entity represented by the start symbol. The start symbol is then rewritten, using a BNF rule replacing the symbol with the right-hand side sequence of symbols.

For example, in parsing a *j--* program, our initial goal is to parse the start symbol, compilationUnit. The compilationUnit is defined by a single extended-BNF rule.

$$\text{compilationUnit} ::= [\texttt{package qualifiedIdentifier ;}] \qquad (3.15)$$
$$\{\texttt{import qualifiedIdentifier ;}\}$$
$$\{\text{typeDeclaration}\}\ \texttt{EOF}$$

So, the goal of parsing a compilationUnit in the input can be rewritten as a number of sub-goals:

1. If there is a **package** statement in the input sentence, then we parse that.

2. If there are **import** statements in the input, then we parse them.

3. If there are any type declarations, then we parse them.

4. Finally, we parse the terminating **EOF** token.

Parsing a token, like `package`, is simple enough. The token should be the first token in the as-yet unscanned input sentence. If it is, we simply scan it; otherwise we raise an error.

Parsing a non-terminal is treated as another parsing (sub-) goal. For example, in the `package` statement, once we have scanned the `package` token, we are left with parsing a qualifiedIdentifier. This is defined by yet another BNF rule

$$\text{qualifiedIdentifier} ::= \texttt{<identifier>} \{ \texttt{. <identifier>} \} \tag{3.16}$$

Here we scan an `<identifier>` (treated by the lexical scanner as a token). And, long as we see another period in the input, we scan that period and scan another `<identifier>`.

That we start at the start symbol and continually rewrite non-terminals using BNF rules until we eventually reach leaves (the tokens are the leaves) makes this a *top-down* parsing technique. Because we, at each step in parsing a non-terminal, replace a parsing goal with a sequence of sub-goals, we often call this a *goal-oriented* parsing technique.

How do we decide what next step to take? For example, how do we decide whether or not there are more `import` statements to parse? We decide by looking at the next unscanned input token. If it is an `import`, we have another `import` statement to parse; otherwise we go on to parsing type declarations. This kind of decision is more explicitly illustrated by the definition for the statement non-terminal (3.17).

$$
\begin{aligned}
\text{statement} ::= \;& \text{block} \\
| \;& \texttt{<identifier>} : \text{statement} \\
| \;& \texttt{if} \text{ parExpression statement } [\texttt{else} \text{ statement}] \\
| \;& \texttt{while} \text{ parExpression statement} \\
| \;& \texttt{return} \text{ [expression] ;} \\
| \;& \texttt{;} \\
| \;& \text{statementExpression ;}
\end{aligned}
\tag{3.17}
$$

When faced with the goal of parsing a statement, we have six alternatives to choose from, depending on the next unscanned input token:

1. If the next token is a {, we parse a block;

2. If the next token is an `if`, we parse an `if` statement;

3. If the next token is a `while`, we parse a `while` statement;

4. If the next token is a `return`, we parse a `return` statement;

5. If the next token is a semicolon, we parse an empty statement;

6. Otherwise (or based on the next token being one of a set of tokens, any one of which may begin an expression) we parse a statementExpression.

In this instance, the decision may be made looking at the next single unscanned input token; when this is the case, we say that our grammar is LL(1). In some cases, one must look ahead several tokens in the input to decide which alternative to take. In all cases, because we can predict which of several alternative right-hand sides of a BNF rule to apply, based on the next input token(s), we say this is a predictive parsing technique.

There are two principal top-down (goal-oriented, or predictive) parsing techniques available to us:

1. Parsing by recursive descent; and

2. LL(1) or LL(k) parsing.

3.3.1   Parsing by Recursive Descent

Parsing by recursive descent involves writing a method (or procedure) for parsing each non-terminal according to the production rules that define that non-terminal. Depending on the next unscanned input symbol, the method decides which rule to apply and then scans any terminals (tokens) in the right-hand side by calling upon the `Scanner`, and parses any non-terminals by recursively invoking the methods that parse them.

This is a strategy we use in parsing *j--* programs. We already saw the method `compilationUnit()` that parses a *j--* compilationUnit in Section 1.4.3.

As another example, consider the rule defining qualifiedIdentifier:

qualifiedIdentifier ::= <identifier> {. <identifier>}   (3.18)

As we saw above, parsing a qualifiedIdentifier such as `java.lang.Class` is straightforward:

1. One looks at the next incoming token and if it is an identifier, scans it. If it is not an identifier, then one raises an error.

2. Repeatedly, as long as the next incoming token is a period:

   a. One scans the period.
   b. One looks at the next incoming token and if it is an identifier, scans it. Otherwise, one raises an error.

The method for implementing this not only parses the input, but also constructs an AST node for recording the fully qualified name as a string. The method is quite simple but introduces two helper methods: `have()` and `mustBe()`, which we use frequently throughout the parser.

```
private TypeName qualifiedIdentifier() {
    int line = scanner.token().line();
    mustBe(IDENTIFIER);
    String qualifiedIdentifier = scanner.previousToken().image();
    while (have(DOT)) {
        mustBe(IDENTIFIER);
        qualifiedIdentifier += "."
            + scanner.previousToken().image();
    }
    return new TypeName(line, qualifiedIdentifier);
}
```

The method `have()` is a predicate. It looks at the next incoming token (supplied by the `Scanner`), and if that token matches its argument, then it scans the token in the input and returns `true`. Otherwise, it scans nothing and returns `false`.

The method `mustBe()` requires that the next incoming token match its argument. It looks at the next token and if it matches its argument, it scans that token in the input. Otherwise, it raises an error[3].

As another example, consider our syntax for statements.

statement ::= block   (3.19)
            | <identifier> : statement
            | if parExpression statement [else statement]
            | while parExpression statement
            | return [expression] ;
            | ;
            | statementExpression ;

---

[3] As we shall see below, `mustBe()` provides a certain amount of error recovery.

As we saw above, the problem parsing a sentence against the grammar is deciding which rule to apply when looking at the next unscanned token.

```
private JStatement statement() {
    int line = scanner.token().line();
    if (see(LCURLY)) {
        return block();
    } else if (have(IF)) {
        JExpression test = parExpression();
        JStatement consequent = statement();
        JStatement alternate = have(ELSE) ? statement() : null;
        return new JIfStatement(line, test, consequent, alternate);
    } else if (have(WHILE)) {
        JExpression test = parExpression();
        JStatement statement = statement();
        return new JWhileStatement(line, test, statement);
    } else if (have(RETURN)) {
        if (have(SEMI)) {
            return new JReturnStatement(line, null);
        } else {
            JExpression expr = expression();
            mustBe(SEMI);
            return new JReturnStatement(line, expr);
        }
    } else if (have(SEMI)) {
        return new JEmptyStatement(line);
    } else { // Must be a statementExpression
        JStatement statement = statementExpression();
        mustBe(SEMI);
        return statement;
    }
}
```

Notice the use of `see()` in looking to see if the next token is a left curly bracket {, the start of a block. Method `see()` is a predicate that simply looks at the incoming token to see if it is the token that we are looking for; in no case is anything scanned. The method `block()` scans the {. On the other hand, `have()` is used to look for (and scan, if it finds it) either an `if`, a `while`, a `return`, or a `;`. If none of these particular tokens are found, then we have neither a block, an if-statement, a while-statement, a return-statement, nor an empty statement; we assume the parser is looking at a statement expression. Given that so many tokens may begin a statementExpression, we treat that as a default; if the next token is not one that may start a statementExpression, then `statementExpression()` (or one of the methods to which it delegates the parse) will eventually detect the error. The important thing is that the error is detected before any additional tokens are scanned so its location is accurately located.

Lookahead

The parsing of statements works because we can determine which rule to follow in parsing the statement based only on the next unscanned symbol in the input source. Unfortunately, this is not always the case. Sometimes we must consider the next few symbols in deciding what to do. That is, we must look ahead in the input stream of tokens to decide which rule to apply. For example, consider the syntax for simple unary expression.

$$
\begin{aligned}
\text{simpleUnaryExpression} ::= \ & \text{! unaryExpression} \\
& | \ (\ \text{basicType} \ ) \ \text{unaryExpression} \ //\text{cast} \\
& | \ (\ \text{referenceType} \ ) \ \text{simpleUnaryExpression} \ // \ \text{cast} \\
& | \ \text{postfixExpression}
\end{aligned} \tag{3.20}
$$

For this, we need special machinery. Not only must we differentiate between the two kinds (basic type and reference type) of casts, but we must also distinguish a cast from a postfix expression that is a parenthesized expression, for example (x).

Consider the `Parser` code for parsing a simple unary expression:

```
private JExpression simpleUnaryExpression() {
    int line = scanner.token().line();
    if (have(LNOT)) {
        return new JLogicalNotOp(line, unaryExpression());
    } else if (seeCast()) {
        mustBe(LPAREN);
        boolean isBasicType = seeBasicType();
        Type type = type();
        mustBe(RPAREN);
        JExpression expr = isBasicType
            ? unaryExpression()
            : simpleUnaryExpression();
        return new JCastOp(line, type, expr);
    } else {
        return postfixExpression();
    }
}
```

Here we use the predicate `seeCast()` to distinguish casts from parenthesized expressions, and `seeBasicType()` to distinguish between casts to basic types from casts to reference types. Now, consider the two predicates.

First, the simpler `seeBasicType()`.

```
private boolean seeBasicType() {
    if (see(BOOLEAN) || see(CHAR) || see(INT)) {
        return true;
    } else {
        return false;
    }
}
```

The predicate simply looks at the next token to see whether or not it denotes a basic type. The method `simpleUnaryExpression()` can use this because it has factored out the opening parenthesis, which is common to both kinds of casts.

Now consider the more difficult `seeCast()`.

```
private boolean seeCast() {
    scanner.recordPosition();
    if (!have(LPAREN)) {
        scanner.returnToPosition();
        return false;
    }
    if (seeBasicType()) {
        scanner.returnToPosition();
        return true;
    }
    if (!see(IDENTIFIER)) {
        scanner.returnToPosition();
        return false;
    } else {
        scanner.next(); // Scan the IDENTIFIER
        // A qualified identifier is ok
        while (have(DOT)) {
            if (!have(IDENTIFIER)) {
                scanner.returnToPosition();
                return false;
            }
```

```
            }
        }
        while (have(LBRACK)) {
            if (!have(RBRACK)) {
                scanner.returnToPosition();
                return false;
            }
        }
        if (!have(RPAREN)) {
            scanner.returnToPosition();
            return false;
        }
        scanner.returnToPosition();
        return true;
    }
```

Here, seeCast() must look ahead in the token stream to consider a sequence of tokens in making its decision. But our lexical analyzer Scanner keeps track of only the single incoming symbol. For this reason, we define a second lexical analyzer LookaheadScanner, which encapsulates our Scanner but provides extra machinery that allows one to look ahead in the token stream. It includes a means of remembering tokens (and their images) that have been scanned, a method recordPosition() for marking a position in the token stream, and returnToPosition() for returning the lexical analyzer to that recorded position (that is, for backtracking). Of course, calls to the two methods may be nested, so that one predicate (for example, seeCast()) may make use of another (for example, seeBasicType()). Therefore, all of this information must be recorded on a pushdown stack.

Error Recovery

What happens when the parser detects an error? This will happen when mustBe() comes across a token that it is not expecting. The parser could simply report the error and quit. But we would rather have the parser report the error, and then continue parsing so that it might detect any additional syntax errors. This facility for continuing after an error is detected is called *error recovery*.

Error recovery can be difficult. The parser must not only detect syntax errors but it must sufficiently recover its state so as to continue parsing without introducing additional spurious error messages. Many parser generators[4] provide elaborate machinery for programming effective error recovery.

For the *j--* parser, we provide limited error recovery in the mustBe() method, which was proposed by [Turner, 1977]. First, consider the definitions for see() and have().

```
    private boolean see(TokenKind sought) {
        return (sought == scanner.token().kind());
    }

    private boolean have(TokenKind sought) {
        if (see(sought)) {
            scanner.next();
            return true;
        } else {
            return false;
        }
    }
```

---

[4]A *parser generator* is a program that will take a context-free grammar (of a specified class) as input and produce a parser as output. We discuss the generation of parsers in subsequent sections, and we discuss a particular parser generator, JavaCC, in Section 3.5.

These are defined as one would expect. Method `mustBe()`, defined as follows, makes use of a boolean flag, `isRecovered`, which is `true` if either no error has been detected or if the parser has recovered from a previous syntax error. It takes on the value `false` when it is in a state in which it has not yet recovered from a syntax error.

```
boolean isRecovered = true;

private void mustBe(TokenKind sought) {
    if (scanner.token().kind() == sought) {
        scanner.next();
        isRecovered = true;
    } else if (isRecovered) {
        isRecovered = false;
        reportParserError("%s found where %s sought", scanner
            .token().image(), sought.image());
    } else {
        // Do not report the (possibly spurious) error,
        // but rather attempt to recover by forcing a match.
        while (!see(sought) && !see(EOF)) {
            scanner.next();
        }
        if (see(sought)) {
            scanner.next();
            isRecovered = true;
        }
    }
}
```

When `mustBe()` first comes across an input token that it is not looking for (it is in the recovered state), it reports an error and goes into an unrecovered state. If, in a subsequent use of `mustBe()`, it finds another syntax error, it does not report the error, but rather it attempts to get back into a recovered state by repeatedly scanning tokens until it comes across the one it is seeking. If it succeeds in finding that token, it goes back into a recovered state. It may not succeed but instead scan to the end of the file; in this case, parsing stops. Admittedly, this is a very naïve error recovery scheme, but it works amazingly well for its simplicity[5].

3.3.2   LL(1) Parsing

The recursive invocation of methods in the recursive descent parsing technique depends on the underlying program stack for keeping track of the recursive calls. The LL(1) parsing technique makes this stack explicit. The first L in its name indicates a left-to-right scan of the input token stream; the second L signifies that it produces a left parse, which is a left-most derivation. The (1) indicates we just look ahead at the next 1 symbol in the input to make a decision.

Like recursive descent, the LL(1) technique is top-down, goal oriented, and predictive.

LL(1) Parsing Algorithm

An LL(1) parser works in typical top-down fashion. At the start, the start symbol is pushed onto the stack, as the initial goal. Depending on the first token in the input stream of tokens to be parsed, the start symbol is replaced on the stack by a sequence of symbols from the right-hand side of a rule defining the start symbol. Parsing continues by parsing each symbol as it is removed from the top of the stack:

---

[5]It reminds us of the story of the dancing dog: one does not ask how well the dog dances but is amazed that he dances at all.

- If the top symbol is a terminal, it scans that terminal from the input. If the next incoming token does not match the terminal on the stack, then an error is raised.

- If the top symbol is a non-terminal, the parser looks at the next incoming token in the input stream to decide what production rule to apply to expand the non-terminal taken from the top of the stack.

Every LL(1) parser shares the same basic parsing algorithm, which is table-driven. A unique parsing table is produced for each grammar. This table has a row for each non-terminal that can appear on the stack, and a column for each terminal token, including special terminator # to mark the end of the string. The parser consults this table, given the non-terminal on top of the stack and the next incoming token, to determine which BNF rule to use in rewriting the non-terminal. It is important that the grammar be such that one may always unambiguously decide what BNF rule to apply; equivalently, no table entry may contain more than one rule.

For example, consider the following grammar:

$$
\begin{aligned}
&1.\ E\ ::= T\ E' \\
&2.\ E'\ ::= +\ \mathrm{T}\ E' \\
&3.\ E'\ ::= \epsilon \\
&4.\ T\ ::= F\ T' \\
&5.\ T'\ ::= *\ F\ T' \\
&6.\ T'\ ::= \epsilon \\
&7.\ F\ ::= (E) \\
&8.\ F\ ::= \mathtt{id}
\end{aligned}
\qquad (3.21)
$$

This grammar describes the same language as that described by the grammar (3.8) in Section 3.2.2. Another grammar that, using the extensions described in Section 3.2.1, describes the same language is

$$
\begin{aligned}
&E ::= T\ \{+\ T\} \\
&T ::= F\ \{*\ F\} \\
&F ::= (E)\ |\ \mathtt{id}
\end{aligned}
\qquad (3.22)
$$

Such a grammar lends itself to parsing by recursive descent. But, the advantage of the grammar in (3.21) is that we can define an LL(1) parsing table for it. The parsing table for the grammar in (3.21) is given in Figure 3.5.

	+	*	(	)	id	#
E			1		1	
E'	2			3		3
T			4		4	
T'	6	5		6		6
F			7		8	

FIGURE 3.5 LL(1) parsing table for the grammar in Example 3.21.

The numbers in the table's entries refer to the numbers assigned to BNF rules in the example grammar (3.21). An empty entry indicates an error: when the parser has the given non-terminal on top of the stack and the given token as the next input symbol, the parser is in an erroneous state. The LL(1) parsing algorithm is parameterized by this table. As we mentioned earlier, it makes use of an explicit pushdown stack.

Algorithm 3.1 LL(1) Parsing Algorithm

Input: LL(1) parsing table *table*, production rules *rules*, and a sentence w, where w is a string of terminals followed by a terminator #

Output: a left-parse, which is a left-most derivation for w

Stack *stk* initially contains the terminator # and the start symbol S, with S on top

Symbol *sym* is the first symbol in the sentence w

while true **do**

  Symbol *top* ← *stk*.pop()

  if *top* = *sym* = # **then**

    Halt successfully

  else if *top* is a terminal symbol **then**

    if *top* = *sym* **then**

      Advance *sym* to be the next symbol in w

    else

      Halt with an error: a *sym* found where a *top* was expected

    end if

  else if top is a non-terminal Y **then**

    index ← *table*[Y, *sym*]

    if *index* ≠ *err* **then**

      rule ← *rules*[*index*]

      Say rule is $Y ::= X_1 X_2 \ldots X_n$; push $X_n, \ldots, X_2, X_1$ onto the stack *stk*, with X_1 on top

    end if

  else

    Halt with an error

  end if

end while

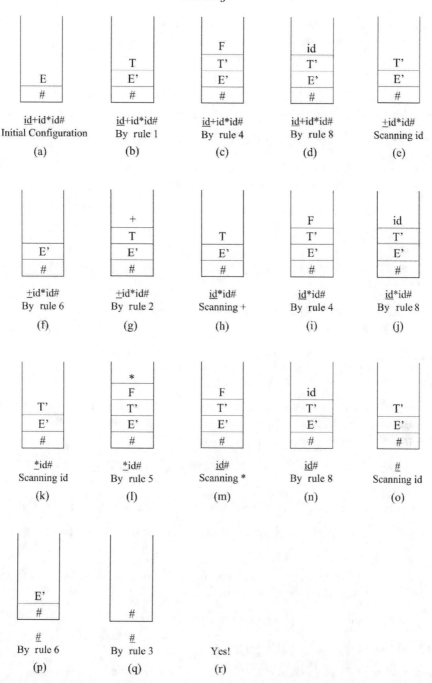

FIGURE 3.6 The steps in parsing `id + id * id` against the LL(1) parsing table in Figure 3.5.

For example, consider parsing the input sentence, `id + id * id #`. The parser would go through the states illustrated in Figure 3.6.

a. Initially, the parser begins with a `#` and E on top of its stack. E represents the goal, the entity that we wish to parse.

b. As an `id` is the incoming token, table[E, `id`] = 1 tells us to apply rule 1, $E ::= TE'$, and replace the E on the stack with the right-hand side, T and E', with T on top. Basically, we plan to accomplish the goal of parsing an E by accomplishing the two sub- goals of parsing a T and then an E'.

c. Seeing an `id` for the incoming token, we apply rule 4, $T ::= FT'$, and replace the T on the stack with the right-hand side F and T', with F on top.

d. We apply rule 8, $F ::= $ `id`, replacing the top goal of F with an `id`.

e. The goal of parsing the `id` on top of the stack is trivially satisfied: we pop the `id` from the stack and scan the `id` in the input; the next incoming symbol is now a +.

f. Now we have the goal of parsing a T' when looking at the input, +. We apply rule 6, T' $::= \epsilon$, replacing the T' on the stack with the empty string (that is, nothing at all). Just the goal E' remains on the stack.

g. Now the goal E', when looking at the input token +, is replaced using rule 2 by +, T and E', with the + on top.

h. The + is trivially parsed; it is popped from the stack and scanned from the input. The next incoming token is `id`.

i. Applying rule 4, $T ::= FT'$, we replace the T on the stack by F and T', with F on top.

j. Applying rule 8, we replace the F on the stack with `id`.

k. The goal of parsing the `id` on top of the stack is trivially satisfied: we pop the `id` from the stack and scan the `id` in the input; the goal on the stack is now a T' and the next incoming symbol is now a *.

l. We apply rule 5 to replace the T' on the stack with a *, F, and another T'. The * is on top.

m. The * on top is easily parsed: it is popped from the stack and scanned from the input.

n. The goal F, atop the stack, when looking at an incoming `id`, is replaced by an id using rule 8.

o. The `id` is popped from the stack and scanned from the input, leaving a T' on the stack and the terminator # as the incoming token.

p. We apply rule 6, replacing T' by the empty string.

q. We apply rule 3, replacing E' by the empty string.

r. There is a # on top of the stack and a # as input. We are done!

An alternative to Figure 3.6 for illustrating the states that the parser goes through is as follows:

Stack	Input	Output
#E	<u>id</u>+id*id#	
#$E'T$	<u>id</u>+id*id#	1
#$E'T'F$	<u>id</u>+id*id#	4
#$E'T'$id	<u>id</u>+id*id#	8
#$E'T'$	<u>+</u>id*id#	
#E'	<u>+</u>id*id#	6
#$E'T$+	<u>+</u>id*id#	2
#$E'T$	<u>id</u> *id#	
#$E'T'F$	<u>id</u>*id#	4
#$E'T'$id	<u>id</u> *id#	8
#$E'T'$	<u>*</u>id#	
#$E'T'F$*	<u>*</u>id#	5
#$E'T'F$	<u>id</u>#	
#$E'T'$id	<u>id</u>#	8
#$E'T'$	<u>#</u>	
#E'	<u>#</u>	6
#	<u>#</u>	3

We are left with the question: How do we construct the parsing table in Figure 3.5? Consider what the entries in the table tell us:

table[Y, a] = i, where i is the number of the rule $Y ::= X_1 X_2 \ldots X_n$

says that if there is the goal Y on the stack and the next unscanned symbol is a, then we can rewrite Y on the stack as the sequence of sub-goals $X_1 X_2 \ldots X_n$. So, the question becomes: When do we replace Y with $X_1 X_2 \ldots X_n$ as opposed to something else? If we consider the last two rules of our grammar (3.21),

$\quad$ 7. $F ::= (E)$
$\quad$ 8. $F ::= $ id

and we have the non-terminal F on top of our parsing stack, the choice is simple. If the next unscanned symbol is an open parenthesis (, we apply rule 7; and if it is an id, we apply rule 8. That is,

$\quad$ table[F, (] = 7
$\quad$ table[F, id] = 8

The problem becomes slightly more complicated when the right-hand side of the rule either starts with a non-terminal or is simply ϵ.

In general, assuming both α and β are (possibly empty) strings of terminals and non-terminals, table[Y, a] = i, where i is the number of the rule $Y ::= X_1 X_2 \ldots X_n$, if either

1. $X_1 X_2 \ldots X_n \overset{*}{\Rightarrow} a\alpha$, or

2. $X_1 X_2 \ldots X_n \overset{*}{\Rightarrow} \epsilon$, and there is a derivation S# $\overset{*}{\Rightarrow} \alpha Y a \beta$, that is, a can follow Y in a derivation.

For this we need two helper functions: *first* and *follow*.

First and Follow

We define, first$(X_1 X_2 \ldots X_n) = \{a | X_1 X_2 \ldots X_n \overset{*}{\Rightarrow} a\alpha, a \in T\}$, that is, the set of all terminals that can start strings derivable from $X_1 X_2 \ldots X_n$. Also, if $X_1 X_2 \ldots X_n \overset{*}{\Rightarrow} \epsilon$, then we say that first$(X_1 X_2 \ldots X_n)$ includes ϵ.

We define two algorithms for computing first: one for a single symbol and the other for a sequence of symbols. The two algorithms are both mutually recursive and make use of *memoization*, that is, what was previously computed for each set is remembered upon each iteration.

Algorithm 3.2 Compute first(X) for all symbols X in a Grammar G

Input: a context-free grammar $G = (N, T, S, P)$
Output: first(X) for all symbols X in G
  for each terminal x **do**
    first$(x) \leftarrow \{x\}$
  end for
  for each non-terminal X **do**
    first$(X) \leftarrow \{\}$
  end for
  if $X ::= \epsilon \in P$ **then**
    add ϵ to first(X)
  end if
  repeat
    for each $Y ::= X_1 X_2 \ldots X_n \in P$ **do**
      add all symbols from first$(X_1 X_2 \ldots X_n)$ to first(Y)
    end for
  until no new symbols are added to any set

Notice that the computation of first in Algorithm 3.2 might take account of the entirety of every right-hand side of each rule defining the symbol we are interested in.

Algorithm 3.3 Compute first$(X_1 X_2 \ldots X_n)$ for a Grammar G

Input: a context-free grammar $G = (N, T, S, P)$ and a sequence $X_1 X_2 \ldots X_n$
Output: first$(X_1 X_2 \ldots X_n)$
  Set $S \leftarrow$ first(X_1)
  $i \leftarrow 2$
  while $\epsilon \in S$ and $i \leq n$ **do**
    $S \leftarrow S - \epsilon$
    Add first(X_i) to S
    $i \leftarrow i + 1$
  end while
  return S

Algorithm 3.3 says that to compute first for a sequence of symbols $X_1 X_2 \ldots X_n$, we start with first(X_1). If $X_1 \overset{*}{\Rightarrow} \epsilon$, then we must also include first(X_2). If $X_2 \overset{*}{\Rightarrow} \epsilon$, then we must also include first(X_3). And so on.

Example. Consider our example grammar in (3.21). The computation of first for the terminals, by step 1 of Algorithm 3.2 is trivial:

$$\begin{aligned}
\text{first}(\texttt{+}) &= \{\texttt{+}\} \\
\text{first}(\texttt{*}) &= \{\texttt{*}\} \\
\text{first}(\texttt{(}) &= \{\texttt{(}\} \\
\text{first}(\texttt{)}) &= \{\texttt{)}\} \\
\text{first}(\texttt{id}) &= \{\texttt{id}\}
\end{aligned}$$

Step 2 of Algorithm 3.2 gives

$$\begin{aligned}
\text{first}(E) &= \{\} \\
\text{first}(E') &= \{\} \\
\text{first}(T) &= \{\} \\
\text{first}(T') &= \{\} \\
\text{first}(F) &= \{\}
\end{aligned}$$

Step 3 of Algorithm 3.2 yields

$$\begin{aligned}
\text{first}(E') &= \{\epsilon\} \\
\text{first}(T') &= \{\epsilon\}
\end{aligned}$$

Step 4 of Algorithm 3.2 is repeatedly executed until no further symbols are added. The first round yields

$$\begin{aligned}
\text{first}(E) &= \{\} \\
\text{first}(E') &= \{\texttt{+}, \epsilon\} \\
\text{first}(T) &= \{\} \\
\text{first}(T') &= \{\texttt{*}, \epsilon\} \\
\text{first}(F) &= \{\texttt{(}, \texttt{id}\}
\end{aligned}$$

The second round of step 4 yields

$$\begin{aligned}
\text{first}(E) &= \{\} \\
\text{first}(E') &= \{\texttt{+}, \epsilon\} \\
\text{first}(T) &= \{\texttt{(}, \texttt{id}\} \\
\text{first}(T') &= \{\texttt{*}, \epsilon\} \\
\text{first}(F) &= \{\texttt{(}, \texttt{id}\}
\end{aligned}$$

The third round of step 4 yields

$$\begin{aligned}
\text{first}(E) &= \{\texttt{(}, \texttt{id}\} \\
\text{first}(E') &= \{\texttt{+}, \epsilon\} \\
\text{first}(T) &= \{\texttt{(}, \texttt{id}\} \\
\text{first}(T') &= \{\texttt{*}, \epsilon\} \\
\text{first}(F) &= \{\texttt{(}, \texttt{id}\}
\end{aligned} \tag{3.23}$$

The fourth round of step 4 adds no symbols to any set, leaving us with (3.23).

We are left with the question as to when is a rule, $X ::= \epsilon$ applicable? For this we need the notion of follow.

We define follow$(X) = \{a | S \overset{*}{\Rightarrow} wX\alpha \text{ and } \alpha \overset{*}{\Rightarrow} a \ldots\}$, that is, all terminal symbols that start terminal strings derivable from what can follow X in a derivation. Another definition is as follows.

1. Follow(S) contains **#**, that is, the terminator follows the start symbol.

2. If there is a rule $Y ::= \alpha X \beta$ in P, follow(X) contains first(β) $- \{\epsilon\}$.

3. If there is a rule $Y ::= \alpha X \beta$ in P and either $\beta = \epsilon$ or first(β) contains ϵ, follow(X) contains follow(Y).

This definition suggests a straightforward algorithm.

Algorithm 3.4 Compute follow(X) for all non-terminals X in a Grammar G

Input: a context-free grammar $G = (N, T, S, P)$
Output: follow(X) for all non-terminals X in G
  follow(S) $\leftarrow \{$**#**$\}$
  for each non-terminal $X \in S$ **do**
    follow(X) $\leftarrow \{\}$
  end for
  repeat
    for each rule $Y ::= X_1 X_2 \ldots X_n \in P$ **do**
      for each non-terminal X_i **do**
        Add first($X_{i+1} X_{i+2} \ldots X_n$) $- \{\epsilon\}$ to follow(X_i), and if X_i is the right-most symbol
        or first($X_{i+1} X_{i+2} \ldots X_n$) contains ϵ, add follow(Y) to follow(X_i)
      end for
    end for
  until no new symbols are added to any set

Example. Again, consider our example grammar (3.21). By steps 1 and 2 of Algorithm 3.4,

$$
\begin{aligned}
\text{follow}(E) &= \{\text{\#}\} \\
\text{follow}(E') &= \{\} \\
\text{follow}(T) &= \{\} \\
\text{follow}(T') &= \{\} \\
\text{follow}(F) &= \{\}
\end{aligned}
$$

From rule 1, $E ::= T E'$, follow(T) contains first(E')$-\{\epsilon\} = \{$**+**$\}$, and because first(E') contains ϵ, it also contains follow(E). Also, follow(E') contains follow(E). So, round 1 of step 3 yields,

$$
\begin{aligned}
\text{follow}(E') &= \{\text{\#}\} \\
\text{follow}(T) &= \{\text{+}, \text{\#}\}
\end{aligned}
$$

We get nothing additional from rule 2, $E' ::= \text{+} T E'$. follow(T) contains first(E')$-\{\epsilon\}$, but we saw that in rule 1. Also, follow(E') contains follow(E'), but that says nothing.

We get nothing additional from rule 3, $E' ::= \epsilon$.

From rule 4, $T ::= F T'$, follow(F) contains first(T') - $\{\}$ = $\{$*****$\}$, and because first(T') contains ϵ, it also contains follow(T). Also, follow(T') contains follow(T). So, we have

$$
\begin{aligned}
\text{follow}(T') &= \{\text{+}, \text{\#}\} \\
\text{follow}(F) &= \{\text{+}, \text{*}, \text{\#}\}
\end{aligned}
$$

Rules 5 and 6 give us nothing new (for the same reasons rules 2 and 3 did not).

Rule 7, $F ::= (E)$, adds) to follow(E), so

$$\text{follow}(E) = \{), \#\}$$

Rule 8 gives us nothing.
Summarizing round 1 of step 3, gives

$$\begin{aligned}
\text{follow}(E) &= \{), \#\} \\
\text{follow}(E') &= \{\#\} \\
\text{follow}(T) &= \{+, \#\} \\
\text{follow}(T') &= \{+, \#\} \\
\text{follow}(F) &= \{+, *, \#\}
\end{aligned}$$

Now, in round 2 of step 3, the) that was added to follow(E) trickles down into the other follow sets:

From rule 1, $E ::= TE'$, because first(E') contains ϵ, follow(T) contains follow(E). Also, follow(E') contains follow(E). So, we have

$$\begin{aligned}
\text{follow}(E') &= \{), \#\} \\
\text{follow}(T) &= \{+,), \#\}
\end{aligned}$$

From rule 4, $T ::= FT'$, because first(T') contains ϵ, follow(F) contains follow(T). Also, follow(T') contains follow(T). So, we have

$$\begin{aligned}
\text{follow}(T') &= \{+,), \#\} \\
\text{follow}(F) &= \{+, *,), \#\}
\end{aligned}$$

So round 2 produces

$$\begin{aligned}
\text{follow}(E) &= \{), \#\} \\
\text{follow}(E') &= \{), \#\} \\
\text{follow}(T) &= \{+,), \#\} \\
\text{follow}(T') &= \{+,), \#\} \\
\text{follow}(F) &= \{+, *,), \#\}
\end{aligned} \qquad (3.24)$$

Round 3 of step 3 adds no new symbols to any set, so we are left with (3.24).

Constructing the LL(1) Parse Table

Recall, assuming both α and β are (possibly empty) strings of terminals and non-terminals, table$[Y, a] = i$, where i is the number of the rule $Y ::= X_1 X_2 \ldots X_n$ if either

1. $X_1 X_2 \ldots X_n \overset{*}{\Rightarrow} a\alpha$, or

2. $X_1 X_2 \ldots X_n \overset{*}{\Rightarrow} \epsilon$, and there is a derivation $S\# \overset{*}{\Rightarrow} \alpha Y a\beta$, that is, a can follow Y in a derivation.

This definition, together with the functions, first and follow, suggests Algorithm 3.5.

Algorithm 3.5 Construct an LL(1) Parse Table for a Grammar $G = (N, T, S, P)$

Input: a context-free grammar $G = (N, T, S, P)$
Output: LL(1) Parse Table for G
  for each non-terminal $Y \in G$ **do**
    for each rule $Y ::= X_1 X_2 \ldots X_n \in P$ with number i **do**
      for each terminal $a \in \text{first}(X_1 X_2 \ldots X_n) - \{\epsilon\}$ **do**
        $\text{table}[Y, a] \leftarrow i$
        if $\text{first}(X_1 X_2 \ldots X_n)$ contains ϵ **then**
          for each terminal a (or **#**) in $\text{follow}(Y)$ **do**
            $\text{table}[Y, a] \leftarrow i$
          end for
        end if
      end for
    end for
  end for

Example. Let as construct the parse table for (3.21). For the non-terminal E, we consider rule 1: $E ::= T E'$. $\text{first}(T E') = \{(, \text{id}\}$. So,

$$\text{table}[E, (] = 1$$
$$\text{table}[E, \text{id}] = 1$$

Because $\text{first}(T E')$ does not contain ϵ, we need not consider $\text{follow}(E)$.
For the non-terminal E', we first consider rule 2: $E' ::= + T E'$; $\text{first}(+ T E') = \{+\}$ so

$$\text{table}[E', +] = 2$$

Rule 3: $E' ::= \epsilon$ is applicable for symbols in $\text{follow}(E') = \{), \#\}$, so

$$\text{table}[E',)] = 3$$
$$\text{table}[E', \#] = 3$$

For the non-terminal T, we consider rule 4: $T ::= F T'$. $\text{first}(F T') = \{(, \text{id}\}$, so

$$\text{table}[T, (] = 4$$
$$\text{table}[T, \text{id}] = 4$$

Because $\text{first}(F T')$ does not contain ϵ, we need not consider $\text{follow}(T)$.
For non-terminal T', we first consider rule: 5: $T' ::= * F T'$; $\text{first}(* F T')$, so

$$\text{table}[T', *] = 5$$

Rule 6: $T' ::= \epsilon$ is applicable for symbols in $\text{follow}(T') = \{+,), \#\}$, so

$$\text{table}[T', +] = 6$$
$$\text{table}[T',)] = 6$$
$$\text{table}[T', \#] = 6$$

For the non-terminal F, we have two rules. First, given rule 7: $F ::= (E)$, and that first$((E)) = \{(\}$,

$$\text{table}[F, (] = 7$$

Second, given rule 8: $F ::= \text{id}$, and since (obviously) first$(\text{id}) = \{\text{id}\}$,

$$\text{table}[F, \text{id}] = 8$$

LL(1) and LL(k) Grammars

We say a grammar is LL(1) if the parsing table produced by Algorithm 3.5 produces no conflicts, that is, no entries having more than one rule. If there were more than one rule, then the parser would no longer be deterministic.

Furthermore, if a grammar is shown to be LL(1) then it is unambiguous. This is easy to show. An ambiguous grammar would lead to two left-most derivations that differ in at least one step, meaning at least two rules are applicable at that step. So an ambiguous grammar cannot be LL(1).

It is possible for a grammar not to be LL(1) but LL(k) for some $k > 1$. That is, the parser should be able to determine its moves looking k symbols ahead. In principle, this would mean a table having columns for each combination of k symbols. But this would lead to very large tables; indeed, the table size grows exponentially with k and would be unwieldy even for $k = 2$. On the other hand, an LL(1) parser generator based on the table construction in Algorithm 3.5 might allow one to specify a k-symbol lookahead for specific non-terminals or rules. These special cases can be handled specially by the parser and so need not lead to overly large (and, most likely sparse) tables. The JavaCC parser generator, which we discuss in Section 3.5, makes use of this focused k-symbol lookahead strategy.

Removing Left Recursion and Left-Factoring Grammars

Not all context-free grammars are LL(1). But for many that are not, one may define equivalent grammars (that is, grammars describing the same language) that are LL(1).

Left Recursion

One class of grammar that is not LL(1) is a grammar having a rule with left recursion, for example *direct, left recursion,*

$$
\begin{aligned}
&\text{Y} ::= \text{Y } \alpha \\
&\text{Y} ::= \beta
\end{aligned}
\tag{3.25}
$$

Clearly, a grammar having these two rules is not LL(1), because, by definition, first$(\text{Y}\alpha)$ must include first(β) making it impossible to discern which rule to apply for expanding Y. But introducing an extra non-terminal, an extra rule, and replacing the left recursion with right recursion easily removes the direct left recursion:

$$
\begin{aligned}
&\text{Y } ::= \beta \text{ Y}' \\
&\text{Y}' ::= \alpha \text{ Y}' \\
&\text{Y}' ::= \epsilon
\end{aligned}
\tag{3.26}
$$

Example. Such grammars are not unusual. For example, the first context-free grammar we saw (3.8) describes the same language as does the (LL(1)) grammar (3.21). We repeat this grammar as (3.27).

$$
\begin{aligned}
E &::= E + T \\
E &::= T \\
T &::= T * \mathrm{F} \\
T &::= F \\
F &::= (E) \\
F &::= \mathtt{id}
\end{aligned}
\tag{3.27}
$$

The left recursion captures the left-associative nature of the operators + and *. But because the grammar has left-recursive rules, it is not LL(1). We may apply the left-recursion removal rule (3.26) to this grammar.

First, applying the rule to E to produce

$$
\begin{aligned}
E &::= T\ E' \\
E' &::= +\ T\ E' \\
E' &::= \epsilon
\end{aligned}
$$

Applying the rule to T yields

$$
\begin{aligned}
T &::= F\ T' \\
T' &::= *\ F\ T' \\
T' &::= \epsilon
\end{aligned}
$$

Giving us the LL(1) grammar

$$
\begin{aligned}
E &::= T\ E' \\
E' &::= +\ T\ E' \\
E' &::= \epsilon \\
T &::= F\ T' \\
T' &::= *\ F\ T' \\
T' &::= \epsilon \\
F &::= (E) \\
F &::= \mathtt{id}
\end{aligned}
\tag{3.28}
$$

Where have we seen this grammar before?

Much less common, particularly in grammars describing programming languages, is *indirect left recursion*. Algorithm 3.6 deals with these rare cases.

Algorithm 3.6 Left Recursion Removal for a Grammar $G = (N, T, S, P)$

Input: a context-free grammar $G = (N, T, S, P)$
Output: G with left recursion eliminated
  Arbitrarily enumerate the non-terminals of $G : X_1, X_2, \ldots, X_n$
  for $i := 1$ to n **do**
    for $j := 1$ to $i - 1$ **do**
      Replace each rule in P of the form $X_i ::= X_j \alpha$ by the rules $X_i ::= \beta_1 \alpha | \beta_2 \alpha | \ldots | \beta_k \alpha$
      where $X_j ::= \beta_1 | \beta_2 | \ldots | \beta_k$ are the current rules defining X_i
      Eliminate any immediate left recursion using (3.25)
    end for
  end for

Example. Consider the following grammar.

$$S ::= A\mathsf{a} \mid \mathsf{b}$$
$$A ::= S\mathsf{c} \mid \mathsf{d}$$

In step 1, we can enumerate the non-terminals using subscripts to record the numbering: S_1 and A_2. This gives us a new set of rules:

$$S_1 ::= A_2\mathsf{a} \mid \mathsf{b}$$
$$A_2 ::= S_1\mathsf{c} \mid \mathsf{d}$$

In the first iteration of step 2 ($i = 1$), no rules apply. In the second iteration ($i = 1, j = 2$), the rule

$$A_2 ::= S_1\mathsf{c}$$

applies. We replace it with two rules, expanding S_1, to yield

$$S_1 ::= A_2\mathsf{a} \mid \mathsf{b}$$
$$A_2 ::= A_2\mathsf{ac} \mid \mathsf{bc} \mid \mathsf{d}$$

We then use the transformation (3.26) to produce the grammar

$$S_1 \;::= A_2\mathsf{a} \mid \mathsf{b}$$
$$A_2 \;::= \mathsf{bc}A'_2 \mid \mathsf{d}A'_2$$
$$A'_2 ::= \mathsf{ac}A'_2$$
$$A'_2 ::= \epsilon$$

Or, removing the subscripts,

$$S \;::= A\mathsf{a} \mid \mathsf{b}$$
$$A \;::= \mathsf{bc}A' \mid \mathsf{d}A'$$
$$A' ::= \mathsf{ac}A'$$
$$A' ::= \epsilon$$

Left factoring

Another common property of grammars that violates the LL(1) property is when two or more rules defining a non-terminal share a common prefix:

$$Y ::= \alpha\,\beta$$
$$Y ::= \alpha\,\gamma$$

The common α violates the LL(1) property. But, as long as first(β) and first(γ) are disjoint, this is easily solved by introducing a new non-terminal:

$$Y \ ::= \alpha Y'$$
$$Y' ::= \beta \tag{3.29}$$
$$Y' ::= \gamma$$

Example. Reconsider (3.14).

$$S ::= \text{if } E \text{ do } S$$
$$\quad | \text{ if } E \text{ then } S \text{ else } S$$
$$\quad | \text{ s}$$
$$E ::= \text{ e}$$

Following the rewriting rule (3.29), we can reformulate the grammar as

$$S \ ::= \text{if } E \ S'$$
$$\quad | \text{ s}$$
$$S' ::= \text{do } S$$
$$\quad | \text{ then } S \text{ else } S$$
$$E \ ::= \text{ e}$$

3.4 Bottom-Up Deterministic Parsing

In bottom-up parsing, one begins with the input sentence and scanning it from left-to-right, recognizes sub-trees at the leaves and builds a complete parse tree from the leaves up to the start symbol at the root.

3.4.1 Shift-Reduce Parsing Algorithm

For example, consider our old friend, the grammar (3.8) repeated here as (3.30):

1. $E ::= E + T$
2. $E \ ::= T$
3. $T \ ::= T * F$ (3.30)
4. $T \ ::= F$
5. $F ::= (E)$
6. $F ::= \text{id}$

And say we want to parse the input string, `id + id * id`. We would start off with the initial configuration:

Stack	Input	Action
#	id+id*id#	

At the start, the terminator is on the stack, and the input consists of the entire input sentence followed by the terminator. The first action is to *shift* the first unscanned input symbol (it is underlined) onto the stack.

Stack	Input	Action
#	<u>id</u>+id*id#	shift
#id	<u>+</u>id*id#	

From this configuration, the next action is to *reduce* the id on top of the stack to an F using rule 6.

Stack	Input	Action
#	<u>id</u>+id*id#	shift
#id	<u>+</u>id*id#	reduce (6)
#F	<u>+</u>id*id#	

From this configuration, the next two actions involve reducing the F to a T (by rule 4), and then to an E (by rule 2).

Stack	Input	Action
#	<u>id</u>+id*id#	shift
#id	<u>+</u>id*id#	reduce (6)
#F	<u>+</u>id*id#	reduce (4)
#T	<u>+</u>id*id#	reduce (2)
#E	<u>+</u>id*id#	

The parser continues in this fashion, by a sequence of shifts and reductions, until we reach a configuration where #E is on the stack (E on top) and the sole unscanned symbol in the input is the terminator #. At this point, we have reduced the entire input string to the grammar's start symbol E, so we can say the input is *accepted*.

Stack	Input	Action
#	<u>id</u>+id*id#	shift
#id	<u>+</u>id*id#	reduce (6)
#F	<u>+</u>id*id#	reduce (4)
#T	<u>+</u>id*id#	reduce (2)
#E	<u>+</u>id*id#	shift
#E+	<u>id</u>*id#	shift
#E+id	<u>*</u>id#	reduce (6)
#E+F	<u>*</u>id#	reduce (4)
#E+T	<u>*</u>id#	shift
#E+T*	<u>id</u>#	shift
#E+T*id	<u>#</u>	reduce (6)
#E+T*F	<u>#</u>	reduce (3)
#E+T	<u>#</u>	reduce (1)
#E	<u>#</u>	accept

Notice that the sequence of reductions 6, 4, 2, 6, 4, 6, 3, 1 represents the right-most derivation of the input string but in reverse:

$$\begin{aligned}
\underline{E} &\Rightarrow E + \underline{T} \\
&\Rightarrow E + T * \underline{F} \\
&\Rightarrow E + \underline{T} * \text{id} \\
&\Rightarrow E + \underline{F} * \text{id} \\
&\Rightarrow \underline{E} + \text{id} * \text{id} \\
&\Rightarrow \underline{T} + \text{id} * \text{id} \\
&\Rightarrow \underline{F} + \text{id} * \text{id} \\
&\Rightarrow \text{id} + \text{id} * \text{id}
\end{aligned}$$

That it is in reverse makes sense because this is a bottom-up parse. The question arises: How does the parser know when to *shift* and when to *reduce*? When reducing, how many symbols on top of the stack play a role in the reduction? And, when reducing, by *which rule* does it make its reduction?

For example, in the derivation above, when the stack contains #E+T and the next incoming token is a $*$, how do we know that we are to *shift* (the $*$ onto the stack) rather than *reduce* either the E+T to an E or the T to an E?

Notice two things:

1. Ignoring the terminator #, the stack configuration combined with the unscanned input stream represents a sentential form in a right-most derivation of the input.

2. The part of the sentential form that is reduced to a non-terminal is always on top of the stack. So all actions take place at the *top* of the stack. We either shift a token onto the stack, or we reduce what is already there.

We call the sequence of terminals on top of the stack that are reduced to a single non-terminal at each reduction step the *handle*. More formally, in a right-most derivation,

$$S \overset{*}{\Rightarrow} \alpha Y w \Rightarrow \alpha \beta w \overset{*}{\Rightarrow} uw, \text{ where } uw \text{ is the sentence,}$$

the handle is the rule $Y ::= \beta$ and a position in the right sentential form $\alpha\beta w$ where β may be replaced by Y to produce the previous right sentential form $\alpha Y w$ in a right-most derivation from the start symbol S. Fortunately, there are a finite number of possible handles that may appear on top of the stack.

So, when a handle appears on top of the stack,

Stack	Input
#$\alpha\beta$	w

we reduce that handle (β to Y in this case).

Now if β is the sequence $X_1, X_2, \ldots, X_n$, then we call any subsequence, $X_1, X_2, \ldots, X_i$, for $i \le n$ a *viable prefix*. Only viable prefixes may appear on the top of the parse stack. If there is not a handle on top of the stack and shifting the first unscanned input token from the input to the stack results in a viable prefix, a shift is called for.

3.4.2   LR(1) Parsing

One way to drive the shift/reduce parser is by a kind of DFA that recognizes viable prefixes and handles. The tables that drive our LR(1) parser are derived from this DFA.

The LR(1) Parsing Algorithm

Before showing how the tables are constructed, let us see how they are used to parse a sentence. The LR(1) parser algorithm is common to all LR(1) grammars and is driven by two tables, constructed for particular grammars: an Action table and a Goto table.

The algorithm is a state machine with a pushdown stack, driven by two tables: Action and Goto. A configuration of the parser is a pair, consisting of the state of the stack and the state of the input:

Stack	Input
$s_0 X_1 s_1 X_2 s_2 \ldots X_m s_m$	$a_k a_{k+1} \ldots a_n$

where the s_i are states, the X_i are (terminal or non-terminal) symbols, and $a_k a_{k+1} \ldots a_n$ are the unscanned input symbols. This configuration represents a right sentential form in a right-most derivation of the input sentence,

$$X_1 X_2 \ldots X_m a_k a_{k+1} \ldots a_n$$

Algorithm 3.7 The LR(1) Parsing Algorithm

Input: Action and Goto tables, and the input sentence w to be parsed, followed by the terminator **#**

Output: a right-most derivation in reverse

Initially, the parser has the configuration

Stack	Input
s_0	$a_1 a_2 \ldots a_n$**#**

where $a_1 a_2 \ldots a_n$ is the input sentence

repeat

If Action$[s_m, a_k] = ss_i$, the parser executes a *shift* (the s stands for "shift") and goes into state s_i, going into the configuration

Stack	Input
$s_0 X_1 s_1 X_2 s_2 \ldots X_m s_m a_k s_i$	$a_{k+1} \ldots a_n$**#**

Otherwise, if Action$[s_m, a_k] = ri$ (the r stands for "reduce"), where i is the number of the production rule $Y ::= X_j X_{j+1} \ldots X_m$, then replace the symbols and states $X_j s_j X_{j+1} s_{j+1} \ldots X_m s_m$ by Ys, where $s = $ Goto$[s_{j-1}, Y]$. The parser then outputs production number i. The parser goes into the configuration

Stack	Input
$s_0 X_1 s_1 X_2 s_2 \ldots X_{j-1} s_{j-1} Y s$	$a_{k+1} \ldots a_n$**#**

Otherwise, if Action$[s_m, a_k] = $ accept, then the parser halts and the input has been successfully parsed

Otherwise, if Action$[s_m, a_k] = $ error, then the parser raises an error. The input is not in the language

until either the sentence is parsed or an error is raised

Example. Consider (again) our grammar for simple expressions, now in (3.31).

1. $E ::= E + T$　　　　　　　　　　　　　　　　　　　　　　(3.31)
2. $E ::= T$
3. $T ::= T * F$
4. $T ::= F$
5. $F ::= (E)$
6. $F ::= \texttt{id}$

The Action and Goto tables are given in Figure 3.7.

	Action						Goto		
	+	*	(	)	id	#	E	T	F
0			s4		s5		1	2	3
1	s6					accept			
2	r2	s7				r2			
3	r4	r4				r4			
4			s11		s12		8	9	10
5	r6	r6				r6			
6			s4		s5			13	3
7			s4		s5				14
8	s16			s15					
9	r2	s17		r2					
10	r4	r4		r4					
11			s11		s12		18	9	10
12	r6	r6		r6					
13	r1	s7				r1			
14	r3	r3				r3			
15	r5	r5				r5			
16			s11		s12			19	10
17			s11		s12				20
18	s16			s21					
19	r1	s17		r1					
20	r3	r3		r3					
21	r5	r5		r5					

FIGURE 3.7 The Action and Goto tables for the grammar in (3.31) (blank implies error).

Consider the steps for parsing `id + id * id`. Initially, the parser is in state 0, so a 0 is pushed onto the stack.

Stack	Input	Action
0	id+id*id#	

The next incoming symbol is an id, so we consult the Action table Action[0, id] to determine what to do in state 0 with an incoming token id. The entry is $s5$, so we shift the id onto the stack and go into state 5 (pushing the new state onto the stack above the id).

Stack	Input	Action
0	id+id*id#	shift 5
0id5	+id*id#	

Now, the 5 on top of the stack indicates we are in state 5 and the incoming token is +, so we consult Action[5, +]; the $r6$ indicates a reduction using rule 6: $F ::= $ id. To make the reduction, we pop $2k$ items off the stack, where k is the number of symbols in the rule's right-hand side; in our example, $k = 1$ so we pop both the 5 and the id.

Stack	Input	Action
0	id+id*id#	shift 5
0id5	+id*id#	reduce 6, output a 6
0	+id*id#	

Because we are reducing the right-hand side to an F in this example, we push the F onto the stack.

Stack	Input	Action
0	id+id*id#	shift 5
0id5	+id*id#	reduce 6, output a 6
0F	+id*id#	

And finally, we consult Goto[0, F] to determine which state the parser, initially in state 0, should go into after parsing an F. Because Goto[0, F] = 3, this is state 3. We push the 3 onto the stack to indicate the parser's new state.

Stack	Input	Action
0	id+id*id#	shift 5
0id5	+id*id#	reduce 6, output a 6
0F3	+id*id#	

From state 3 and looking at the incoming token +, Action[3, +] tells us to reduce using rule 4: $T ::= F$.

Stack	Input	Action
0	id+id*id#	shift 5
0id5	+id*id#	reduce 6, output a 6
0F3	+id*id#	reduce 4, output a 4
0T2	+id*id#	

From state 2 and looking at the incoming token +, Action[2, +] tells us to reduce using rule 2: $E ::= T$.

Stack	Input	Action
0	<u>id</u>+id*id#	shift 5
0id5	<u>+</u>id*id#	reduce 6, output a 6
0F3	<u>+</u>id*id#	reduce 4, output a 4
0T2	<u>+</u>id*id#	reduce 2, output a 2
0E1	<u>+</u>id*id#	

From state 1 and looking the incoming token +, Action[3, +] = $s6$ tells us to shift (the + onto the stack and go into state 6.

Stack	Input	Action
0	<u>id</u>+id*id#	shift 5
0id5	<u>+</u>id*id#	reduce 6, output a 6
0F3	<u>+</u>id*id#	reduce 4, output a 4
0T2	<u>+</u>id*id#	reduce 2, output a 2
0E1	<u>+</u>id*id#	shift 6
0E1+6	<u>id</u>*id#	

Continuing in this fashion, the parser goes through the following sequence of configurations and actions:

Stack	Input	Action
0	<u>id</u>+id*id#	shift 5
0id5	<u>+</u>id*id#	reduce 6, output a 6
0F3	<u>+</u>id*id#	reduce 4, output a 4
0T2	<u>+</u>id*id#	reduce 2, output a 2
0E1	<u>+</u>id*id#	shift 6
0E1+6	<u>id</u>*id#	shift 5
0E1+6id5	<u>*</u>id#	reduce 6, output 6
0E1+6F3	<u>*</u>id#	reduce 4, output 4
0E1+6T13	<u>*</u>id#	shift 7
0E1+6T13*7	<u>id</u>#	shift 5
0E1+6T13*7id5	<u>#</u>	reduce 6, output 6
0E1+6T13*7F14	<u>#</u>	reduce 3, output 3
0E1+6T13	<u>#</u>	reduce 1, output 1
0E1	#	accept

In the last step, the parser is in state 1 and the incoming token is the terminator #; Action[1, #] says we accept the input sentence; the sentence has been successfully parsed.

Moreover, the parser has output 6, 4, 2, 6, 4, 6, 3, 1, which is a right-most derivation of the input string in reverse: 1, 3, 6, 4, 6, 2, 4, 6, that is

$$\underline{E} \Rightarrow E + \underline{T}$$
$$\Rightarrow E + T * \underline{F}$$
$$\Rightarrow E + \underline{T} * \text{id}$$
$$\Rightarrow E + \underline{F} * \text{id}$$
$$\Rightarrow \underline{E} + \text{id} * \text{id}$$
$$\Rightarrow \underline{T} + \text{id} * \text{id}$$
$$\Rightarrow \underline{F} + \text{id} * \text{id}$$
$$\Rightarrow \text{id} + \text{id} * \text{id}$$

A careful reading of the preceding steps suggests that explicitly pushing the symbols onto the stack is unnecessary because the symbols are implied by the states themselves. An industrial-strength LR(1) parser will simply maintain a stack of states. We include the symbols only for illustrative purposes.

For all of this to work, we must go about constructing the tables Action and Goto. To do this, we must first construct the grammar's LR(1) canonical collection.

The LR(1) Canonical Collection

The LR(1) parsing tables, Action and Goto, for a grammar G are derived from a DFA for recognizing the possible handles for a parse in G. This DFA is constructed from what is called an *LR(1) canonical collection*, in turn a collection of sets of *items* of the form

$$[Y ::= \alpha \cdot \beta, \, \mathsf{a}] \tag{3.32}$$

where $Y ::= \alpha\beta$ is a production rule in the set of productions P, α and β are (possibly empty) strings of symbols, and a is a *lookahead*. The item represents a potential handle. The $\cdot$ is a position marker that marks the top of the stack, indicating that we have parsed the α and still have the β ahead of us in satisfying the Y. The lookahead symbol, a, is a token that can follow Y (and so, $\alpha\beta$) in a legal right-most derivation of some sentence.

- If the position marker comes at the start of the right-hand side in an item,

$$[Y ::= \cdot \, \alpha \, \beta, \, \mathsf{a}]$$

the item is called a *possibility*. One way of parsing the Y is to first parse the α and then parse the β, after which point the next incoming token will be an a. The parse might be in the following configuration:

Stack	Input
$\#\gamma$	$u\mathsf{a}\ldots$

where $\alpha\beta \overset{*}{\Rightarrow} u$, where u is a string of terminals.

- If the position marker comes after a string of symbols α but before a string of symbols β in the right-hand side in an item,

$$[Y ::= \alpha \cdot \beta, \, \mathsf{a}]$$

the item indicates that α has been parsed (and so is on the stack) but that there is still β to parse from the input:

Stack	Input
#$\gamma\alpha$	va...

where $\beta \overset{*}{\Rightarrow} v$, where v is a string of terminals.

- If the position marker comes at the end of the right-hand side in an item,

$$[Y ::= \alpha \; \beta \cdot, \, \mathsf{a}]$$

the item indicates that the parser has successfully parsed $\alpha\beta$ in a context where Ya would be valid, the $\alpha\beta$ can be reduced to a Y, and so $\alpha\beta$ is a handle. That is, the parse is in the configuration

Stack	Input
#$\gamma\alpha\beta$	a...

and the reduction of $\alpha\beta$ would cause the parser to go into the configuration

Stack	Input
#γY	a...

A non-deterministic finite-state automaton (NFA) that recognizes viable prefixes and handles can be constructed from items like that in (3.32). The items record the progress in parsing various language fragments. We also know that, given this NFA, we can construct an equivalent DFA using the powerset construction that we saw in Section 2.7. The states of the DFA for recognizing viable prefixes are derived from sets of these items, which record the current state of an LR(1) parser. Here, we shall construct these sets, and so construct the DFA, directly (instead of first constructing a NFA).

So, the states in our DFA will be constructed from sets of items like that in (3.32). We call the set of states the *canonical collection*.

To construct the canonical collection of states, we first must augment our grammar G with an additional start symbol S' and an additional rule,

$$S' ::= S$$

so as to yield a grammar G', which describes the same language as does G, but which does not have its start symbol on the right-hand side of any rule. For example, augmenting our grammar (3.31) for simple expressions gives us the augmented grammar in (3.33).

$$(3.33)$$

0. $E' ::= E$
1. $E\ \ ::= E + T$
2. $E\ \ ::= T$
3. $T\ \ ::= T * F$
4. $T\ \ ::= F$
5. $F\ \ ::= (E)$
6. $F\ \ ::= \mathtt{id}$

We then start constructing item sets from this augmented grammar. The first set, representing the initial state in our DFA, will contain the LR(1) item:

$$\{[E' ::= \cdot\ E, \#]\} \tag{3.34}$$

which says that parsing an E' means parsing an E from the input, after which point the next (and last) remaining unscanned token should be the terminator #. But at this point, we have not yet parsed the E; the $\cdot$ in front of it indicates that it is still ahead of us.

Now that we must parse an E at this point means that we might be parsing either an $E + T$ (by rule 1 of 3.33) or a T (by rule 2). So the initial set would also contain

$$[E\ \ ::= \cdot\ E + T, \#] \tag{3.35}$$
$$[E\ \ ::= \cdot\ T, \#]$$

In fact, the initial set will contain additional items implied by (3.34). We call the initial set (3.34) the *kernel*. From the kernel, we can then compute the *closure*, that is, all items implied by the kernel. Algorithm 3.8 computes the closure for any set of items.

Algorithm 3.8 Computing the Closure of a Set of Items

Input: a set of items, s
Output: closure(s)
  add s to closure(s)
  repeat
    if closure(s) contains an item of the form

$$[Y ::= \alpha \cdot X\ \beta, \mathsf{a}]$$

    add the item

$$[X ::= \cdot\ \gamma, \mathsf{b}]$$

    for every rule $X ::= \gamma$ in P and for every token b in first($\beta\mathsf{a}$).
  until no new items may be added

Example. To compute the closure of our kernel (3.34), that is, closure($\{[E' ::= \cdot E, \#]\}$), by step 1 is initially

$$\{[E' ::= \cdot\ E, \#]\} \tag{3.36}$$

We then invoke step 2. Because the $\cdot$ comes before the E, and because we have the rule $E ::= E + T$ and $E ::= T$, we add $[E ::= \cdot\ E + T, \#]$ and $[E ::= \cdot\ T, \#]$ to get

$$\{[E' ::= \cdot\ E, \#], \tag{3.37}$$
$$[E\ \ ::= \cdot\ E + T, \#],$$
$$[E\ \ ::= \cdot\ T, \#]\}$$

The item $[E ::= \cdot\ E + T,\ \#]$ implies

$$[E\ ::=\ \cdot\ E + T,\ +]$$
$$[E\ ::=\ \cdot\ T,\ +]$$

because $\text{first}(+T\#) = \{+\}$. Now, given that these items differ from previous items only in the lookaheads, we can use the more compact notation $[E ::= \cdot\ E + T,\ +/\#]$ for representing the two items

$$[E\ ::=\ \cdot\ E + T,\ +]\ \text{and}$$
$$[E\ ::=\ \cdot\ E + T,\ \#]$$

So we get

$$\{[E'\ ::=\ \cdot\ E,\ \#],$$
$$[E\ ::=\ \cdot\ E + T,\ +/\#],$$
$$[E\ ::=\ \cdot\ T,\ +/\#]\} \tag{3.38}$$

The items $[E ::= \cdot\ T,\ +/\#]$ imply additional items (by similar logic), leading to

$$\{[E'\ ::=\ \cdot\ E,\ \#],$$
$$[E\ ::=\ \cdot\ E + T,\ +/\#],$$
$$[E\ ::=\ \cdot\ T,\ +/\#],$$
$$[T\ ::=\ \cdot\ T\ *\ F,\ +/*/\#],$$
$$[T\ ::=\ \cdot\ F,\ +/*/\#]\} \tag{3.39}$$

And finally the items $[T ::= \cdot\ F,\ +/*/\#]$ imply additional items (by similar logic), leading to

$$s_0 = \{[E'\ ::=\ \cdot\ E,\ \#],$$
$$[E\ ::=\ \cdot\ E + T,\ +/\#],$$
$$[E\ ::=\ \cdot\ T,\ +/\#],$$
$$[T\ ::=\ \cdot\ T\ *\ F,\ +/*/\#],$$
$$[T\ ::=\ \cdot\ F,\ +/*/\#],$$
$$[F\ ::=\ \cdot(E),\ +/*/\#],$$
$$[F\ ::=\ \cdot\ \text{id},\ +/*/\#]\} \tag{3.40}$$

The item set (3.40) represents the initial state s_0 in our canonical LR(1) collection.

As an aside, notice that the closure of $\{[E ::= \cdot E,\ \#]\}$ represents all of the states in an NFA, that could be reached from the initial item by ϵ-moves alone, that is, without scanning anything from the input. That portion of the NFA that is equivalent to the initial state s_0 in our DFA is illustrated in Figure 3.8.

We now need to compute all of the states and transitions of the DFA that recognizes viable prefixes and handles. For any item set s, and any symbol $X \in (T \cup N)$,

$$\text{goto}(s, X) = \text{closure}(r),$$

where $r = \{[Y ::= \alpha X \cdot \beta, a] | [Y ::= \alpha \cdot X\beta, a]\}$[6]. That is, to compute $\text{goto}(s, X)$, we take all items from s with a $\cdot$ before the X and move them after the X; we then take the closure of that. Algorithm 3.9 does this.

---

[6]The | operator is being used as the set notation "for all", not the BNF notation for alternation.

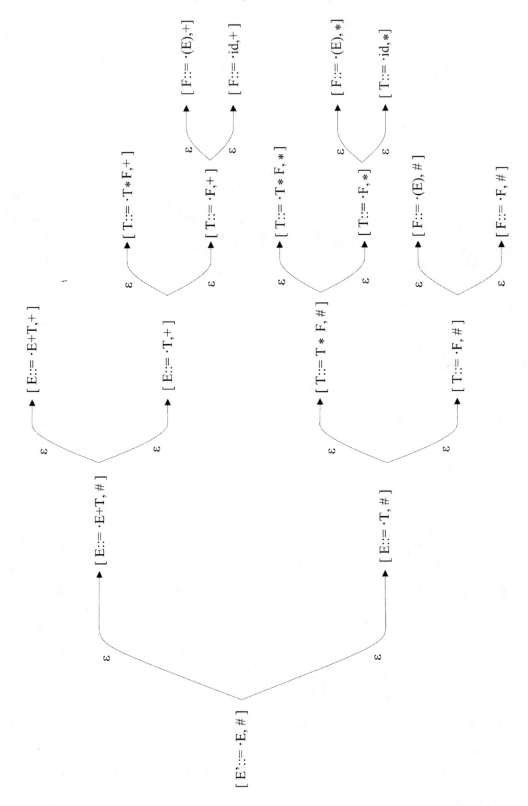

FIGURE 3.8 The NFA corresponding to s_0.

Algorithm 3.9 Computing goto

Input: a state s, and a symbol $X \in T \cup N$

Output: the state, goto(s, X)

$\quad r \leftarrow \{\}$
$\quad$**for** each item $[Y ::= \alpha \cdot X\beta, \text{a}]$ in s **do**
$\quad\quad$add $[Y ::= \alpha X \cdot \beta, \text{a}]$ to r
$\quad$**end for**
$\quad$**return** closure(r)

Example. Consider the computation of goto(s_0, E), where s_0 is in (3.40). The relevant items in s_0 are $[E' ::= \cdot E, \#]$ and $[E ::= \cdot E + T, +/\#]$. Moving the $\cdot$ to the right of the E in the items gives us $\{[E' ::= E\cdot, \#], [E ::= E\cdot + T, +/\#]\}$. The closure of this set is the set itself; let us call this state s_1.

$$
\begin{aligned}
\text{goto}(s_0, E) &= s_1 \\
&= \{[E' ::= E \cdot, \#], \\
&\quad\ [E \ ::= E \cdot + T, +/\#]\}
\end{aligned}
$$

In a similar manner, we can compute

$$
\begin{aligned}
\text{goto}(s_0, T) &= s_2 \\
&= \{[E \ ::= T \cdot, +/\#], \\
&\quad\ [T \ ::= T \cdot * F, +/*/\#]\}
\end{aligned}
$$

$$
\begin{aligned}
\text{goto}(s_0, F) &= s_3 \\
&= \{[T \ ::= F \cdot, +/*/\#]\}
\end{aligned}
$$

goto$(s_0, ()$ involves a closure because moving the $\cdot$ across the $($ puts it in front of the E.

$$
\begin{aligned}
\text{goto}(s_0, () &= s_4 \\
&= \{[F \ ::= (\cdot E), \ +/*/\#], \\
&\quad\ [E \ ::= \cdot E + T, +/)], \\
&\quad\ [E \ ::= \cdot T, +/)], \\
&\quad\ [T \ ::= \cdot T * F, +/*/)], \\
&\quad\ [T \ ::= \cdot F, +/*/)], \\
&\quad\ [F \ ::= \cdot (E), +/*/)], \\
&\quad\ [F \ ::= \cdot \text{id}, +/*/)]\}
\end{aligned}
$$

$$
\begin{aligned}
\text{goto}(s_0, \text{id}) &= s_5 \\
&= \{[F \ ::= \text{id} \cdot, +/*/\#]\}
\end{aligned}
$$

We continue in this manner, computing goto for the states we have, and then for any new states repeatedly until we have defined no more new states. This gives us the canonical LR(1) collection.

Algorithm 3.10 Computing the LR(1) Collection

Input: a context-free grammar $G = (N, T, S, P)$
Output: the canonical LR(1) collection of states $c = \{s_0, s_1, \ldots, s_n\}$

Define an augmented grammar G', which is G with the added non-terminal S' and added production rule $S' ::= S$, where S is G's start symbol. The following steps apply to G'. Enumerate the production rules beginning at 0 for the newly added production.

$c \leftarrow \{s_0\}$ where $s_0 = \text{closure}(\{[S' ::= \cdot S, \#]\})$
repeat
    for each s in c, and for each symbol $X \in T \cup N$ **do**
        if $\text{goto}(s, X) \neq \emptyset$ and $\text{goto}(s, X) \notin c$ **then**
            add $\text{goto}(s, X)$ to c.
        end if
    end for
until no new states are added to c

Example. We can now resume computing the LR(1) canonical collection for the simple expression grammar, beginning from state s_1:

$$\text{goto}(s_1, \texttt{+}) = s_6$$
$$= \{[E ::= E + \cdot T, \texttt{+/\#}],$$
$$[T ::= \cdot T * F, \texttt{+/*/\#}],$$
$$[T ::= \cdot F, \texttt{+/*/\#}],$$
$$[F ::= \cdot (E), \texttt{+/*/\#}],$$
$$[F ::= \cdot \texttt{id}, \texttt{+/*/\#}]\}$$

There are no more moves from s_1. Similarly, from s_2,

$$\text{goto}(s_2, \texttt{*}) = s_7$$
$$= \{[T ::= T * \cdot F, \texttt{+/*/\#}],$$
$$[F ::= \cdot (E), \texttt{+/*/\#}],$$
$$[F ::= \cdot \texttt{id}, \texttt{+/*/\#}]\}$$

Notice that the closure of $\{[T ::= T * \cdot F, \texttt{+/*/\#}]\}$ carries along the same lookaheads because no symbol follows the F in the right-hand side. There are no gotos from s_3, but several from s_4.

$$\text{goto}(s_4, E) = s_8$$
$$= \{[F ::= (E \cdot), \texttt{+/*/\#}],$$
$$[E ::= E \cdot + T, \texttt{+/)}]\}$$

$$\text{goto}(s_4, T) = s_9$$
$$= \{[E ::= T \cdot, \texttt{+/)}],$$
$$[T ::= T \cdot * F, \texttt{+/*/)}]\}$$

$$\text{goto}(s_4, F) = s_{10}$$
$$= \{[T ::= F \cdot, \texttt{+/*/)}]\}$$

$$\begin{aligned}
\text{goto}(s_4, \text{ (}) = s_{11}\\
= \{&[F \ ::= \ (\ \cdot \ E), \ \text{+/*/)}],\\
&[E \ ::= \ \cdot \ E + T, \ \text{+/)}],\\
&[E \ ::= \ \cdot \ T, \ \text{+/)}],\\
&[T \ ::= \ \cdot \ T * F, \ \text{+/*/)}],\\
&[T \ ::= \ \cdot \ F, \ \text{+/*/)}],\\
&[F \ ::= \ \cdot \ (E), \ \text{+/*/)}],\\
&[F \ ::= \ \cdot \ \text{id}, \ \text{+/*/)}]\}
\end{aligned}$$

Notice that s_{11} differs from s_4 in only the lookaheads for the first item.

$$\begin{aligned}
\text{goto}(s_4, \text{ id}) = s_{12}\\
= \{&[F \ ::= \ \text{id} \ \cdot, \ \text{+/*/)}]\}
\end{aligned}$$

There are no moves from s_5, so consider s_6:

$$\begin{aligned}
\text{goto}(s_6, T) = s_{13}\\
= \{&[E \ ::= \ E + T \ \cdot, \ \text{+/\#}],\\
&[T \ ::= \ T \cdot * F, \ \text{+/*/\#}]\}
\end{aligned}$$

Now, $\text{goto}(s_6, F) = \{[T ::= F\cdot, \ \text{+/*/\#}]\}$ but that is s_3. $\text{goto}(s_6, \text{ (})$ is closure($\{[F ::= (\cdot E), \ \text{+/*/\#}]\}$), but that is s_4. And $\text{goto}(s_6, \text{ id})$ is s_5.

$$\begin{aligned}
\text{goto}(s_6, F) &= s_3\\
\text{goto}(s_6, \text{ (}) &= s_4\\
\text{goto}(s_6, \text{ id}) &= s_5
\end{aligned}$$

Consider s_7, s_8, and s_9.

$$\begin{aligned}
\text{goto}(s_7, F) &= s_{14}\\
&= \{[T \ ::= \ T * F \ \cdot, \ \text{+/*/\#}]\}\\
\text{goto}(s_7, \text{ (}) &= s_4\\
\text{goto}(s_7, \text{ id}) &= s_5\\
\text{goto}(s_8, \text{)}) &= s_{15}\\
&= \{[F \ ::= \ (E) \ \cdot, \ \text{+/*/\#}]\}\\
\text{goto}(s_8, \text{ +}) &= s_{16}\\
&= \{[E \ ::= \ E + \cdot \ T, \ \text{+/)}],\\
&\quad\ [T \ ::= \ \cdot \ T * F, \ \text{+/*/)}],\\
&\quad\ [T \ ::= \ \cdot \ F, \ \text{+/*/)}],\\
&\quad\ [F \ ::= \ \cdot \ (E), \ \text{+/*/)}],\\
&\quad\ [F \ ::= \ \cdot \ \text{id}, \ \text{+/*/)}]\}\\
\text{goto}(s_9, \text{ *}) &= s_{17}\\
&= \{[T \ ::= \ T * \cdot \ F, \ \text{+/*/)}],\\
&\quad\ [F \ ::= \ \cdot \ (E), \ \text{+/*/)}],\\
&\quad\ [F \ ::= \ \cdot \ \text{id}, \ \text{+/*/)}]\}
\end{aligned}$$

There are no moves from s_{10}, but several from s_{11}:

$$\text{goto}(s_{11}, E) = s_{18}$$
$$= \{[F \ ::= (E \cdot), \text{+/*/}]\},$$
$$[E \ ::= E \cdot + T, \text{+/}]\}$$
$$\text{goto}(s_{11}, T) = s_9$$
$$\text{goto}(s_{11}, F) = s_{10}$$
$$\text{goto}(s_{11}, () = s_{11}$$
$$\text{goto}(s_{11}, \text{id}) = s_{12}$$

There are no moves from s_{12}, but there is a move from s_{13}:

$$\text{goto}(s_{13}, *) = s_7$$

There are no moves from s_{14} or s_{15}, but there are moves from s_{16}, s_{17}, s_{18}, and s_{19}:

$$\text{goto}(s_{16}, T) = s_{19}$$
$$= \{[E \ ::= E + T \cdot, \text{+/}]\},$$
$$[T \ ::= T \cdot * F, \text{+/*/}]\}$$
$$\text{goto}(s_{16}, F) = s_{10}$$
$$\text{goto}(s_{16}, () = s_{11}$$
$$\text{goto}(s_{16}, \text{id}) = s_{12}$$
$$\text{goto}(s_{17}, F) = s_{20}$$
$$= \{[T \ ::= T * F \cdot, \text{+/*/}]\}$$
$$\text{goto}(s_{17}, () = s_{11}$$
$$\text{goto}(s_{17}, \text{id}) = s_{12}$$
$$\text{goto}(s_{18},)) = s_{21}$$
$$= \{[F \ ::= (E) \cdot, \text{+/*/}]\}$$
$$\text{goto}(s_{18}, +) = s_{16}$$
$$\text{goto}(s_{19}, *) = s_{17}$$

There are no moves from s_{20} or s_{21}, so we are done. The LR(1) canonical collection consists of twenty-two states $s_0 \ldots s_{21}$. The entire collection is summarized in the table below.

$s_0 = \{[E' ::= \cdot \ E, \text{\#}],$
　　$[E \ ::= \cdot \ E + T, \text{+/\#}],$
　　$[E \ ::= \cdot \ T, \text{+/\#}],$
　　$[T \ ::= \cdot \ T * F, \text{+/*/\#}],$
　　$[T \ ::= \cdot \ F, \text{+/*/\#}],$
　　$[F \ ::= \cdot (E), \text{+/*/\#}],$
　　$[F \ ::= \cdot \ \text{id}, \text{+/*/\#}]\}$

$\text{goto}(s_0, E) = s_1$
$\text{goto}(s_0, T) = s_2$
$\text{goto}(s_0, F) = s_3$
$\text{goto}(s_0, () = s_4$
$\text{goto}(s_0, \text{id}) = s_5$

$s_{11} = \{[F \ ::= (\cdot \ E), \text{+/*/}],$
　　$[E \ ::= \cdot \ E + T, \text{+/}],$
　　$[E \ ::= \cdot \ T, \text{+/}],$
　　$[T \ ::= \cdot \ T * F, \text{+/*/}],$
　　$[T \ ::= \cdot \ F, \text{+/*/}],$
　　$[F \ ::= \cdot \ (E), \text{+/*/}],$
　　$[F \ ::= \cdot \ \text{id}, \text{+/*/}]\}$

$\text{goto}(s_{11}, E) = s_{18}$
$\text{goto}(s_{11}, T) = s_9$
$\text{goto}(s_{11}, F) = s_{10}$
$\text{goto}(s_{11}, () = s_{11}$
$\text{goto}(s_{11}, \text{id}) = s_{12}$

$s_1 = \{[E' ::= E \cdot, \text{\#}],$
　　$[E \ ::= E \cdot + T, \text{+/\#}]\}$

$\text{goto}(s_1, +) = s_6$

$s_{12} = \{[F \ ::= \text{id} \cdot, \text{+/*/}]\}$

$s_2 = \{[E ::= T \cdot, +/\#],$ $\text{goto}(s_2, *) = s_7$
     $[T ::= T \cdot * F, +/*/\#]\}$

$s_3 = \{[T ::= F \cdot, +/*/\#]\}$

$s_4 = \{[F ::= (\cdot E), +/*/\#],$ $\text{goto}(s_4, E) = s_8$
     $[E ::= \cdot E + T, +/)],$ $\text{goto}(s_4, T) = s_9$
     $[E ::= \cdot T, +/)],$ $\text{goto}(s_4, F) = s_{10}$
     $[T ::= \cdot T * F, +/*/)],$ $\text{goto}(s_4, () = s_{11}$
     $[T ::= \cdot F, +/*/)],$ $\text{goto}(s_4, \text{id}) = s_{12}$
     $[F ::= \cdot (E), +/*/)],$
     $[F ::= \cdot \text{id}, +/*/)]\}$

$s_5 = \{[F ::= \text{id} \cdot, +/*/\#]\}$

$s_6 = \{[E ::= E + \cdot T, +/\#],$ $\text{goto}(s_6, T) = s_{13}$
     $[T ::= \cdot T * F, +/*/\#],$ $\text{goto}(s_6, F) = s_3$
     $[T ::= \cdot F, +/*/\#],$ $\text{goto}(s_6, () = s_4$
     $[F ::= \cdot (E), +/*/\#],$ $\text{goto}(s_6, \text{id}) = s_5$
     $[F ::= \cdot \text{id}, +/*/\#]\}$

$s_7 = \{[T ::= T * \cdot F, +/*/\#],$ $\text{goto}(s_7, F) = s_{14}$
     $[F ::= \cdot (E), +/*/\#],$ $\text{goto}(s_7, () = s_4$
     $[F ::= \cdot \text{id}, +/*/\#]\}$ $\text{goto}(s_7, \text{id}) = s_5$

$s_8 = \{[F ::= (E \cdot), +/*/\#],$ $\text{goto}(s_8,)) = s_{15}$
     $[E ::= E \cdot + T, +/)]\}$ $\text{goto}(s_8, +) = s_{16}$

$s_9 = \{[E ::= T \cdot, +/)],$ $\text{goto}(s_9, *) = s_{17}$
     $[T ::= T \cdot * F, +/*/)]\}$

$s_{10} = \{[T ::= F \cdot, +/*/)]\}$

$s_{13} = \{[E ::= E + T \cdot, +/\#],$ $\text{goto}(s_{13}, *) = s_7$
     $[T ::= T \cdot * F, +/*/\#]\}$

$s_{14} = \{[T ::= T * F \cdot, +/*/\#]\}$

$s_{15} = \{[F ::= (E) \cdot, +/*/\#]\}$

$s_{16} = \{[E ::= E + \cdot T, +/)],$ $\text{goto}(s_{16}, T) = s_{19}$
     $[T ::= \cdot T * F, +/*/)],$ $\text{goto}(s_{16}, F) = s_{10}$
     $[T ::= \cdot F, +/*/)],$ $\text{goto}(s_{16}, () = s_{11}$
     $[F ::= \cdot (E), +/*/)],$ $\text{goto}(s_{16}, \text{id}) = s_{12}$
     $[F ::= \cdot \text{id}, +/*/)]\}$

$s_{17} = \{[T ::= T * \cdot F, +/*/)],$ $\text{goto}(s_{17}, F) = s_{20}$
     $[F ::= \cdot (E), +/*/)],$ $\text{goto}(s_{17}, () = s_{11}$
     $[F ::= \cdot \text{id}, +/*/)]\}$ $\text{goto}(s_{17}, \text{id}) = s_{12}$

$s_{18} = \{[F ::= (E \cdot), +/*/)],$ $\text{goto}(s_{18},)) = s_{21}$
     $[E ::= E \cdot + T, +/)]\}$ $\text{goto}(s_{18}, +) = s_{16}$

$s_{19} = \{[E ::= E + T \cdot, +/)],$ $\text{goto}(s_{19}, *) = s_{17}$
     $[T ::= T \cdot * F, +/*/)]\}$

$s_{20} = \{[T ::= T * F \cdot, +/*/)]\}$

$s_{21} = \{[F ::= (E) \cdot, +/*/)]\}$

We may now go about constructing the tables Action and Goto.

Constructing the LR(1) Parsing Tables

The LR(1) parsing tables Action and Goto are constructed from the LR(1) canonical collection, as prescribed in Algorithm 3.11.

Algorithm 3.11 Constructing the LR(1) Parsing Tables for a Context-Free Grammar

---

Input: a context-free grammar $G = (N, T, S, P)$
Output: the LF(1) tables Action and Goto

1. Compute the LR(1) canonical collection $c = \{s_0, s_1, \ldots, s_n\}$. State i of the parser corresponds to the item set s_i. State 0, corresponding to the item set s_0, which contains the item $[S' ::= \cdot S, \#]$, is the parser's initial state

2. The Action table is constructed as follows:

   a. For each transition, $\text{goto}(s_i, a) = s_j$, where a is a terminal, set $\text{Action}[i, a] = sj$. The s stands for "shift"

   b. If the item set s_k contains the item $[S' ::= S \cdot, \#]$, set $\text{Action}[k, \#] = \text{accept}$

   c. For all item sets s_i, if s_i contains an item of the form $[Y ::= \alpha \cdot, a]$, set $\text{Action}[i, a] = rp$, where p is the number corresponding to the rule $Y ::= \alpha$. The r stands for "reduce"

   d. All undefined entries in Action are set to *error*

3. The Goto table is constructed as follows:

   a. For each transition, $\text{goto}(s_i, Y) = s_j$, where Y is a non-terminal, set $\text{Goto}[i, Y] = j$.

   b. All undefined entries in Goto are set to *error*

If all entries in the Action table are unique, then the grammar G is said to be LR(1).

Example. Let us say we are computing the Action and Goto tables for the arithmetic expression grammar in (3.31). We apply Algorithm 3.10 for computing the LR(1) canonical collection. This produces the twenty-two item sets shown in the table before. Adding the extra production rule and enumerating the production rules gives us the augmented grammar in (3.41).

0. $E' ::= E$
1. $E ::= E + T$
2. $E ::= T$ (3.41)
3. $T ::= T * F$
4. $T ::= F$
5. $F ::= (E)$
6. $F ::= \text{id}$

We must now apply steps 2 and 3 of Algorithm 3.11 for constructing the tables Action and Goto. Both tables will each have twenty-two rows for the twenty-two states, derived from the twenty-two item sets in the LR(1) canonical collection: 0 to 21. The Action table will have six columns, one for each terminal symbol: +, *, (,), id, and the terminator #. The Goto table will have three columns, one for each of the original non-terminal symbols: E, T, and F. The newly added non-terminal E' does not play a role in the parsing process. The tables are illustrated in Figure 3.7. To see how these tables are constructed, let us derive the entries for several states.

First, let us consider the first four states of the Action table:

- The row of entries for state 0 is derived from item set s_0.

  - By step 2a of Algorithm 3.11, the transition $goto(s_0, () = s_4$ implies Action[0, (] = $s4$, and $goto(s_0, \text{id}) = s_5$ implies Action[0, id] = $s5$. The $s4$ means "shift the next input symbol (onto the stack and go into state 4"; the $s5$ means "shift the next input symbol id onto the stack and go into state 5."

- The row of entries for state 1 is derived from item set s_1:

  - By step 2a, the transition $goto(s_1, +) = s_6$ implies Action[1, +] = $s6$. Remember, the $s6$ means "shift the next input symbol + onto the stack and go into state 6".

  - By step 2b, because item set s_1 contains $[E' ::= E\cdot, \#]$, Action[1, #] = accept. This says, that if the parser is in state 1 and the next input symbol is the terminator #, the parser accepts the input string as being in the language.

- The row of entries for state 2 is derived from item set s_2:

  - By step 2a, the transition $goto(s_2, *) = s_7$ implies Action[2, *] = $s7$.

  - By step 2c, the items[7] $[E ::= T\cdot, +/\#]$ imply two entries: Action[2, #] = $r2$ and Action[2, +] = $r2$. These entries say that if the parser is in state 2 and the next incoming symbol is either a # or a +, reduce the T on the stack to a E using production rule 2: $E ::= T$.

- The row of entries for state 3 is derived from item set s_3:

  - By step 2c, the items $[T ::= F\cdot, +/*/\#]$ imply three entries: Action[3, #]= $r4$, Action[3, +] = $r4$, and Action[3, *] = $r4$. These entries say that if the parser is in state 3 and the next incoming symbol is either a #, a +, or a * reduce the F on the stack to a T using production rule 4: $T ::= F$.

All other entries in rows 0, 1, 2, and 3 are left blank to indicate an error. If, for example, the parser is in state 0 and the next incoming symbol is a +, the parser raises an error. The derivations of the entries in rows 4 to 21 in the Action table (see Figure 3.7) are left as an exercise.

Now let us consider the first four states of the Goto table:

- The row of entries for state 0 is derived from item set s_0:

  - By step 3a of Algorithm 3.11, the $goto(s_0, E) = 1$ implies Goto[0, E] = 1, $goto(s_0, T) = 2$ implies Goto[0, E] = 4, and $goto(s_0, F) = 3$ implies Goto[0, E] = 3. The entry Goto[0, E] = 1 says that in state 0, once the parser scans and parses an E, the parser goes into state 1.

- The row of entries for state 1 is derived from item set s_1. Because there are no transitions on a non-terminal from item set s_1, no entries are indicated for state 1 in the Goto table.

- The row of entries for state 2 is derived from item set s_2. Because there are no transitions on a non-terminal from item set s_2, no entries are indicated for state 2 in the Goto table.

---

[7]Recall that the $[E ::= T\cdot, +/\#]$ denotes two items: $[E ::= T\cdot, +]$ and $[E ::= T\cdot, \#]$.

- The row of entries for state 3 is derived from item set s_3. Because there are no transitions on a non-terminal from item set s_3, no entries are indicated for state 3 in the Goto table.

All other entries in rows 0, 1, 2, and 3 are left blank to indicate an error. The derivations of the entries in rows 4 to 21 in the Goto table (see Figure 3.7) are left as an exercise.

Conflicts in the Action Table

There are two different kinds of conflicts possible for an entry in the Action table:

1. The first is the shift-reduce conflict, which can occur when there are items of the forms

$$[Y ::= \alpha \,\cdot, \text{a}] \text{ and}$$
$$[Y ::= \alpha \,\cdot \text{a}\beta, \text{b}]$$

The first item suggests a reduce if the next unscanned token is an a; the second suggests a shift of the a onto the stack.

Although such conflicts may occur for unambiguous grammars, a common cause is ambiguous constructs such as

$$S ::= \text{if } (E) \; S$$
$$S ::= \text{if } (E) \; S \text{ else } S$$

As we saw in Section 3.2.3, language designers will not give up such ambiguous constructs for the sake of parser writers. Most parser generators that are based on LR grammars permit one to supply an extra disambiguating rule. For example, the rule in this case would be to favor a shift of the else over a reduce of the "if (E) S" to an S.

2. The second kind of conflict that we can have is the reduce-reduce conflict. This can happen when we have a state containing two items of the form

$$[X ::= \alpha \,\cdot, \text{a}] \text{ and}$$
$$[Y ::= \beta \,\cdot, \text{a}]$$

Here, the parser cannot distinguish which production rule to apply in the reduction.

Of course, we will never have a shift-shift conflict, because of the definition of goto for terminals. Usually, a certain amount of tinkering with the grammar is sufficient for removing bona fide conflicts in the Action table for most programming languages.

3.4.3 LALR(1) Parsing

Merging LR(1) States

An LR(1) parsing table for a typical programming language such as Java can have thousands of states, and so thousands of rows. One could argue that, given the inexpensive memory nowadays, this is not a problem. On the other hand, smaller programs and data make for faster running programs so it would be advantageous if we might be able to reduce the number of states. LALR(1) is a parsing method that does just this.

If you look at the LR(1) canonical collection of states in Figure 3.8, which we computed for our example grammar (3.31), you will find that many states are virtually identical—they differ only in their lookahead tokens. Their cores—the core of an item is just the rule and position marker portion—are identical. For example, consider states s_2 and s_9:

$$s_2 = \{[E ::= T \cdot, \, +/\#],$$
$$[T ::= T \cdot * F, \, +/*/\#]\}$$

$$s_9 = \{[E ::= T \cdot, \, +/)],$$
$$[T ::= T \cdot * F, \, +/*/)]\}$$

They differ only in the lookaheads # and). Their cores are the same:

$$s_2 = \{[E ::= T \cdot],$$
$$[T ::= T \cdot * F]\}$$

$$s_9 = \{[E ::= T \cdot],$$
$$[T ::= T \cdot * F]\}$$

What happens if we merge them, taking a union of the items, into a single state, $s_{2.9}$? Because the cores are identical, taking a union of the items merges the lookaheads:

$$s_{2.9} = \{[E ::= T \cdot, \, +/)/\#],$$
$$[T ::= T \cdot * F, \, +/*/)/\#]\}$$

Will this cause the parser to carry out actions that it is not supposed to? Notice that the lookaheads play a role only in reductions and never in shifts. It is true that the new state may call for a reduction that was not called for by the original LR(1) states; yet an error will be detected before any progress is made in scanning the input.

Similarly, looking at the states in Figure 3.8, one can merge states s_3 and s_{10}, s_2 and s_9, s_4 and s_{11}, s_5 and s_{12}, s_6 and s_{16}, s_2 and s_{17}, s_8 and s_{18}, s_{13} and s_{19}, s_{14} and s_{20}, and s_{15} and s_{21}. This allows us to reduce the number of states by ten. In general, for bona fide programming languages, one can reduce the number of states by an order of magnitude.

LALR(1) Table Construction

There are two approaches to computing the LALR(1) states, and so the LALR(1) parsing tables.

LALR(1) table construction from the LR(1) states

In the first approach, we first compute the full LR(1) canonical collection of states, and then perform the state merging operation illustrated above for producing what we call the LALR(1) canonical collection of states.

Algorithm 3.12 Constructing the LALR(1) Parsing Tables for a Context-Free Grammar

Input: a context-free grammar $G = (N, T, S, P)$
Output: the LALR(1) tables Action and Goto

1. Compute the LR(1) canonical collection $c = \{s_0, s_1, \ldots, s_n\}$

2. Merge those states whose item cores are identical. The items in the merged state are a union of the items from the states being merged. This produces an LALR(1) canonical collection of states

3. The goto function for each new merged state is the union of the goto for the individual merged states

4. The entries in the Action and Goto tables are constructed from the LALR(1) states in the same way as for the LR(1) parser in Algorithm 3.11

If all entries in the Action table are unique, then the grammar G is said to be LALR(1).

Example. Reconsider our grammar for simple expressions from (3.31), and (again) repeated here as (3.42).

0. $E' ::= E$ (3.42)
1. $E\ ::= E + T$
2. $E\ ::= T$
3. $T\ ::= T * F$
4. $T\ ::= F$
5. $F\ ::= (E)$
6. $F\ ::= \text{id}$

Step 1 of Algorithm 3.12 has us compute the LR(1) canonical collection that was shown in a table above, and is repeated here.

$s_0 = \{[E' ::= \cdot\, E, \#],$ $\text{goto}(s_0, E) = s_1$
$\quad [E\ ::= \cdot\, E + T, +/\#],$ $\text{goto}(s_0, T) = s_2$
$\quad [E\ ::= \cdot\, T, +/\#],$ $\text{goto}(s_0, F) = s_3$
$\quad [T\ ::= \cdot\, T * F, +/*/\#],$ $\text{goto}(s_0, () = s_4$
$\quad [T\ ::= \cdot\, F, +/*/\#],$ $\text{goto}(s_0, \text{id}) = s_5$
$\quad [F\ ::= \cdot\, (E), +/*/\#],$
$\quad [F\ ::= \cdot\, \text{id}, +/*/\#]\}$

$s_{11} = \{[F\ ::= (\,\cdot\, E), +/*/)],$ $\text{goto}(s_{11}, E) = s_{18}$
$\quad [E\ ::= \cdot\, E + T, +/)],$ $\text{goto}(s_{11}, T) = s_9$
$\quad [E\ ::= \cdot\, T, +/)],$ $\text{goto}(s_{11}, F) = s_{10}$
$\quad [T\ ::= \cdot\, T * F, +/*/)],$ $\text{goto}(s_{11}, () = s_{11}$
$\quad [T\ ::= \cdot\, F, +/*/)],$ $\text{goto}(s_{11}, \text{id}) = s_{12}$
$\quad [F\ ::= \cdot\, (E), +/*/)],$
$\quad [F\ ::= \cdot\, \text{id}, +/*/)]\}$

$s_1 = \{[E' ::= E\, \cdot, \#],$ $\text{goto}(s_1, +) = s_6$
$\quad [E\ ::= E\, \cdot\, + T, +/\#]\}$

$s_{12} = \{[F\ ::= \text{id}\, \cdot, +/*/)]\}$

$s_2 = \{[E\ ::= T\, \cdot, +/\#],$ $\text{goto}(s_2, *) = s_7$
$\quad [T\ ::= T\, \cdot\, * F, +/*/\#]\}$

$s_{13} = \{[E\ ::= E + T\, \cdot, +/\#],$ $\text{goto}(s_{13}, *) = s_7$
$\quad [T\ ::= T\, \cdot\, * F, +/*/\#]\}$

$s_3 = \{[T\ ::= F\, \cdot, +/*/\#]\}$

$s_{14} = \{[T\ ::= T * F\, \cdot, +/*/\#]\}$

$s_4 = \{[F ::= (\cdot\ E),\ \textbf{+/*/\#}],$ $goto(s_4, E) = s_8$ $\quad$ $s_{15} = \{[F ::= (E)\ \cdot,\ \textbf{+/*/\#}]\}$
$\quad\quad [E ::= \cdot\ E + T,\ \textbf{+/)}],$ $goto(s_4, T) = s_9$
$\quad\quad [E ::= \cdot\ T,\ \textbf{+/)}],$ $goto(s_4, F) = s_{10}$
$\quad\quad [T ::= \cdot\ T * F,\ \textbf{+/*/)}],$ $goto(s_4, \textbf{(}) = s_{11}$
$\quad\quad [T ::= \cdot\ F,\ \textbf{+/*/)}],$ $goto(s_4, \textbf{id}) = s_{12}$
$\quad\quad [F ::= \cdot\ (E),\ \textbf{+/*/)}],$
$\quad\quad [F ::= \cdot\ \textbf{id},\ \textbf{+/*/)}]\}$

$s_5 = \{[F ::= \textbf{id}\ \cdot,\ \textbf{+/*/\#}]\}$

$s_{16} = \{[E ::= E + \cdot\ T,\ \textbf{+/)}],$ $goto(s_{16}, T) = s_{19}$
$\quad\quad [T ::= \cdot\ T * F,\ \textbf{+/*/)}],$ $goto(s_{16}, F) = s_{10}$
$\quad\quad [T ::= \cdot\ F,\ \textbf{+/*/)}],$ $goto(s_{16}, \textbf{(}) = s_{11}$
$\quad\quad [F ::= \cdot\ (E),\ \textbf{+/*/)}],$ $goto(s_{16}, \textbf{id}) = s_{12}$
$\quad\quad [F ::= \cdot\ \textbf{id},\ \textbf{+/*/)}]\}$

$s_6 = \{[E ::= E + \cdot\ T,\ \textbf{+/\#}],$ $goto(s_6, T) = s_{13}$
$\quad\quad [T ::= \cdot\ T * F,\ \textbf{+/*/\#}],$ $goto(s_6, F) = s_3$
$\quad\quad [T ::= \cdot\ F,\ \textbf{+/*/\#}],$ $goto(s_6, \textbf{(}) = s_4$
$\quad\quad [F ::= \cdot\ (E),\ \textbf{+/*/\#}],$ $goto(s_6, \textbf{id}) = s_5$
$\quad\quad [F ::= \cdot\ \textbf{id},\ \textbf{+/*/\#}]\}$

$s_{17} = \{[T ::= T * \cdot\ F,\ \textbf{+/*/)}],$ $goto(s_{17}, F) = s_{20}$
$\quad\quad [F ::= \cdot\ (E),\ \textbf{+/*/)}],$ $goto(s_{17}, \textbf{(}) = s_{11}$
$\quad\quad [F ::= \cdot\ \textbf{id},\ \textbf{+/*/)}]\}$ $goto(s_{17}, \textbf{id}) = s_{12}$

$s_7 = \{[T ::= T * \cdot\ F,\ \textbf{+/*/\#}],$ $goto(s_7, F) = s_{14}$
$\quad\quad [F ::= \cdot\ (E),\ \textbf{+/*/\#}],$ $goto(s_7, \textbf{(}) = s_4$
$\quad\quad [F ::= \cdot\ \textbf{id},\ \textbf{+/*/\#}]\}$ $goto(s_7, \textbf{id}) = s_5$

$s_{18} = \{[F ::= (E\ \cdot),\ \textbf{+/*/)}],$ $goto(s_{18}, \textbf{)}) = s_{21}$
$\quad\quad [E ::= E\ \cdot + T,\ \textbf{+/)}]\}$ $goto(s_{18}, \textbf{+}) = s_{16}$

$s_8 = \{[F ::= (E\ \cdot),\ \textbf{+/*/\#}],$ $goto(s_8, \textbf{)}) = s_{15}$
$\quad\quad [E ::= E\ \cdot + T,\ \textbf{+/)}]\}$ $goto(s_8, \textbf{+}) = s_{16}$

$s_{19} = \{[E ::= E + T\ \cdot,\ \textbf{+/)}],$ $goto(s_{19}, *) = s_{17}$
$\quad\quad [T ::= T\ \cdot * F,\ \textbf{+/*/)}]\}$

$s_9 = \{[E ::= T\ \cdot,\ \textbf{+/)}],$ $goto(s_9, *) = s_{17}$
$\quad\quad [T ::= T\ \cdot * F,\ \textbf{+/*/)}]\}$

$s_{20} = \{[T ::= T * F\ \cdot,\ \textbf{+/*/)}]\}$

$s_{10} = \{[T ::= F\ \cdot,\ \textbf{+/*/)}]\}$

$s_{21} = \{[F ::= (E)\ \cdot,\ \textbf{+/*/)}]\}$

Merging the states and re-computing the gotos gives us the LALR(1) canonical collection illustrated in the table below.

$s_0 =$

$\{[E' ::= \cdot\ E,\ \textbf{\#}],$ $goto(s_0, E) = s_1$
$\quad [E ::= \cdot\ E + T,\ \textbf{+/\#}],$ $goto(s_0, T) = s_{2.9}$
$\quad [E ::= \cdot\ T,\ \textbf{+/\#}],$ $goto(s_0, F) = s_{3.10}$
$\quad [T ::= \cdot\ T * F,\ \textbf{+/*/\#}],$ $goto(s_0, \textbf{(}) = s_{4.11}$
$\quad [T ::= \cdot\ F,\ \textbf{+/*/\#}],$ $goto(s_0, \textbf{id}) = s_{5.12}$
$\quad [F ::= \cdot (E),\ \textbf{+/*/\#}],$
$\quad [F ::= \cdot\ \textbf{id},\ \textbf{+/*/\#}]\}$

$s_{6.16} =$

$\{[E ::= E + \cdot\ T,\ \textbf{+/)/\#}],$ $goto(s_{6.16}, T) = s_{13.19}$
$\quad [T ::= \cdot\ T * F,\ \textbf{+/*/)/\#}],$ $goto(s_{6.16}, F) = s_{3.10}$
$\quad [T ::= \cdot\ F,\ \textbf{+/*/)/\#}],$ $goto(s_{6.16}, \textbf{(}) = s_{4.11}$
$\quad [F ::= \cdot\ (E),\ \textbf{+/*/)/\#}],$ $goto(s_{6.16}, \textbf{id}) = s_{5.12}$
$\quad [F ::= \cdot\ \textbf{id},\ \textbf{+/*/)/\#}]\}$

$s_1 =$

$\{[E' ::= E\ \cdot,\ \textbf{\#}],$ $goto(s_1, \textbf{+}) = s_{6.16}$
$\quad [E ::= E\ \cdot + T,\ \textbf{+/\#}]\}$

$s_{7.17} =$

$\{[T ::= T * \cdot\ F,\ \textbf{+/*/)/\#}],$ $goto(s_{7.17}, F) = s_{14.20}$
$\quad [F ::= \cdot\ (E),\ \textbf{+/*/)/\#}],$ $goto(s_{7.17}, \textbf{(}) = s_{4.11}$
$\quad [F ::= \cdot\ \textbf{id},\ \textbf{+/*/)/\#}]\}$ $goto(s_{7.17}, \textbf{id}) = s_{5.12}$

$s_{2.9} =$

$\{[E ::= T \cdot, +/)/\#],$ $\quad$ goto$(s_{2.9}, *) = s_{7.17}$
$[T ::= T \cdot * F, +/*/)/\#]\}$

$s_{3.10} =$

$\{[T ::= F \cdot, +/*/)/\#]\}$

$s_{4.11} =$

$\{[F ::= (\cdot E), +/*/)/\#],$ $\quad$ goto$(s_{4.11}, E) = s_{8.18}$
$[E ::= \cdot E + T, +/)],$ $\quad\quad$ goto$(s_{4.11}, T) = s_{2.9}$
$[E ::= \cdot T, +/)],$ $\quad\quad\quad$ goto$(s_{4.11}, F) = s_{3.10}$
$[T ::= \cdot T * F, +/*/)],$ $\quad$ goto$(s_{4.11}, () = s_{4.11}$
$[T ::= \cdot F, +/*/)],$ $\quad\quad$ goto$(s_{4.11}, \text{id}) = s_{5.12}$
$[F ::= \cdot (E), +/*/)],$
$[F ::= \cdot \text{id}, +/*/)]\}$

$s_{5.12} = \{[F ::= \text{id} \cdot, +/*/)/\#]\}$

$s_{8.18} =$

$\{[F ::= (E \cdot), +/*/)/\#],$ goto$(s_{8.18},)) = s_{15.21}$
$[E ::= E \cdot + T, +/)]\}$ $\quad$ goto$(s_{8.18}, +) = s_{6.16}$

$s_{13.19} =$

$\{[E ::= E + T \cdot, +/)/\#],$ $\quad$ goto$(s_{13.19}, *) = s_{7.17}$
$[T ::= T \cdot * F, +/*/)/\#]\}$

$s_{14.20} = \{[T ::= T * F \cdot, +/*/)/\#]\}$

$s_{15.21} = \{[F ::= (E) \cdot, +/*/)]\}$

The LALR(1) parsing tables are given in Figure 3.9.

	Action						Goto		
	+	*	(	)	id	#	E	T	F
0			s4		s5		1	2	3
1	s6					accept			
2.9	r2	s7.17		r2		r2			
3.10	r4	r4		r4		r4			
4.11			s11		s12		8.18	2.9	3.10
5.12	r6	r6		r6		r6			
6.16			s4		s5			13.19	3.10
7.17			s4		s5				14.20
8.18	s16			s15					
13.19	r1	s7				r1			
14.20	r3	r3				r3			
15.21	r5	r5				r5			

FIGURE 3.9 The LALR(1) parsing tables for the Grammar in (3.42)

Of course, this approach of first generating the LR(1) canonical collection of states and then merging states to produce the LALR(1) collection consumes a great deal of space. But once the tables are constructed, they are small and workable. An alternative approach, which does not consume so much space, is to do the merging as the LR(1) states are produced.

Merging the states as they are constructed

Our algorithm for computing the LALR(1) canonical collection is a slight variation on Algorithm 3.10; it is Algorithm 3.13.

Algorithm 3.13 Computing the LALR(1) Collection of States

Input: a context-free grammar $G = (N, T, S, P)$
Output: the canonical LALR(1) collection of states $c = \{s_0, s_1, \ldots, s_n\}$

Define an augmented grammar G' which is G with the added non-terminal S' and added production rule $S' ::= S$, where S is G's start symbol. The following steps apply to G'. Enumerate the production rules beginning at 0 for the newly added production

$c \leftarrow \{s_0\}$, where $s_0 = \text{closure}(\{[S' ::= \cdot S, \#]\})$
repeat
    for each s in c, and for each symbol $X \in T \cup N$ **do**
        if $\text{goto}(s, X) \neq \emptyset$ and $\text{goto}(s, X) \notin c$ **then**
            Add $s = \text{goto}(s, X)$ to c
            Check to see if the cores of an existing state in c are equivalent to the cores of s
            If so, merge s with that state
            Otherwise, add s to the collection c
        end if
    end for
until no new states are added to c

There are other enhancements we can make to Algorithm 3.13 to conserve even more space. For example, as the states are being constructed, it is enough to store their kernels. The closures may be computed when necessary, and even these may be cached for each non-terminal symbol.

LALR(1) Conflicts

There is the possibility that the LALR(1) table for a grammar may have conflicts where the LR(1) table does not. Therefore, while it should be obvious that every LALR(1) grammar is an LR(1) grammar, not every LR(1) grammar is an LALR(1) grammar.

How can these conflicts arise? A shift-reduce conflict cannot be introduced by merging two states, because we merge two states only if they have the same core items. If the merged state has an item that suggests a shift on a terminal a and another item that suggests a reduce on the lookahead a, then at least one of the two original states must have contained both items, and so caused a conflict.

On the other hand, merging states can introduce reduce-reduce conflicts. An example arises in grammar given in Exercise 3.20. Even though LALR(1) grammars are not as powerful as LR(1) grammars, they are sufficiently powerful to describe most programming languages. This, together with their small (relative to LR) table size, makes the LALR(1) family of grammars an excellent candidate for the automatic generation of parsers. Stephen C. Johnson's YACC, for "Yet Another Compiler-Compiler" [Johnson, 1975], based on LALR(1) techniques, was probably the first practical bottom-up parser generator. GNU has developed an open-source version called Bison [Donnelly and Stallman, 2011].

3.4.4   LL or LR?

Figure 3.10 illustrates the relationships among the various categories of grammars we have been discussing.

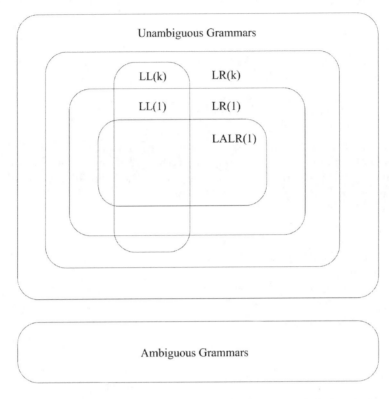

FIGURE 3.10 Categories of context-free grammars and their relationship.

Theoretically, LR(1) grammars are the largest category of grammars that can be parsed deterministically while looking ahead just one token. Of course, LR(k) grammars for $k > 1$ are even more powerful, but one must look ahead k tokens, more importantly, the parsing tables must (in principle) keep track of all possible token strings of length k. So, in principle, the tables can grow exponentially with k.

LALR(1) grammars make for parsers that are almost as powerful as LR(1) grammars but result in much more space-efficient parsing tables. This goes some way in explaining the popularity of parser generators such as YACC and Bison.

LL(1) grammars are the least powerful category of grammars that we have looked at. Every LL(1) grammar is an LR(1) grammar and every LL(k) grammar is an LR(k) grammar.

Also, every LR(k) grammar is an unambiguous grammar. Indeed, the LR(k) category is the largest category of grammars for which we have a means for testing membership. There is no general algorithm for telling us whether an arbitrary context-free grammar is unambiguous. But we can test for LR(1) or LR(k) for some k; and if it is LR(1) or LR(k), then it is unambiguous.

In principle, recursive descent parsers work only when based on LL(1) grammars. But as we have seen, one may program the recursive descent parser to look ahead a few symbols, in those places in the grammar where the LL(1) condition does not hold.

LL(1), LR(1), LALR(1), and recursive descent parsers have all been used to parse one programming language or another. LL(1) and recursive descent parsers have been applied to most of Nicklaus Wirth's languages, for example, Algol-W, Pascal, and Modula. Recursive descent was used to produce the parser for the first implementations of C but then using YACC; that and the fact that YACC was distributed with Unix popularized it. YACC was the first LALR(1) parser generator with a reasonable execution time.

Interestingly, LL(1) and recursive descent parsers are enjoying greater popularity, for example for the parsing of Java. Perhaps it is the simplicity of the predictive top-down approach. It is now possible to come up with (mostly LL) predictive grammars for most programming languages. True, none of these grammars are strictly LL(1); indeed, they are not even unambiguous, look at the if-else statement in Java and almost every other programming language. But these special cases may be handled specially, for example by selective looking ahead k symbols in rules where it is necessary, and favoring the scanning of an *else* when it is part of an if-statement. There are parser generators that allow the parser developer to assert these special conditions in the grammatical specifications. One of these is JavaCC, which we discuss in the next section.

3.5 Parser Generation Using JavaCC

In Chapter 2 we saw how JavaCC can be used to generate a lexical analyzer for j-- from an input file (j--.jj) specifying the lexical structure of the language as regular expressions. In this section, we will see how JavaCC can be used to generate an LL(k) recursive descent parser for j-- from a file specifying its syntactic structure as EBNF (extended BNF) rules.

In addition to containing the regular expressions for the lexical structure for j--, the j--.jj file also contains the syntactic rules for the language. The Java code between the PARSER_BEGIN(JavaCCParser) and PARSER_END(JavaCCParser) block in the j--.jj file is copied verbatim to the generated JavaCCParser.java file in the jminusminus package. This code defines helper functions, which are available for use within the generated parser. Some of the helpers include reportParserError() for reporting errors and recoverFromError() for recovering from errors. Following this block is the specification for the scanner for j--, and following that is the specification for the parser for j--.

We now describe the JavaCC syntactic specification. The general layout is this: we define a start symbol, which is a high-level non-terminal (compilationUnit in case of j--) that references lower-level non-terminals. These lower-level non-terminals in turn reference the tokens defined in the lexical specification.

When building a syntactic specification, we are not limited to literals and simple token references. We can use the following EBNF syntax:

- [a] for "zero or one", or an "optional" occurrence of a

- (a)∗ for "zero or more" occurrences of a

- a|b for alternation, that is, either a or b

- () for grouping

The syntax for a non-terminal declaration (or, production rule) in the input file almost resembles that of a java method declaration; it has a return type (could be void), a name, can accept arguments, and has a body that specifies the extended BNF rules along with any actions that we want performed as the production rule is parsed. It also has a block preceding the body; this block declares any local variables used within the body. Syntactic

actions, such as creating an AST node, are java code embedded within blocks. JavaCC turns
the specification for each non-terminal into a java method within the generated parser.

As an example, let us look at how we specify the following rule:

qualifiedIdentifier ::= <identifier> { . <identifier>}

for parsing a qualified identifier using JavaCC.

```
private TypeName qualifiedIdentifier(): {
    int line = 0;
    String qualifiedIdentifier = "";
}
{
    try {
        <IDENTIFIER>
        {
            line = token.beginLine;
            qualifiedIdentifier = token.image;
        }
        (
            <DOT> <IDENTIFIER>
            { qualifiedIdentifier += "." + token.image; }
        )*
    }
    catch (ParseException e) {
        recoverFromError(new int[] { SEMI, EOF }, e);
    }
    { return
        new TypeName(line, qualifiedIdentifier); }
}
```

Let us walk through the above method in order to make sense out of it.

- The `qualifiedIdentifier` non-terminal, as in the case of the hand-written parser,
  is `private` method and returns an instance of `TypeName`. The method does not take
  any arguments.

- The local variable block defines two variables, line and `qualifiedIdentifier`; the
  former is for tracking the line number in the source file, and the latter is for accumu-
  lating the individual identifiers (x, y, and z in x.y.z, for example) into a qualified
  identifier (x.y.z, for example). The variables line and `qualifiedIdentifier` are used
  within the body that actually parses a qualified identifier.

- In the body, all parsing is done within the try-catch block. When an identifier token[8]
  `<IDENTIFIER>` is encountered, its line number in the source file and its image are
  recorded in the respective variables; this is an action and hence is within a block. We
  then look for zero or more occurrences of the tokens `<DOT>` `<IDENTIFIER>` within the
  `()*` EBNF construct. For each such occurrence, we append the image of the identifier
  to the `qualifiedIdentifier` variable; this is also an action and hence is java code
  within a block. Once a qualified identifer has been parsed, we return (again an action
  and hence is java code within a block) an instance of `TypeName`.

- JavaCC raises a `ParseException` when encountering a parsing error. The instance of
  `ParseException` stores information about the token that was found and the token
  that was sought. When such an exception occurs, we invoke our own error recovery

---

[8]The token variable stores the current token information: `token.beginLine` stores the line number
in which the token occurs in the source file and `token.image` stores the token's image (for example, the
identifier name in case of the `<IDENTIFIER>` token).

method `recoverFromError()` and try to recover to the nearest semicolon (`SEMI`) or to the end of file (`EOF`). We pass to this method the instance of `ParseException` so that the method can report a meaningful error message.

As another example, let us see how we specify the non-terminal statement,

statement ::= block
         | <identifier> : statement
         | if parExpression statement [else statement]
         | while parExpression statement
         | return [expression] ;
         | ;
         | statementExpression ;

for parsing statements in *j--*.

```
private JStatement statement(): {
    int line = 0;
    JStatement statement = null;
    JExpression test = null;
    JStatement consequent = null;
    JStatement alternate = null;
    JStatement body = null;
    JExpression expr = null;
}
{
    try {
        statement = block() |
        <IF> { line = token.beginLine; }
        test = parExpression()
        consequent = statement()

        // Even without the lookahead below, which is added to
        // suppress JavaCC warnings, dangling if-else problem is
        // resolved by binding the alternate to the closest
        // consequent.
        [
            LOOKAHEAD(<ELSE>)
            <ELSE> alternate = statement()
        ]
        { statement =
            new JIfStatement(line, test, consequent, alternate); } |
        <WHILE> { line = token.beginLine; }
        test = parExpression()
        body = statEOT()
        { statement = new JWhileStatement(line, test, body); } |
        <RETURN> { line = token.beginLine; }
        [
            expr = expression()
        ]
        <SEMI>
        { statement = new JReturnStatement(line, expr); } |
        <SEMI>
        { statement = new JEmptyStatement(line); } |
        // Must be a statementExpression
        statement = statementExpression()
        <SEMI>
    }
    catch (ParseException e) {
        recoverFromError(new int[] { SEMI, EOF }, e);
    }
    { return statement; }
```

```
}
```

We will jump right into the try block to see what is going on.

- If the current token is <LCURLY>, which marks the beginning of a block, the lower-level non-terminal block is invoked to parse block-statement. The value returned by block is assigned to the local variable statement.

- If the token is <IF>, we get its line number and parse an if statement; we delegate to the lower-level non-terminals parExpression and statement to parse the test expression, the consequent, and the (optional) alternate statements. Note the use of the | and [] JavaCC constructs for alternation and option. Once we have successfully parsed an if statement, we create an instance of the AST node for an if statement and assign it to the local variable statement.

- If the token is <WHILE>, <RETURN>, or <SEMI>, we parse a while, return, or an empty statement.

- Otherwise, it must be a statement expression, which we parse by simply delegating to the lower-level non-terminal statementExpression. In each case we set the local variable statement to the appropriate AST node instance.

Finally, we return the local variable statement and this completes the parsing of a *j--* statement.

Lookahead

As in the case of a recursive descent parser, we cannot always decide which production rule to use in parsing a non-terminal just by looking at the current token; we have to look ahead at the next few symbols to decide. JavaCC offers a function called LOOKAHEAD that we can use for this purpose. Here is an example in which we parse a simple unary expression in *j--*, expressed by the BNF rule:

simpleUnaryExpression ::= ! unaryExpression
                  | (basicType) unaryExpression //cast
                  | (referenceType) simpleUnaryExpression // cast
                  | postfixExpression

```
private JExpression simpleUnaryExpression(): {
    int line = 0;
    Type type = null;
    JExpression expr = null, unaryExpr = null, simpleUnaryExpr = null;
}
{
    try {
        <LNOT> { line = token.beginLine; }
        unaryExpr = unaryExpression()
        { expr = new JLogicalNotOp(line, unaryExpr); } |
        LOOKAHEAD(<LPAREN> basicType() <RPAREN>)
        <LPAREN> { line = token.beginLine; }
        type = basicType()
        <RPAREN>
        unaryExpr = unaryExpression()
        { expr = new JCastOp(line, type, unaryExpr); } |
        LOOKAHEAD(<LPAREN> referenceType() <RPAREN>)
        <LPAREN> { line = token.beginLine; }
        type = referenceType()
```

```
        <RPAREN>
        simpleUnaryExpr = simpleUnaryExpression()
        { expr = new JCastOp(line, type, simpleUnaryExpr); } |
        expr = postfixExpression()
    }
    catch (ParseException e) {
        recoverFromError(new int[] { SEMI, EOF }, e);
    }
    { return expr ; }
}
```

We use the `LOOKAHEAD` function to decide between a cast expression involving a basic type, a cast expression involving a reference type, and a postfix expression, which could also begin with an `LPAREN` (`(x)` for example). Notice how we are spared the chore of writing our own non-terminal-specific lookahead functions. Instead, we simply invoke the JavaCC `LOOKAHEAD` function by passing in tokens we want to look ahead. Thus, we do not have to worry about backtracking either; `LOOKAHEAD` does it for us behind the scenes. Also notice how, as in `LOOKAHEAD(<LPAREN> basicType()<RPAREN>)`, we can pass both terminals and non-terminals to `LOOKAHEAD`.

Error Recovery

JavaCC offers two error recovery mechanisms, namely *shallow* and *deep* error recovery. We employ the latter in our implementation of a JavaCC parser for *j--*. This involves catching within the body of a non-terminal, the `ParseException` that is raised in the event of a parsing error. The exception instance `e` along with skip-to tokens are passed to our `recoverFromError()` error recovery function. The exception instance has information about the erroneous token that was found and the token that was expected, and `skipTo` is an array of tokens that we would like to skip to in order to recover from the error. Here is the function:

```
private void recoverFromError(int[] skipTo, ParseException e) {
    // Get the possible expected tokens
    StringBuffer expected = new StringBuffer();
    for (int i = 0; i < e.expectedTokenSequences.length; i++) {
        for (int j = 0; j < e.expectedTokenSequences[ i ].length;
            j++) {
            expected.append("\n");
            expected.append("     ");
            expected.append(tokenImage[
                e.expectedTokenSequences[ i ][ j ] ]);
            expected.append("...");
        }
    }

    // Print error message
    if (e.expectedTokenSequences.length == 1) {
        reportParserError("\"%s\" found where %s sought",
            getToken(1), expected);
    }
    else {
        reportParserError("\"%s\" found where one of %s sought",
            getToken(1), expected);
    }

    // Recover
    boolean loop = true;
    do {
        token = getNextToken();
        for (int i = 0; i < skipTo.length; i++) {
```

```
            if (token.kind == skipTo[ i ]) {
                loop = false;
                break;
            }
        }
    } while(loop);
}
```

First, the function, from the token that was found and the token that was sought, constructs and displays an appropriate error message. Second, it recovers by skipping to the nearest token in the skipto list of tokens.

In the current implementation of the parser for *j--*, all non-terminals specify SEMI and EOF as skipTo tokens. This error recovery scheme could be made more sophisticated by specifying the follow of the non-terminal as skipTo tokens.

Note that when ParseException is raised, control is transferred to the calling non-terminal. Thus when an error occurs within higher non-terminals, the lower non-terminals go unparsed.

Generating a Parser Versus Hand-Writing a Parser

If you compare the JavaCC specification for the parser for *j--* with the hand-written parser, you will notice that they are very much alike. This would make you wonder whether we are gaining anything by using JavaCC. The answer is, yes we are. Here are some of the benefits:

- Lexical structure is much more easily specified using regular expressions.

- EBNF constructs are allowed.

- Lookahead is easier; it is given as a function and takes care of backtracking.

- Choice conflicts are reported when lookahead is insufficient

- Sophisticated error recovery mechanisms are available.

Other parser generators, including ANTLR[9] for Java, also offer the above advantages over hand-written parsers.

3.6 Further Readings

For a thorough and classic overview of context-free parsing, see [Aho et al., 2007].

The context-free syntax for Java may be found in [Gosling et al., 2005]; see chapters 2 and 18. This book is also published online at http://docs.oracle.com/javase/specs/.

LL(1) parsing was introduced in [Lewis and Stearns, 1968] and [Knuth, 1971b]. Recursive descent was introduced in [Lewis et al., 1976]. The simple error-recovery scheme used in our parser comes from [Turner, 1977].

See Chapter 5 in [Copeland, 2007] for more on how to generate parsers using JavaCC. See Chapter 7 for more information on error recovery. See Chapter 8 for a case study—parser for JavaCC grammar. JavaCC itself is open-source software, which may be obtained from https://javacc.dev.java.net/. Also, see [van der Spek et al., 2005] for a discussion of error recovery in JavaCC.

---

[9]A parser generator; http://www.antlr.org/.

See [Aho et al., 1975] for an introduction to YACC. The canonical open-source implementation of the LALR(1) approach to parser generation is given by [Donnelly and Stallman, 2011]. See [Burke and Fisher, 1987] for a nice approach to LALR(1) parser error recovery.

Other shift-reduce parsing strategies include both simple-precedence and operator-precedence parsing. These are nicely discussed in [Gries, 1971].

3.7 Exercises

Exercise 3.1. Consult Chapter 18 of the *Java Language Specification* [Gosling et al., 2005]. There you will find a complete specification of Java's context-free syntax.

a. Make a list of all the expressions that are in Java but not in *j--*.

b. Make a list of all statements that are in Java but not in *j--*.

c. Make a list of all type declarations that are in Java but not in *j--*.

d. What other linguistic constructs are in Java but not in *j--*?

Exercise 3.2. Consider the following grammar:

$S ::= (L) \mid$ a
$L ::= L\ S \mid \epsilon$

a. What language does this grammar describe?

b. Show the parse tree for the string (a ( ) (a (a))).

c. Derive an equivalent LL(1) grammar.

Exercise 3.3. Show that the following grammar is ambiguous.

$S ::=$ a S b $S \mid$ b S a $S \mid \epsilon$

Exercise 3.4. Show that the following grammar is ambiguous. Come up with an equivalent grammar that is not ambiguous.

$E ::= E$ and $E \mid E$ or $E \mid$ true $\mid$ false

Exercise 3.5. Write a grammar that describes the language of Roman numerals.

Exercise 3.6. Write a grammar that describes Lisp *s*-expressions.

Exercise 3.7. Write a grammar that describes a number of (zero or more) a's followed by an equal number of b's.

Exercise 3.8. Show that the following grammar is not LL(1).

$S ::=$ a b $\mid A$ b
$A ::=$ a a $\mid$ c d

Exercise 3.9. Consider the following context-free grammar:

$S ::= B \text{ a} \mid \text{a}$
$B ::= \text{c} \mid \text{b } C \text{ } B$
$C ::= \text{c } C \mid \epsilon$

a. Compute first and follow for S, B, and C.

b. Construct the LL(1) parsing table for this grammar.

c. Is this grammar LL(1)? Why or why not?

Exercise 3.10. Consider the following context-free grammar:

$S ::= A \text{ a} \mid \text{a}$
$A ::= \text{c} \mid \text{b } B$
$B ::= \text{c } B \mid \epsilon$

a. Compute first and follow for S, A, and B.

b. Construct the LL(1) parsing table for the grammar.

c. Is this grammar LL(1)? Why or why not?

Exercise 3.11. Consider the following context-free grammar:

$S ::= A \text{ a}$
$A ::= \text{b d } B \mid \text{e } B$
$B ::= \text{c } A \mid \text{d } B \mid \epsilon$

a. Compute first and follow for S, A, and B.

b. Construct an LL(1) parsing table for this grammar.

c. Show the steps in parsing b d c e a.

Exercise 3.12. Consider the following context-free grammar:

$S ::= AS \mid \text{b}$
$A ::= SA \mid \text{a}$

a. Compute first and follow for S and A.

b. Construct the LL(1) parsing table for the grammar.

c. Is this grammar LL(1)? Why or why not?

Exercise 3.13. Show that the following grammar is LL(1).

$S ::= A \text{ a } A \text{ b}$
$S ::= B \text{ b } B \text{ a}$
$A ::= \epsilon$
$B ::= \epsilon$

Exercise 3.14. Consider the following grammar:

$E ::= E \text{ or } T \mid T$
$T ::= T \text{ and } F \mid F$
$F ::= \text{not } F \mid (E) \mid \text{i}$

a. Is this grammar LL(1)? If not, derive an equivalent grammar that is LL(1).

b. Construct the LL(1) parsing table for the LL(1) grammar.

c. Show the steps in parsing not i and i or i.

Exercise 3.15. Consider the following grammar:

$S ::= L = R$
$S ::= R$
$L ::= * R$
$L ::= i$
$R ::= L$

a. Construct the canonical LR(1) collection.

b. Construct the Action and Goto tables.

c. Show the steps in the parse for * i = i.

Exercise 3.16. Consider the following grammar:

$S ::= (L) \mid a$
$L ::= L , S \mid S$

a. Compute the canonical collection of LR(1) items for this grammar.

b. Construct the LR(1) parsing table for this grammar.

c. Show the steps in parsing the input string ((a , a), a).

d. Is this an LALR(1) grammar?

Exercise 3.17. Consider the following grammar.

$S ::= A a \mid b A c \mid d c \mid b d a$
$A ::= d$

a. What is the language described by this grammar?

b. Compute first and follow for all non-terminals.

c. Construct the LL(1) parsing table for this grammar. Is it LL(1)? Why?

d. Construct the LR(1) canonical collection, and the Action and Goto tables for this grammar. Is it LR(1)? Why or why not?

Exercise 3.18. Is the following grammar LR(1)? LALR(1)?

$S ::= C C$
$C ::= a C \mid b$

Exercise 3.19. Consider the following context-free grammar:

$S ::= A a \mid b A c \mid d c \mid b d a$
$A ::= d$

a. Compute the canonical LR(1) collection for this grammar.

b. Compute the Action and Goto tables for this grammar.

c. Show the steps in parsing b d c.

d. Is this an LALR(1) grammar?

Exercise 3.20. Show that the following grammar is LR(1) but not LALR(1).

$S ::= $ a B c $ | $ b C d $ | $ a C d b B d
$B ::= $ e
$C ::= $ e

Exercise 3.21. Modify the Parser to parse and return nodes for the double literal and the float literal.

Exercise 3.22. Modify the Parser to parse and return nodes for the long literal.

Exercise 3.23. Modify the Parser to parse and return nodes for all the additional operators that are defined in Java but not yet in *j--*.

Exercise 3.24. Modify the Parser to parse and return nodes for conditional expressions, for example, (a > b)? a : b.

Exercise 3.25. Modify the Parser to parse and return nodes for the for-statement, including both the basic for-statement and the enhanced for-statement.

Exercise 3.26. Modify the Parser to parse and return nodes for the switch-statement.

Exercise 3.27. Modify the Parser to parse and return nodes for the try-catch-finally statement.

Exercise 3.28. Modify the Parser to parse and return nodes for the throw-statement.

Exercise 3.29. Modify the Parser to deal with a throws-clause in method declarations.

Exercise 3.30. Modify the Parser to deal with methods and constructors having variable arity, that is, a variable number of arguments.

Exercise 3.31. Modify the Parser to deal with both static blocks and instance blocks in type declarations.

Exercise 3.32. Although we do not describe the syntax of generics in Appendix C, it is described in Chapter 18 of the *Java Language Specification* [Gosling et al., 2005]. Modify the Parser to parse generic type definitions and generic types.

Exercise 3.33. Modify the j--.jj file in the compiler's code tree for adding the above (3.22 through 3.31) syntactic constructs to *j--*.

Exercise 3.34. Say we wish to add a do-until statement to *j--*. For example,

```
do {
    x = x * x;
}
until (x > 1000);
```

a. Write a grammar rule for defining the context-free syntax for a new do-until statement.

b. Modify the Scanner to deal with any necessary new tokens.

c. Modify the Parser to parse and return nodes for the do-until statement.

Chapter 4

Type Checking

4.1 Introduction

Type checking, or more formally *semantic analysis*, is the final step in the analysis phase. It is the compiler's last chance to collect information necessary to begin the synthesis phase. Semantic analysis includes the following:

- Determining the types of all names and expressions.

- Type checking: insuring that all expressions are properly typed, for example, that the operands of an operator have the proper types.

- A certain amount of storage analysis, for example determining the amount of storage that is required in the current stack frame to store a local variable (one word for ints, two words for longs). This information is used to allocate locations (at offsets from the base of the current stack frame) for parameters and local variables.

- A certain amount of AST tree rewriting, usually to make implicit constructs more explicit.

Semantic analysis of *j--* programs involves all of the following operations.

- Like Java, *j--* is *strictly-typed*; that is, we want to determine the types of all names and expressions at compile time.

- A *j--* program must be well-typed; that is, the operands to all operations must have appropriate types.

- All *j--* local variables (including formal parameters) must be allocated storage and assigned locations within a method's stack frame.

- The AST for *j--* requires a certain amount of sub-tree rewriting. For example, field references using simple names must be rewritten as explicit field selection operations. And declared variable initializations must be rewritten as explicit assignment statements.

4.2 *j--* Types

4.2.1 Introduction to *j--* Types

A type in *j--* is either a primitive type or a reference type.

j-- **primitive types:**

- `int` - 32 bit two's complement integers
- `boolean` - taking the value `true` or `false`
- `char` - 16 bit Unicode (but many systems deal only with the lower 8 bits)

j-- **reference types:**

- Arrays
- Objects of a type described by a class declaration
- Built-in objects `java.lang.Object` and `java.lang.String`

j-- code may interact with classes from the Java library but it must be able to do so using only these types.

4.2.2   Type Representation Problem

The question arises: How do we represent a type in our compiler? For example, how do we represent the types `int`, `int[]`, `Factorial`, `String[][]`? The question must be asked in light of two desires:

1. We want a simple, but extensible representation. We want no more complexity than is necessary for representing all of the types in *j*-- and for representing any (Java) types that we may add in exercises.

2. We want the ability to interact with the existing Java class libraries.

Two solutions come immediately to mind:

1. Java types are represented by objects of (Java) type `java.lang.Class`. `Class` is a class defining the interface necessary for representing Java types. Because *j*-- is a subset of Java, why not use `Class` objects to represent its types? Unfortunately, the interface is not as convenient as we might like.

2. A home-grown representation may be simpler. One defines an abstract class (or interface) `Type`, and concrete sub-classes (or implementations) `PrimitiveType`, `ReferenceType`, and `ArrayType`.

4.2.3   Type Representation and Class Objects

Our solution is to define our own class `Type` for representing types, with a simple interface but also encapsulating the `java.lang.Class` object that corresponds to the Java representation for that same type.

But the parser does not know anything about types. It knows neither what types have been declared nor which types have been imported. For this reason we define two placeholder type representations:

1. `TypeName` - for representing named types recognized by the parser like user-defined classes or imported classes until such time as they may be resolved to their proper `Type` representation.

2. `ArrayTypeName` - for representing array types recognized by the parser like `String[]`, until such time that they may resolved to their proper `Type` representation.

During analysis, `TypeNames` and `ArrayTypeNames` are resolved to the `Types` that they represent. Type resolution involves looking up the type names in the symbol table to determine which defined type or imported type they name. More specifically,

- A `TypeName` is resolved by looking it up in the current context, our representation of our symbol table. The `Type` found replaces the `TypeName`.[1] Finally, the `Type`'s accessibility from the place the `TypeName` is encountered is checked.

- An `ArrayTypeName` has a base type. First the base type is resolved to a `Type`, whose `Class` representation becomes the base type for a new `Class` object for representing the array type[2]. Our new `Type` encapsulates this `Class` object.

- A `Type` resolves to itself.

So that `ArrayTypeNames` and `TypeNames` may stand in for `Types` in the compiler, both are sub-classes of `Type`.

One might ask why the *j--* compiler does not simply use *Java's* `Class` objects for representing types. The answer is twofold:

1. Our `Type` defines just the interface we need.

2. Our `Type` permits the `Parser` to use its sub-types `TypeName` and `ArrayTypeName` in its place when denoting types that have not yet been resolved.

4.3 *j--* Symbol Tables

In general, a symbol table maps names to the things they name, for example, types, formal parameters, and local variables. These mappings are established in a declaration and consulted each time a declared name is encountered.

4.3.1 Contexts and Idefns: Declaring and Looking Up Types and Local Variables

In the *j--* compiler, the symbol table is a tree of `Context` objects, which spans the abstract syntax tree. Each `Context` corresponds to a region of scope in the *j--* source program and contains a map of names to the things they name.

For example, reconsider the simple `Factorial` program. In this version, we mark two locations in the program using comments: `position 1` and `position 2`.

```
package pass;

import java.lang.System;

public class Factorial {
    // Two methods and a field

    public static int factorial(int n) {
```

---

[1] Even though externally defined types must be explicitly imported, if the compiler does not find the name in the symbol table, it attempts to load a class file of the given name and, if successful, declares it.

[2] Actually, because Java does not provide the necessary API for creating `Class` objects that represent array types, we create an instance of that array type and use `getClass()` to get its type's `Class` representation.

```
        // position 1:
        if (n <= 0) {
            return 1;
        } else {
            return n * factorial(n - 1);
        }
    }

    public static void main(String[] args) {
        // position 2:
        int x = n;
        System.out.println(n + "! = " + factorial(x));
    }

    static int n = 5;
}
```

The symbol table for this program, and its relationship to the AST, is illustrated in Figure 4.1.

In its entirety, the symbol table takes the form of a *tree* that corresponds to the shape of the AST. A `context`—that is, a node in this tree—captures the region of scope corresponding to the AST node that points to it. For example, in Figure 4.1,

1. The context pointer from the AST's `JCompilationUnit` node points to the `JCompilationUnitContext` that is at the root of the symbol table.

2. The context pointer from the AST's `JClassDeclaration` points to a `ClassContext`,

3. The context pointer from the AST's two `JMethodDeclarations` each point to a `MethodContext`

4. The context pointer from the AST's two `JBlocks` each point to a `LocalContext`.

On the other hand, from any particular location in the program, looking back toward the root `CompilationUnitContext`, the symbol table looks like a stack of contexts. Each (`surroundingContext`) link back toward the `CompilationUnitContext` points to the context representing the surrounding lexical scope.

For example,

1. The context for `Position 2` pointer in Figure 4.1 points to a `LocalContext`, which declares the local variable x in the body of the `main()` method of the `Factorial` program.

2. Its `surroundingContext` pointer points to a `MethodContext` in which the formal parameter, `args` is declared.

3. Its `surroundingContext` pointer points to a `ClassContext` in which nothing is declared.

4. Its `surroundingContext` pointer points to a `CompilationUnitContext` at the base of the stack, which contains the declared types for the entire program.

During analysis, when the compiler encounters a variable, it looks up that variable in the symbol table by name, beginning at the `LocalContext` most recently created in the symbol table, for example, that pointed to by the pointer labeled "context for `Position 1`." Type names are looked up in the `CompilationUnitContext`. To make this easier, each context maintains three pointers to surrounding contexts, as illustrated in Figure 4.2.

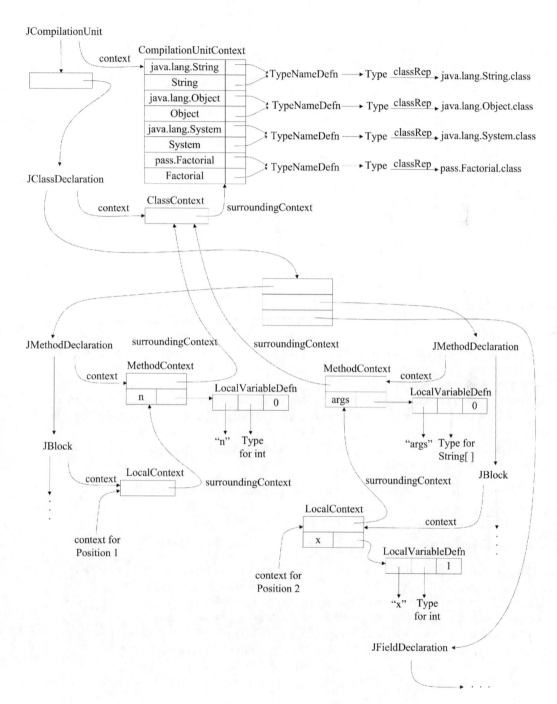

FIGURE 4.1 The symbol table for the Factorial program.

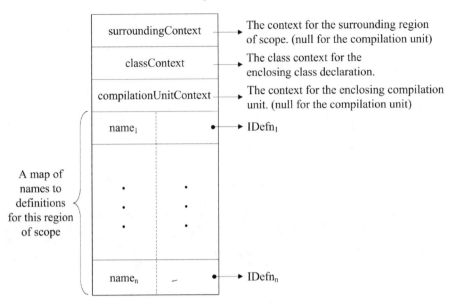

FIGURE 4.2 The structure of a context.

The pointer `surroundingContext` points to the context for the surrounding region of scope; the chain of these pointers is the stack that captures the nested scope in *j--* programs. The pointer `compilationUnitContext` points to the context for the enclosing compilation unit, that is, a `CompilationUnitContext`. In the current definition of *j--*, there is just one compilation unit but one might imagine an extension to *j--* permitting the compilation of several files, each of which defines a compilation unit. The pointer `classContext` points to the context (a `ClassContext`) for the enclosing class. As we shall see below, no names are declared in a `ClassContext` but this could change if we were to add nested type declarations to *j--*.

A `CompilationUnitContext` represents the scope of the entire program and contains a mapping from names to types:

- The implicitly declared types, `java.lang.Object`, and `java.lang.String`

- Imported types

- User-defined types, that is, types introduced in class declarations

A `ClassContext` represents the scope within a class declaration. In the *j--* symbol table, no names are declared here. All members, that is all constructors, methods and fields are recorded in the `Class` object that represents the type; we discuss this in the next section. If we were to add nested type declarations to *j--*, they might be declared here.

A `MethodContext` represents the scope within a method declaration. A method's formal parameters are declared here. A `MethodContext` is a kind of `LocalContext`.

A `LocalContext` represents the scope within a block, that is, the region between two curly brackets { and }. This includes the block defining the body to a method. Local variables are declared here.

Each kind of context derives from (extends) the class `Context`, which supplies the mapping from names to definitions (`IDefns`). Because a method defines a local context, `MethodContext` extends `LocalContext`. The inheritance tree for contexts is illustrated in Figure 4.3

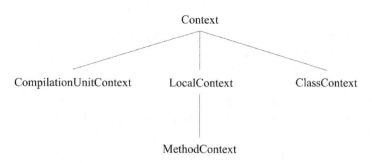

FIGURE 4.3 The inheritance tree for contexts.

An `IDefn` is the interface type for symbol table definitions, which has two implementations:

1. A `TypeNameDefn`, which defines a type name. An `IDefn` of this sort encapsulates the `Type` that it denotes.

2. A `LocalVariableDefn` defines a local variable and encapsulates the name, its `Type`, and an offset in the current run-time stack frame. (We discuss stack frame offsets in Section 4.5.2.)

4.3.2 Finding Method and Field Names in Type Objects

As we discussed in Section 4.2.3, the types defined by classes are represented in the same way as they are in the Java library, so that we may consistently manipulate types that we define and types that we import.

Class member (field and method in j--) names are not declared in a `ClassContext`, but in the `Types` that they declare. We rely on the encapsulated `Class` object to store the interface information, and we rely on Java reflection to query a type for information about its members.

For example, `Type` supports a method `fieldFor()` that, when given a name, returns a `Field` with the given name that is defined for that type. It delegates the finding of this field to a search of the underlying `Class` object and can be seen from the code that defines `fieldFor()`:

```
public Field fieldFor(String name) {
    Class<?> cls = classRep;
    while (cls != null) {
        java.lang.reflect.Field[] fields =
            cls.getDeclaredFields();
        for (java.lang.reflect.Field field:fields) {
            if (field.getName().equals(name)) {
                return new Field(field);
            }
        }
        cls = cls.getSuperclass();
    }
    return null;
}
```

This code first looks for the named field in the `Class` object for the `Type` being defined. If it does not find it there, it looks in the `Class` object for the `Type`'s super type, and so on until either the field is found or we come to the base of the inheritance tree, in which case `null` is returned to indicate that no such field was found. If we find the `java.lang.reflect.Field`, we encapsulate it within our own locally defined `Field`. More interface for querying a `Type` about its members is implemented, delegating the reflection to the underlying `Class` object that is kept in the `Type`'s `classRep` field.

The `Class` objects are created for declared classes in the `preAnalyze()` phase and are queried during the `analyze()` phase. This is made possible by the `CLEmitter`'s ability to create partial class files—class files that define the headers for methods (but not the bodies) and the types of fields (but not their initializations). These analysis issues are discussed in the next two sections.

4.4 Pre-Analysis of *j--* Programs

4.4.1 An Introduction to Pre-Analysis

The semantic analysis of *j--* (and Java) programs requires two traversals of the AST because a class name or a member name may be referenced before it is declared in the source program. The traversals are accomplished by methods (the method `preAnalyze()` for the first traversal and the method `analyze()` for the second), that invoke themselves at the child nodes for recursively descending the AST to its leaves.

But the first traversal need not traverse the AST as deeply as the second traversal. The only names that may be referred to before they are declared are type names (that is, class names in *j--*) and members.

So `preAnalyze()` must traverse down the AST only far enough for

- Declaring imported type names,

- Declaring user-defined class names,

- Declaring fields,

- Declaring methods (including their signatures the types of their parameters).

For this reason, `preAnalyze()` need be defined only in the following types of AST nodes:

- `JCompilationUnit`

- `JClassDeclaration`

- `JFieldDeclaration`

- `JMethodDeclaration`

- `JConstructorDeclaration`

So the `preAnalyze()` phase descends recursively down the AST only as far as the member declarations, but not into the bodies of methods[3].

---

[3]Implementing nested classes would require recursing more deeply.

4.4.2 JCompilationUnit.preAnalyze()

For the JCompilationUnit node at the top of the AST, preAnalyze() does the following:

1. It creates a CompilationUnitContext.

2. It declares the implicit *j*-- types, java.lang.String and java.lang.Object.

3. It declares any imported types.

4. It declares the types defined by class declarations. Here it creates a Type for each declared class, whose classRep refers to a Class object for an empty class. For example, at this point in the pre-analysis phase of our Factorial program above, the Type for Factorial would have a classRep, the Class object for the class:

```
class Factorial {}
```

Later on, in both analyze() and codeGen(), the class will be further defined.

5. Finally, preAnalyze() invokes itself for each of the type declarations in the compilation unit. As we shall see below (Section 4.4.3), this involves, for a class declaration, creating a new Class object that records the interface information for each member and then overwriting the classRep for the declared Type with this new (more fully defined) Class.

Here is the code for preAnalyze() in JCompilationUnit:

```
public void preAnalyze() {
    context = new CompilationUnitContext();

    // Declare the two implicit types java.lang.Object and
    // java.lang.String
    context.addType(0, Type.OBJECT);
    context.addType(0, Type.STRING);

    // Declare any imported types
    for (TypeName imported: imports) {
        try {
            Class<?> classRep =
                Class.forName(imported.toString());
            context.addType(imported.line(),
                Type.typeFor(classRep));
        }
        catch (Exception e) {
            JAST.compilationUnit.reportSemanticError(
                imported.line(),
                "Unable to find %s", imported.toString());
        }
    }

    // Declare the locally declared type(s)
    CLEmitter.initializeByteClassLoader();
    for (JAST typeDeclaration: typeDeclarations) {
        ((JTypeDecl)
            typeDeclaration).declareThisType(context);
    }

    // Pre-analyze the locally declared type(s). Generate
    // (partial) Class instances, reflecting only the member
    // interface type information
    CLEmitter.initializeByteClassLoader();
    for (JAST typeDeclaration: typeDeclarations) {
```

```
        ((JTypeDecl)
          typeDeclaration).preAnalyze(context);
    }
}
```

4.4.3 JClassDeclaration.preAnalyze()

In a class declaration, `preAnalyze()` does the following:

1. It first creates a new `ClassContext`, whose `surroundingContext` points to the `CompilationUnitContext`.

2. It resolves the class's super type.

3. It creates a new `CLEmitter` instance, which will eventually be converted to the `Class` object for representing the declared class.

4. It adds a class header, defining a name and any modifiers, to this `CLEmitter` instance.

5. It recursively invokes `preAnalyze()` on each of the class' members. This causes field declarations, constructors, and method declarations (but with empty bodies) to be added to the `CLEmitter` instance.

6. If there is no explicit constructor (having no arguments) in the set of members, it adds the implicit constructor to the `CLEmitter` instance. For example, for the `Factorial` program above, the following implicit constructor is added, even though it is never used in the `Factorial` program:

```
public Factorial() {
    super.Factorial();
}
```

7. Finally, the `CLEmitter` instance produces a `Class` object, and that replaces the `classRep` for the `Type` of the declared class name in the (parent) `ClassContext`.

Notice that this pass need not descend into method bodies. If *j--*, like full Java, supported nested classes, then `preAnalyze()` would have to examine all of the statements of every method body to see if it were a nested class needing pre-analysis.

The code for `JClassDeclaration`'s `preAnalyze()` is as follows:

```
public void preAnalyze(Context context) {
    // Construct a class context
    this.context = new ClassContext(this, context);

    // Resolve superclass
    superType = superType.resolve(this.context);

    // Creating a partial class in memory can result in a
    // java.lang.VerifyError if the semantics below are
    // violated, so we can't defer these checks to analyze()
    thisType.checkAccess(line, superType);
    if (superType.isFinal()) {
        JAST.compilationUnit.reportSemanticError(line,
            "Cannot extend a final type: %s",
            superType.toString());
    }

    // Create the (partial) class
```

```
        CLEmitter partial = new CLEmitter();

        // Add the class header to the partial class
        String qualifiedName =
            JAST.compilationUnit.packageName() == "" ? name :
                JAST.compilationUnit.packageName() + "/" + name;
        partial.addClass(mods, qualifiedName, superType.jvmName(),
            null, false);

        // Pre-analyze the members and add them to the partial class
        for (JMember member: classBlock) {
            member.preAnalyze(this.context, partial);
            if (member instanceof JConstructorDeclaration &&
                ((JConstructorDeclaration) member).
                    params.size() == 0) {
                hasExplicitConstructor = true;
            }
        }

        // Add the implicit empty constructor?
        if (!hasExplicitConstructor) {
            codegenPartialImplicitConstructor(partial);
        }

        // Get the Class rep for the (partial) class and make it the
        // representation for this type
        Type id = this.context.lookupType(name);
        if (id != null &&
            !JAST.compilationUnit.errorHasOccurred()) {
            id.setClassRep(partial.toClass());
        }
    }
```

4.4.4 JMethodDeclaration.preAnalyze()

Here is the code for **preAnalyze()** in **JMethodDeclaration**:

```
public void preAnalyze(Context context, CLEmitter partial) {
    // Resolve types of the formal parameters
    for (JFormalParameter param: params) {
        param.setType(param.type().resolve(context));
    }

    // Resolve return type
    returnType = returnType.resolve(context);

    // Check proper local use of abstract
    if (isAbstract && body != null) {
        JAST.compilationUnit.reportSemanticError(line(),
            "abstract method cannot have a body");
    }
    else if (body == null && ! isAbstract) {
        JAST.compilationUnit.reportSemanticError(line(),
            "Method with null body must be abstract");
    }
    else if (isAbstract && isPrivate ) {
        JAST.compilationUnit.reportSemanticError(line(),
            "private method cannot be declared abstract");
    }
    else if (isAbstract && isStatic ) {
        JAST.compilationUnit.reportSemanticError(line(),
            "static method cannot be declared abstract");
```

```
        }

        // Compute descriptor
        descriptor = "(";
        for (JFormalParameter param: params) {
            descriptor += param.type().toDescriptor();
        }
        descriptor += ")" + returnType.toDescriptor();

        // Generate the method with an empty body (for now)
        partialCodegen(context, partial);
}
```

Basically, `preAnalyze()` does the following in a method declaration:

1. It resolves the types of its formal parameters and its return type.

2. It checks that any **abstract** modifier is proper.

3. It computes the *method descriptor*, which codifies the method's signature as a string[4]. For example, in the `Factorial` program above,

   - Method `factorial()` has the descriptor (I)I, which indicates a method taking an **int** for an argument and returning an **int** result, and

   - Method `main()` has the descriptor ([Ljava.lang.String;)V, which indicates a method taking a `String[]` argument and not returning anything (that is, a **void** return type).

4. Finally, it calls upon `partialCodegen()` to generate code for the method, but without the body. So the `Class` object that is generated for the enclosing class declaration has, after pre-analysis, at least the interface information for methods, including the types of parameters and the return type.

The code for `partialCodegen()` is as follows:

```
public void partialCodegen(Context context, CLEmitter partial) {
    // Generate a method with an empty body; need a return to
    // make the class verifier happy.
    partial.addMethod(mods, name, descriptor, null, false);

    // Add implicit RETURN
    if (returnType == Type.VOID) {
        partial.addNoArgInstruction(RETURN);
    }
    else if (returnType == Type.INT ||
             returnType == Type.BOOLEAN ||
             returnType == Type.CHAR) {
        partial.addNoArgInstruction(ICONST_0);
        partial.addNoArgInstruction(IRETURN);
    }
    else {
        // A reference type.
        partial.addNoArgInstruction(ACONST_NULL);
        partial.addNoArgInstruction(ARETURN);
    }
}
```

---

[4]Method descriptors are discussed in Appendix D.

4.4.5 JFieldDeclaration.preAnalyze()

Pre-analysis for a JFieldDeclaration is similar to that for a JMethodDeclaration. In a JFieldDeclaration, preAnalyze() does the following:

1. It enforces the rule that fields may not be declared abstract.

2. It resolves the field's declared type.

3. It generates the JVM code for the field declaration, via the CLEmitter created for the enclosing class declaration.

The code itself is rather simple:

```
public void preAnalyze(Context context, CLEmitter partial) {
    // Fields may not be declared abstract.
    if (mods.contains("abstract")) {
        JAST.compilationUnit.reportSemanticError(line(),
            "Field cannot be declared abstract");
    }

    for (JVariableDeclarator decl: decls) {
        // Add field to (partial) class
        decl.setType(decl.type().resolve( context));
        partial.addField(mods, decl.name(),
            decl.type().toDescriptor(), false);
    }
}
```

4.4.6 Symbol Table Built by preAnalyze()

So, pre-analysis recursively descends only so far as the class members declared in a program and constructs only a CompilationUnitContext (in which all types are declared) and a ClassContext. No local variables are declared in pre-analysis. Figure 4.4 illustrates how much of the symbol table is constructed for our Factorial program once pre-analysis is complete.

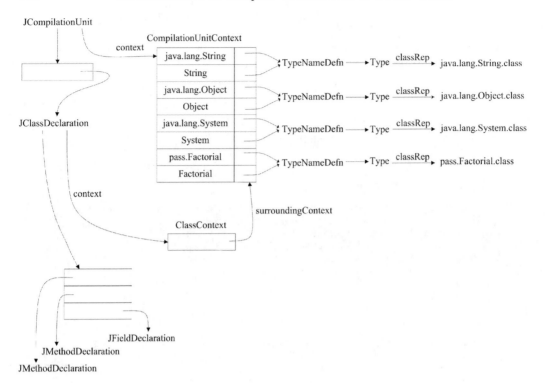

FIGURE 4.4 The symbol table created by the pre-analysis phase for the `Factorial` program.

4.5   Analysis of *j--* Programs

Once we have declared and loaded all imported types, and we have declared the types defined by class declarations, the compiler can execute the analysis phase.

The analysis phase recursively descends throughout the AST all the way to its leaves,

- Rewriting field and local variable initializations as assignments,

- Declaring both formal parameters and local variables,

- Allocating locations in the stack frame for the formal parameters and local variables,

- Computing the types of expressions and enforcing the language type rules,

- Reclassifying ambiguous names, and

- Doing a limited amount of tree surgery.

4.5.1 Top of the AST

Traversing the Top of the AST

At the top of the AST, `analyze()` simply recursively descends into each of the type (class) declarations, delegating analysis to one class declaration at a time:

```
public JAST analyze(Context context) {
    for (JAST typeDeclaration : typeDeclarations) {
        typeDeclaration.analyze(this.context);
    }
    return this;
}
```

Each class declaration in turn iterates through its members, delegating analysis to each of them. The only interesting thing is that, after all of the members have been analyzed, static field initializations are separated from instance field initializations, in preparation for code generation. This is discussed in the next section.

Rewriting Field Initializations as Assignments

In `JFieldDeclaration`, `analyze()` rewrites the field initializer as an explicit assignment statement, analyzes that, and then stores it in the `JFieldDeclaration`'s initializations list.

```
public JFieldDeclaration analyze(Context context) {
    for (JVariableDeclarator decl : decls) {
        // All initializations must be turned into assignment
        // statements and analyzed
        if (decl.initializer() != null) {
            JAssignOp assignOp =
                new JAssignOp(decl.line(),
                              new JVariable(decl.line(),
                                            decl.name()),
                              decl.initializer());
            assignOp.isStatementExpression = true;
            initializations.add(
                new JStatementExpression(decl.line(),
                    assignOp).analyze(context));
        }
    }
    return this;
}
```

Afterward, returning up to `JClassDeclaration`, `analyze()` separates the assignment statements into two lists: one for the static fields and one for the instance fields.

```
// Copy declared fields for purposes of initialization.
for (JMember member : classBlock) {
    if (member instanceof JFieldDeclaration) {
        JFieldDeclaration fieldDecl = (JFieldDeclaration) member;
        if (fieldDecl.mods().contains("static")) {
            staticFieldInitializations.add(fieldDecl);
        } else {
            instanceFieldInitializations.add(fieldDecl);
        }
    }
}
```

Later, `codegen()` will use these lists in deciding whether or not to generate both an instance initializing method and a class initializing method.

For example, consider the static field, declared in the `Factorial` class:

```
static int n = 5;
```

Figure 4.5 illustrates how the sub-tree for the declaration is rewritten. The tree surgery in Figure 4.5 proceeds as follows.

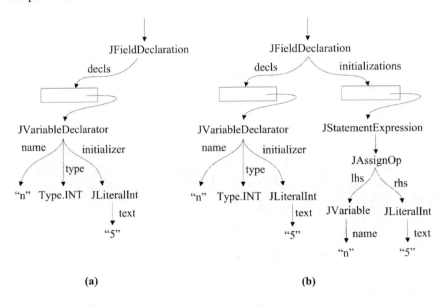

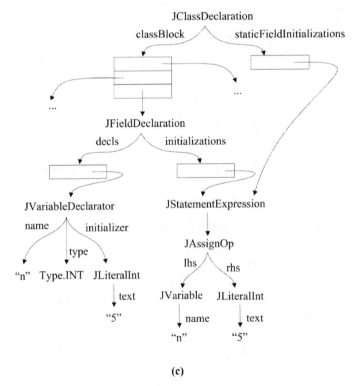

FIGURE 4.5 The rewriting of a field initialization.

1. The original sub-tree, produced by the parser for the static field declaration is shown in (a). The tree for the initial value is stored in `JVariableDeclarator`'s initializer field.

2. During analysis, `JFieldDeclaration`'s `analyze()` rewrites the initializing expression, 5 as an explicit assignment statement, n = 5; producing the sub-tree illustrated in (b).

3. Finally, when analysis returns back up the AST to `JClassDeclaration` (after recursing down the tree, these methods also back out and they may do tasks on the way back up), its `analyze()` copies the initializations to fields in its node. It separates the initializations into two lists: `staticFieldInitializations` for static fields and `instanceFieldInitializations` for instance fields. In the case of our example, n is static, and so the initialization is added to `staticFieldInitializations` as illustrated in (c).

In addition to this tree surgery on field initializations, `analyze()` does little at the top of the AST. The only other significant task is in `JMethodDeclaration`, where `analyze()` declares parameters and allocates locations relative to the base of the current stack frame. We discuss this in the next section.

4.5.2   Declaring Formal Parameters and Local Variables

`MethodContexts` and `LocalContexts`

Both formal parameters and local variables are declared in the symbol table and allocated locations within a method invocation's run-time stack frame. For example, consider the following class declaration:

```
public class Locals {
    public int foo(int t, String u) {
        int v = u.length();
        {
            int w = v + 5, x = w + 7;
            v = w + x;
        }
        {
            int y = 3;
            int z = v + y;
            t = t + y + z;
        }
        return t + v;
    }
}
```

The stack frame allocated for an invocation of `foo()` at run-time by the JVM is illustrated in Figure 4.6. Because `foo()` is an instance method, space is allocated at location 0 for `this`.

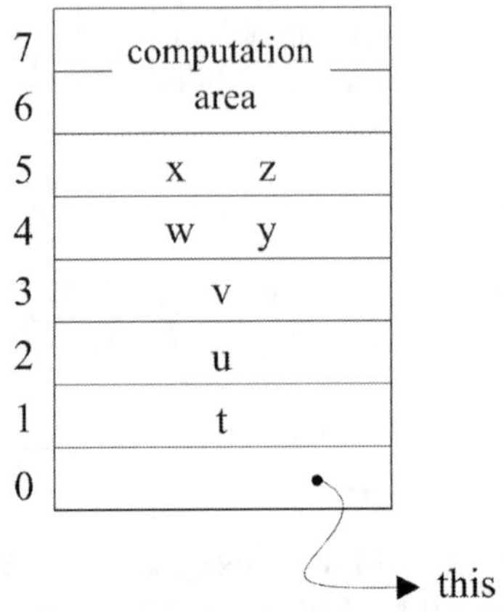

```
7  |___ computation ___|
6  |        area        |
5  |     x       z      |
4  |     w       y      |
3  |         v          |
2  |         u          |
1  |         t          |
0  |                 •  |
```

→ this

FIGURE 4.6 The stack frame for an invocation of `Locals.foo()`.

There are several regions of scope defined within the method `foo()`:

- First, the method itself defines a region of scope, where the formal parameters `t` and `u` are declared.

- The method body defines a nested region of scope, where the local variable `v` is declared.

- Nested within the body are two blocks, each of which defines a region of scope:

   - In the first region, the local variables `w` and `x` are declared.

   - In the second region, the local variables `y` and `z` are declared.

Because these two regions are disjoint, their locally declared variables may share locations on the run-time stack frame. Thus, as illustrated in Figure 4.6, `w` and `y` share the same location, and `x` and `z` share the same location.

During analysis, a context is created for each of these regions of scope: a `MethodContext` for that region defined by the method (and in which the formal parameters `t` and `u` are declared) and a `LocalContext` for each of the blocks.

The code for analyzing a `JMethodDeclaration` performs four steps:

1. It creates a new `MethodContext`, whose `surroundingContext` points back to the previous `ClassContext`.

2. The first stack frame offset is 0; but if this is an instance method, then offset 0 must be allocated to `this`, and the `nextOffset` is incremented to 1.

3. The formal parameters are declared as local variables and allocated consecutive offsets in the stack frame.

4. It analyzes the method's body.

The code is straightforward:

```
public JAST analyze(Context context) {
    this.context = new MethodContext(context, returnType);

    if (!isStatic) {
        // Offset 0 is used to addr "this".
        this.context.nextOffset();
```

```
        }

        // Declare the parameters
        for (JFormalParameter param : params) {
            this.context.addEntry(param.line(), param.name(),
                    new LocalVariableDefn(param.type(), this.context
                        .nextOffset(), null));
        }

            if (body != null) {
            body = body.analyze(this.context);
        }
        return this;
}
```

The code for analyzing a `JBlock` is even simpler; it performs just two steps:

1. It creates a new `LocalContext`, whose `surroundingContext` points back to the previous `MethodContext` (or `LocalContext` in the case of nested blocks). Its `nextOffset` value is copied from the previous context.

2. It analyzes each of the body's statements. Any `JVariableDeclarations` declare their variables in the `LocalContext` created in step 1. Any nested `JBlock` simply invokes this two-step process recursively, creating yet another `LocalContext` for the nested block.

Again, the code is straightforward:

```
public JBlock analyze(Context context) {
    // { ... } defines a new level of scope.
    this.context = new LocalContext(context);

    for (int i = 0; i < statements.size(); i++) {
        statements.set(i, (JStatement) statements.get(i).analyze(
                this.context));
    }
    return this;
}
```

For example, the steps in adding the contexts to the symbol table for the analysis of method `foo()` are illustrated in Figure 4.7. In the contexts, the names should map to objects of type `LocalVariableDefn`, which define the variables and their stack frame offsets. In the figures, the arrows point to the offsets in parentheses.

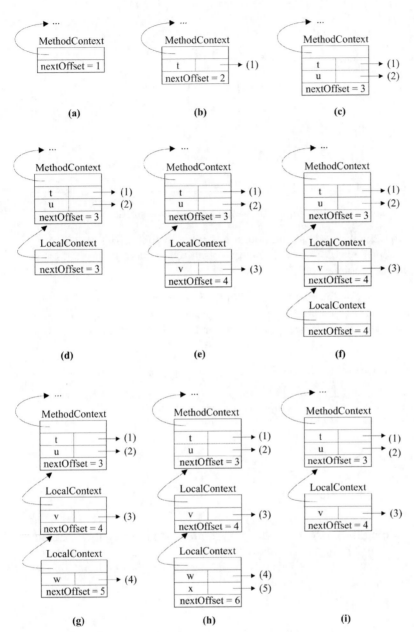

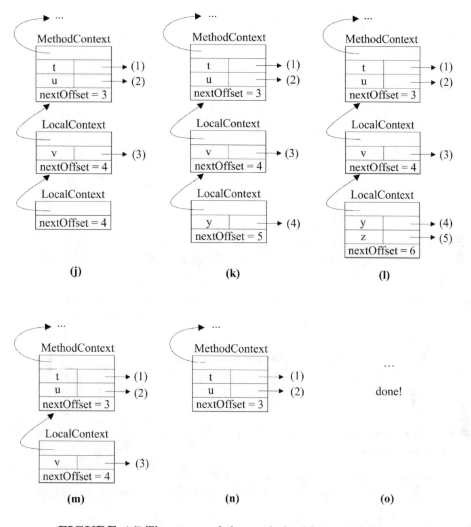

FIGURE 4.7 The stages of the symbol table in analyzing `Locals.foo()`.

Analysis proceeds as follows:

(a) The `analyze()` method for `JMethodDeclaration` creates a new `MethodContext`. Because `foo()` is an instance method, location 0 is allocated to `this`, and the next available stack frame location (`nextOffset`) is 1.

(b) It declares the first formal parameter `t` in this `MethodContext`, allocating it the offset 1 (and incrementing `nextOffset` to 2).

(c) It declares the second formal parameter `u` in this `MethodContext`, allocating it the offset 2 (and incrementing `nextOffset` to 3). Analysis is then delegated to the method's block (a `JBlock`).

(d) The `analyze()` method for `JBlock` creates a new `LocalContext`. Notice how the constructor for the new `LocalContext` copies the value (3) for `nextOffset` from the context for the enclosing method or block:

```
public LocalContext(Context surrounding) {
```

```
            super(surrounding, surrounding.classContext(),
                surrounding.compilationUnitContext());
    offset = (surrounding instanceof LocalContext)
        ? ((LocalContext) surrounding).offset()
        : 0;
    }
```

(e) A `JVariableDeclaration` declares the variable v in the `LocalContext`, allocating it the offset 3.

(f) `analyze()` creates a new `LocalContext` for the nested `JBlock`, copying the `nextOffset` 4 from the context for the surrounding block.

(g) A `JVariableDeclaration` declares the variable w in the new `LocalContext`, for the nested block, allocating it the offset 4.

(h) A second `JVariableDeclaration` declares the variable x in the same `LocalContext`, allocating it the offset 5. The subsequent assignment statement will be analyzed in this context.

(i) When this first nested `JBlock` has been analyzed, `analyze()` returns control to the `analyze()` for the containing `JBlock`, leaving the symbol table in exactly the state that it was in step (e).

(j) `Analyze()` creates a new `LocalContext` for the second nested `JBlock`, copying the `nextOffset` 4 from the context for the surrounding block. Notice the similarity to the state in step (f). In this way, variables y and z will be allocated the same offsets in steps (k) and (l) as w and x were in steps (g) and (h).

(k) A `JVariableDeclaration` declares the variable y in the new `LocalContext`, for the nested block, allocating it the offset 4.

(l) A second `JVariableDeclaration` declares the variable z in the same `LocalContext`, allocating it the offset 5. The subsequent assignment statement will be analyzed in this context.

(m) When this second nested `JBlock` has been analyzed, `analyze()` returns control to the `analyze()` for the containing `JBlock`, leaving the symbol table in exactly the state that it was in steps (e) and (i).

(n) When the method body's `JBlock` has been analyzed, `analyze()` returns control to the `analyze()` for the containing `JMethodDeclaration`, leaving the symbol table in exactly the state that it was in step (c).

(o) When the `JMethodDeclaration` has been analyzed, `analyze()` returns control to the `analyze()` for the containing `JClassDeclaration`, leaving the symbol table in exactly the state that it was before step (a).

We address the details of analyzing local variable declarations in the next section.

Analyzing Local Variable Declarations and Their Initializations

A local variable declaration is represented in the AST with a `JVariableDeclaration`. For example, consider the local variable declaration from `Locals`:

```
int w = v + 5, x = w + 7;
```

Before the `JVariableDeclaration` is analyzed, it appears exactly as it was created by the parser, as is illustrated in Figure 4.8.

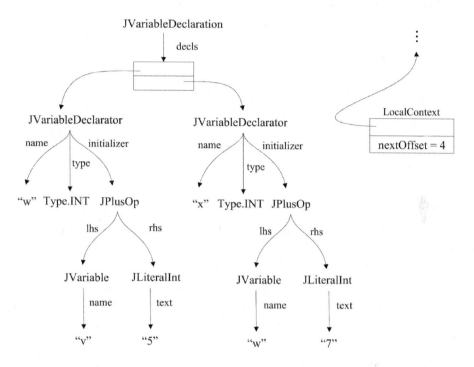

FIGURE 4.8 The sub-tree for `int w = v + 5, x = w + 7;` before analysis.

Figure 4.8 also pictures the `LocalContext` created for the nested block in which the declaration occurs, but before the declaration is analyzed.

The code for analyzing a `JVariableDeclaration` is as follows:

```
public JStatement analyze(Context context) {
    for (JVariableDeclarator decl : decls) {
        // Local variables are declared here (fields are
        // declaredin preAnalyze())
        int offset = ((LocalContext) context).nextOffset();
        LocalVariableDefn defn = new LocalVariableDefn(decl
            .type().resolve(context), offset);

        // First, check for shadowing
        IDefn previousDefn = context.lookup(decl.name());
        if (previousDefn != null
            && previousDefn instanceof LocalVariableDefn) {
            JAST.compilationUnit.reportSemanticError(decl.line(),
                "The name " + decl.name()
                    + " overshadows another local variable.");
        }

        // Then declare it in the local context
        context.addEntry(decl.line(), decl.name(), defn);

        // All initializations must be turned into assignment
        // statements and analyzed
        if (decl.initializer() != null) {
            defn.initialize();
            JAssignOp assignOp = new JAssignOp(decl.line(),
                new JVariable(decl.line(), decl.name()), decl
```

```
                    .initializer());
        assignOp.isStatementExpression = true;
        initializations.add(new JStatementExpression(decl
            .line(), assignOp).analyze(context));
    }
  }
  return this;
}
```

Analysis of a `JVariableDeclaration` such as that in Figure 4.8 involves the following:

1. `LocalVariableDefns` and their corresponding stack frame offsets are allocated for each of the declared variables.

2. The code checks to make sure that the declared variables do not shadow existing local variables.

3. The variables are declared in the local context.

4. Any initializations are rewritten as explicit assignment statements; those assignments are re-analyzed and stored in an `initializations` list. Later, code generation will generate code for any assignments in this list.

Figure 4.9 illustrates the result of analyzing the `JVariableDeclaration` in Figure 4.8. Notice that re-analyzing the assignment statements attaches types to each node in the sub-trees; more on that below.

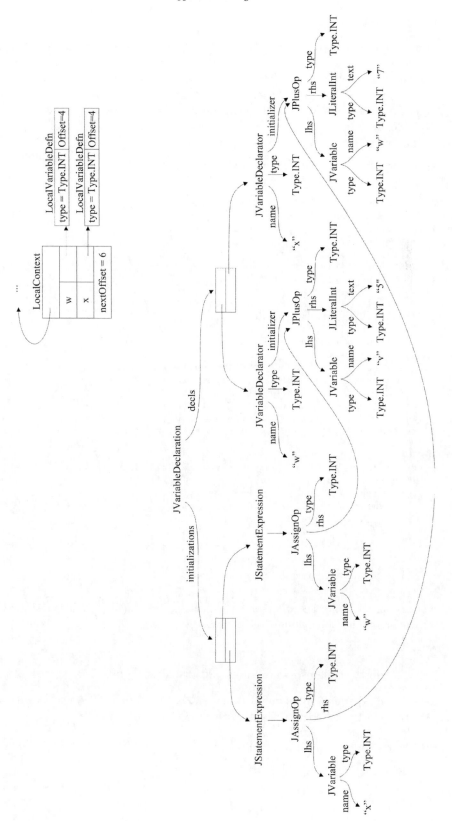

FIGURE 4.9 The sub-tree for `int w = v + 5, x = w + 7;` after analysis.

4.5.3 Simple Variables

Simple variables are represented in the AST as `JVariable` nodes. A simple variable could denote a local variable, a field or a type. Analysis of simple variables involves looking up their names in the symbol table to find their types. If a variable is not found in the symbol table, then we examine the `Type` for the surrounding class (in which the variable appears) to see if it is a field. If it is a field, then the field selection is made explicit by rewriting the tree as a `JFieldSelection`.

The code for `analyze()` in `JVariable` is as follows:

```
public JExpression analyze(Context context) {
    iDefn = context.lookup(name);
    if (iDefn == null) {
        // Not a local, but is it a field?
        Type definingType = context.definingType();
        Field field = definingType.fieldFor(name);
        if (field == null) {
            type = Type.ANY;
            JAST.compilationUnit.reportSemanticError(line,
                "Cannot find name: " + name);
        } else {
            // Rewrite a variable denoting a field as an
            // explicit field selection
            type = field.type();
            JExpression newTree = new JFieldSelection(line(),
                field.isStatic() ||
                (context.methodContext() != null &&
                context.methodContext().isStatic()) ?
                    new JVariable(line(),
                        definingType.toString()) :
                    new JThis(line), name);
            return (JExpression) newTree.analyze(context);
        }
    } else {
        if (!analyzeLhs && iDefn instanceof LocalVariableDefn &&
            !((LocalVariableDefn) iDefn).isInitialized()) {
            JAST.compilationUnit.reportSemanticError(line,
                "Variable " + name + " might not have been
                    initialized");
        }
        type = iDefn.type();
    }
    return this;
}
```

For example, consider a simple case where a variable is declared locally, such as the variable v in our `Locals` class. When analyzing the `return` statement,

```
return t + v;
```

the analysis of v is pretty straightforward and is illustrated in Figure 4.10.

1. Its name is looked up in the symbol table and is found to be associated with the `LocalVariableDefn` of a local variable with type `Type.INT` and offset 3.

2. The `LocalVariableDefn` is recorded in field `iDefn` for later use by code generation.

3. The type field is copied from the `LocalVariableDefn`.

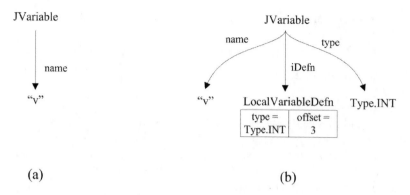

FIGURE 4.10 A locally declared variable (a) before analysis; (b) after analysis.

When the variable denotes a field, analysis is a little more interesting. For example, consider the analysis of the static field n, when it appears in the `main()` method of our `Factorial` class example above.

1. Figure 4.11(a) shows the `JVariable` before it is analyzed.

2. Its name is looked up in the symbol table but is not found. So the defining type (the type declaration enclosing the region where the variable appears) is consulted; n is found to be a static field in class `Factorial`.

3. The implicit static field selection is made explicit by rewriting the tree to represent `pass.Factorial.n`. This produces the `JFieldSelection` illustrated in Figure 4.11 (b).

4. The `JFieldSelection` produced in step 3 is recursively analyzed to determine the types of its target and the result, as illustrated in Figure 4.11(c).

5. It is this sub-tree that is returned to replace the original `JVariable` (a) in the parent AST. This tree rewriting is the reason that `analyze()` everywhere returns a (possibly rewritten) sub-tree.

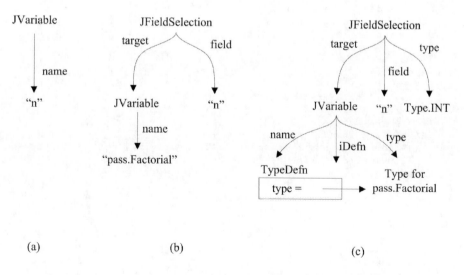

FIGURE 4.11 Analysis of a variable that denotes a static field.

Just how the sub-tree in Figure 4.11(b) is analyzed to produce (c) is discussed in the next section.

4.5.4   Field Selection and Message Expressions

Reclassifying an Ambiguous Target

Both field selections and message expressions have *targets*. In a field selection, the target is either an object or a class from which one wants to select a field. In a message expression, the target is an object or class to which one is sending a message. Unfortunately, the parser cannot always make out the syntactic structure of a target.

For example, consider the field selection

```
w.x.y.z
```

The parser knows this is a field selection of some sort and that z is the field. But, without knowing the types of w, x, and y, the parser cannot know whether

- w is a class name, x is a static field in w, and y is a field of x;

- w is a package containing class x, and y is a static field in x; or

- w.x.y is a fully qualified class name such as java.lang.System.

For this reason, the parser packages up the string "w.x.y" in an AmbiguousName object, attached to either the JFieldSelection or JMessageExpression, so as to put off the decision until analysis.

The first thing that analysis does in either sort of expression is to reclassify the ambiguous target in the context in which it appears. The reclassify() method in AmbiguousName is based on the rules in the *Java Language Specification* [Gosling et al., 2005] for reclassifying an ambiguous name:

```java
public JExpression reclassify(Context context) {
    // Easier because we require all types to be imported.
    JExpression result = null;
    StringTokenizer st = new StringTokenizer(name, ".");

    // Firstly, find a variable or Type.
    String newName = st.nextToken();
    IDefn iDefn = null;

    do {
        iDefn = context.lookup(newName);
        if (iDefn != null) {
            result = new JVariable(line, newName);
            break;
        } else if (!st.hasMoreTokens()) {
            // Nothing found. :(
            JAST.compilationUnit.reportSemanticError(line,
                "Cannot find name " + newName);
            return null;
        } else {
            newName += "." + st.nextToken();
        }
    } while (true);

    // For now we can assume everything else is fields.
    while (st.hasMoreTokens()) {
        result = new JFieldSelection(line, result,
            st.nextToken());
```

```
    }
    return result;
}
```

For example, consider the following message expression:

```
java.lang.System.out.println(...);
```

The parser will have encapsulated the target `java.lang.System.out` in an `AmbiguousName` object. The first thing `analyze()` does for a `JMessageExpression` is to reclassify this `AmbiguousName` to determine the structure of the expression that it denotes. It does this by looking at the ambiguous `java.lang.System.out` from left to right.

1. First, `reclassify()` looks up the simple name `java` in the symbol table.

2. Not finding that, it looks up `java.lang`.

3. Not finding that, it looks up `java.lang.System`, which (assuming `java.lang.System` has been properly imported) it finds to be a class.

4. It then assumes that the rest of the ambiguous part, that is `out`, is a field.

5. Thus, the target is a field selection whose target is `java.lang.System` and whose field name is `out`.

Analyzing a Field Selection

The code for analyzing a field is as follows:

```
public JExpression analyze(Context context) {
    // Reclassify the ambiguous part.

    target = (JExpression) target.analyze(context);
    Type targetType = target.type();

    // We use a workaround for the "length" field of arrays.
    if ((targetType instanceof ArrayTypeName)
        && fieldName.equals("length")) {
        type = Type.INT;
    } else {
        // Other than that, targetType has to be a
        // ReferenceType
        if (targetType.isPrimitive()) {
            JAST.compilationUnit.reportSemanticError(line(),
                "Target of a field selection must "
                    + "be a defined type");
            type = Type.ANY;
            return this;
        }
        field = targetType.fieldFor(fieldName);
        if (field == null) {
            JAST.compilationUnit.reportSemanticError(line(),
                "Cannot find a field: " + fieldName);
            type = Type.ANY;
        } else {
            context.definingType().checkAccess(line,
                (Member) field);
            type = field.type();

            // Non-static field cannot be referenced from a
            // static context.
            if (!field.isStatic()) {
```

```
            if (target instanceof JVariable &&
                ((JVariable) target).iDefn() instanceof
                  TypeNameDefn) {
                JAST.compilationUnit.
                    reportSemanticError(line(),
                    "Non-static field " + fieldName +
                      " cannot be referenced from a static
                      context");
            }
        }
    }
}
    return this;
}
```

After reclassifying any ambiguous part and making that the target, analysis of a JFieldSelection proceeds as follows:

1. It analyzes the target and determines the target's type.

2. It then considers the special case where the target is an array and the field is length. In this case, the type of the "field selection" is Type.INT[5].

3. Otherwise, it ensures that the target is not a primitive and determines whether or not it can find a field of the appropriate name in the target's type. If it cannot, then an error is reported.

4. Otherwise, it checks to make sure the field is accessible to this region, a non-static field is not referenced from a static context, and then returns the analyzed field selection sub-tree.

Analyzing messages expressions is similar, but with the added complication of arguments.

Analyzing a Message Expression

After reclassifying any AmbiguousName, analyzing a JMessageExpression proceeds as follows.

1. It analyzes the arguments to the message and constructs an array of their types.

2. It determines the surrounding, defining class (for determining access).

3. It analyzes the target to which the message is being sent.

4. It takes the message name and the array of argument types and looks for a matching method defined in the target's type. In *j--*, argument types must match exactly. If no such method is found, it reports an error.

5. Otherwise, the target class and method are checked for accessibility, a non-static method is now allowed to be referenced from a static context, and the method's return type becomes the type of the message expression.

```
public JExpression analyze(Context context) {
    // Reclassify the ambiguous part
```

---

[5]This is a Java language hack; length is not really a field but an operation on arrays.

```
    // Then analyze the arguments, collecting
    // their types (in Class form) as argTypes
    argTypes = new Type[arguments.size()];
    for (int i = 0; i < arguments.size(); i++) {
        arguments.set(i, (JExpression) arguments.get(i).analyze(
            context));
        argTypes[i] = arguments.get(i).type();
    }


    // Where are we now? (For access)
    Type thisType = ((JTypeDecl) context.classContext
        .definition()).thisType();


    // Then analyze the target
    if (target == null) {
                // Implied this (or, implied type for statics)
        if (!context.methodContext().isStatic()) {
            target = new JThis(line()).analyze(context);
        }
        else {
            target = new JVariable(line(),
                        context.definingType().toString()).
                            analyze(context);
        }
    } else {
        target = (JExpression) target.analyze(context);
        if (target.type().isPrimitive()) {
            JAST.compilationUnit.reportSemanticError(line(),
                "cannot invoke a message on a primitive type:"
                    + target.type());
        }
    }


    // Find appropriate Method for this message expression
    method = target.type().methodFor(messageName, argTypes);
    if (method == null) {
        JAST.compilationUnit.reportSemanticError(line(),
            "Cannot find method for: "
                + Type.signatureFor(messageName, argTypes));
        type = Type.ANY;
    } else {
        context.definingType().checkAccess(line,
                                        (Member) method);
        type = method.returnType();

        // Non-static method cannot be referenced from a
        // static context.
        if (!method.isStatic()) {
            if (target instanceof JVariable &&
                ((JVariable) target).iDefn() instanceof
                    TypeNameDefn) {
                JAST.compilationUnit.reportSemanticError(line(),
                    "Non-static method " +
                    Type.signatureFor(messageName, argTypes) +
                    "cannot be referenced from a static context");
            }
        }
    }
    return this;
}
```

4.5.5 Typing Expressions and Enforcing the Type Rules

Much of the rest of analysis, as defined for the various kinds of AST nodes, is about computing and checking types and enforcing additional *j--* rules. Indeed, when one reads through
any compiler, one finds lots of code whose only purpose is to enforce a litany of rules.

For most kinds of AST nodes, analysis involves analyzing the sub-trees and checking the
types.

Analyzing a Subtraction Operation

For example, analyzing a `JSubtractOp` is straightforward:

```
public JExpression analyze(Context context) {
    lhs = (JExpression) lhs.analyze(context);
    rhs = (JExpression) rhs.analyze(context);
    lhs.type().mustMatchExpected(line(), Type.INT);
    rhs.type().mustMatchExpected(line(), Type.INT);
    type = Type.INT;
    return this;
}
```

Analyzing a + Operation Causes Tree Rewriting for Strings

On the other hand, analyzing a `JPlusOp` is complicated by the possibility that one of
the operands to the + is a string; in this case, we simply rewrite the `JPlusOp` as a
`JStringConcatenationOp` and analyze that:

```
public JExpression analyze(Context context) {
    lhs = (JExpression) lhs.analyze(context);
    rhs = (JExpression) rhs.analyze(context);
    if (lhs.type() == Type.STRING || rhs.type() == Type.STRING) {
        return (new JStringConcatenationOp(line, lhs, rhs))
            .analyze(context);
    } else if (lhs.type() == Type.INT && rhs.type() == Type.INT){
        type = Type.INT;
    } else {
        type = Type.ANY;
        JAST.compilationUnit.reportSemanticError(line(),
            "Invalid operand types for +");
    }
    return this;
}
```

And, analyzing a `JStringConcatenationOp` is easy because we know at least one of the
operands is a string, so the result has to be a string:

```
public JExpression analyze(Context context) {
    type = Type.STRING;
    return this;
}
```

Analyzing a Literal

Analyzing a literal is trivial. We know its type. For example, the `analyze()` method for
`JLiteralInt` follows:

```
public JExpression analyze(Context context) {
    type = Type.INT;
    return this;
}
```

Analyzing a Control Statement

Analyzing a control statement (for example, if-statement) is pretty straightforward. For example, analysis of the if-statement involves only

1. Analyzing the test and checking that it is a Boolean,

2. Analyzing the consequent (the then part); and finally,

3. If there is an alternate (an else part), analyzing that.

The code for analyzing the `JIfStatement` follows.

```
public JStatement analyze(Context context) {
    test = (JExpression) test.analyze(context);
    test.type().mustMatchExpected(line(), Type.BOOLEAN);
    consequent = (JStatement) consequent.analyze(context);
    if (alternate != null) {
        alternate = (JStatement) alternate.analyze(context);
    }
    return this;
}
```

4.5.6 Analyzing Cast Operations

The *j--* language is stricter than Java when it comes to types. For example, there are no implied conversions in *j--*. When one assigns an expression to a variable, the types must match exactly. The same goes for actual parameters to messages matching the formal parameters of methods.

This does not exclude polymorphism. For example if type `Bar` extends (is a sub-type of) type `Foo`, if `bar` is a variable of type `Bar` and `foo` is a variable of type `Foo`, we can say

```
foo = (Foo) bar;
```

to keep the *j--* compiler happy. Of course, the object that `bar` refers to could be of type `Bar` or any of its sub-types. Polymorphism has not gone away.

Analysis, when encountering a `JCastOp` for an expression such as

$(Type_2)$ expression of $Type_1$

must determine two things:

1. That an expression of type $Type_1$ can be cast to $Type_2$, that is, that the cast is valid.

2. The type of the result. This part is easy: it is simply $Type_2$.

To determine (1), we must consider the possibilities for $Type_1$ and $Type_2$. These are specified in Section 4.5 of the *Java Language Specification* [Gosling et al., 2005].

1. Any type may be cast to itself. Think of this as the *Identity* cast.

2. An arbitrary reference type may be cast to another reference type if and only if either one of the following holds:

   (a) The first type is a sub-type of (extends) the second type. This is called *widening* and requires no action at run-time.

   (b) The second type is a sub-type of the first type. This is called *narrowing* and requires a run-time check to make sure the expression being cast is actually an instance of the type it is being cast to.

3. The following table summarizes other casts. In reading the table, think of the rows as $Type_1$ and the columns as $Type_2$. So the table says whether or not (and how) a type labeling a row may be cast to a type labeling a column.

	boolean	char	int	Boolean	Character	Integer
boolean	Identity	Error	Error	Boxing	Error	Error
char	Error	Identity	Widening	Error	Boxing	Error
int	Error	Narrowing	Identity	Error	Error	Boxing
Boolean	Unboxing	Error	Error	Identity	Error	Error
Character	Error	Unboxing	Error	Error	Identity	Error
Integer	Error	Error	Unboxing	Error	Error	Identity

(a) A `boolean` can always be cast to a `Boolean`. The `Boolean` simply encapsulates the `boolean` value so the operation is called *boxing*. Similarly, a `char` can be boxed as a `Character`, and an `int` can be boxed as an `Integer`.

(b) A `Boolean` can be cast to a `boolean` using *unboxing*, plucking out the encapsulated `boolean` value. Similarly for `Characters` and `Integers`.

(c) One primitive type may often be cast to another. For example, a `char` may be cast to an `int`. This is an example of *widening* and requires no run-time action. An `int` may be cast to a `char`. This is an example of *narrowing* and requires the execution of the `i2c` instruction at run-time.

(d) Some types may not be cast to other types. For example, one may not cast a `boolean` to an `int`. Such casts are invalid.

The code for `analyze()` in `JCastOp` follows:

```
public JExpression analyze(Context context) {
    expr = (JExpression) expr.analyze(context);
    type = cast = cast.resolve(context);
    if (cast.equals(expr.type())) {
        converter = Converter.Identity;
    } else if (cast.isJavaAssignableFrom(expr.type())) {
        converter = Converter.WidenReference;
    } else if (expr.type().isJavaAssignableFrom(cast)) {
        converter = new NarrowReference(cast);
    } else if ((converter =
        conversions.get(expr.type(), cast)) != null) {
    } else {
        JAST.compilationUnit.reportSemanticError(line,
            "Cannot cast a " + expr.type().toString() + " to a "
                + cast.toString());
    }
    return this;
}
```

In the code, `cast` is the type that we want to cast the expression `expr` to. The code not only decides whether or not the cast is valid, but if it is valid, computes the `Converter` that will be used at code generation time to generate any run-time code required for the cast. If no such converter is defined, then the cast is invalid.

One example of a converter is that for narrowing one reference type to another (more specific) reference sub-type; the converter is `NarrowReference` and its code is as follows:

```
class NarrowReference implements Converter {

    private Type target;
```

```
    public NarrowReference(Type target) {
        this.target = target;
    }

    public void codegen(CLEmitter output) {
        output.addReferenceInstruction(CHECKCAST,
                                      target.jvmName());
    }
}
```

The compiler creates a `NarrowReference` instance for each cast of this kind, tailored to the type it is casting to.

The `codegen()` method is used in the next code generation phase. We address code generation in the next chapter.

4.5.7 Java's Definite Assignment Rule

In full Java, every variable (whether it be a local variable or a field) must be definitely assigned before it is accessed in a computation. That is, it must appear on the left-hand side of the = operator before it is accessed. The definite assignment rule, as it applies to Java, is described fully in Chapter 16 of the *Java Language Specification* [Gosling et al., 2005]. We do not have this rule in *j--*, so we need not enforce it in our compiler.

Enforcing the definite assignment rule requires data flow analysis, which determines where in the program variables are defined (assigned values), where in the program the variables' values are used, and so where in the program those values are valid (from assignment to last use).

We discuss data flow analysis in Chapters 6 and 7; our JVM-to-MIPS translator performs data-flow analysis as part of computing live intervals for register allocation.

4.6 Visitor Pattern and the AST Traversal Mechanism

One might ask, "Why does the compiler not use the visitor pattern for traversing the AST in pre-analysis, analysis, and code generation?"

The visitor pattern is one of the design patterns introduced by Erich Gamma et al. [Gamma, 1995]. The visitor pattern serves two purposes:

1. It separates the tree traversal function from the action taken at each node. A separate mechanism, which can be shared by tree-printing, pre-analysis, analysis, and code generation, traverses the tree.

2. It gathers all of the actions for each phase together into one module (or method). Thus, all of the tree-printing code is together, all of the pre-analysis code is together, all of the analysis code is together, and all of the code generation code is together.

The idea is that it separates traversal from actions taken at the nodes (a separation of concerns) and it makes it simpler to add new actions. For example, to add an optimization phase one need simply write a single method, rather than having to write an `optimize()` method for each node type.

Such an organization would be useful if we were planning to add new actions. But we are more likely to be adding new syntax to *j--*, together with its functionality. Making extensions to *j--* would require our modifying each of the phases.

Moreover, often the traversal order for one node type differs from that of another. And sometimes we want to perform actions at a node before, after, and even in between the traversal of the sub-trees. For example, take a look at `codegen()` for the `JIfStatement`:

```
public void codegen(CLEmitter output) {
    String elseLabel = output.createLabel();
    String endLabel = output.createLabel();
    condition.codegen(output, elseLabel, false);
    thenPart.codegen(output);
    if (elsePart != null) {
        output.addBranchInstruction(GOTO, endLabel);
    }
    output.addLabel(elseLabel);
    if (elsePart != null) {
        elsePart.codegen(output);
        output.addLabel(endLabel);
    }
}
```

Mixing the traversal code along with the actions is just so much more flexible. For this reason, the *j--* compiler eschews the visitor pattern.

4.7   Programming Language Design and Symbol Table Structure

Our symbol table for *j--* has two parts:

1. A linked list (or stack) of contexts (hash tables) for capturing the nested scope of type definitions and local variables. Types are maintained in a `CompilationUnitContext` at the base of this list (or stack). Local variables are looked up beginning at the end of this linked list (or top of the stack); this reflects the nested scope of local variables. That names at each level of scope are defined in a hash table speeds up the lookup operation. Keeping a linked list of these hash tables, one for each lexical level of scope, speeds up compile-time block entry and block exit.

2. Classes are represented by `Type` objects, stored in the `CompilationUnitContext`, and built upon Java's `Class` objects—Java's representation for types. This representation reflects the fact that a class' method names and field names are particular to that class, and it allows *j--* programs to engage with existing classes in the Java API.

Thus, the symbol table structure in our compiler is pretty much dictated by the design of the *j--* programming language.

If we were to implement nested class declarations (an exercise in Chapter 5), some of the infrastructure is already in our symbol table design—additional classes can be stored in nested contexts; we may have to modify `addType()` and `lookupType()` somewhat. Other languages that have classes or modules (such as the programming language *Modula*) will want a symbol table whose structure is similar to ours.

But other, simpler procedural languages, such as the C language, in which a name is either external to all functions or local to a function, may have much simpler symbol tables. For C, one could use a single hash table mapping external names to their definitions, and a simple stack of hash tables for dealing with nested blocks inside of functions.

Languages that support dynamic binding (such as some versions of LISP and some functional programming languages) require us to keep the symbol table around at runtime; here, efficiency becomes even more important; hash tables are the usual solution. A radical solution to dealing with the run-time name environments of function programming languages is to do away the names altogether! David Turner [Turner, 1979] proposed a practical scheme where all functional programs might be translated to constant combinators.

4.8 Attribute Grammars

Notice that as we recursively traverse our abstract syntax tree, we are making additional passes over the syntax for enforcing type rules, which are not expressible using the context free syntax. It would be nice if we could somehow attach type rules to our (BNF) grammar rules. An attempt at this takes the form of attribute grammars.

Attribute grammars, first developed by Donald Knuth [Knuth, 1968] as a means of formalizing the semantics of a context-free language, are useful in specifying the syntax and semantics of a programming language. An attribute grammar can be used to specify the context-sensitive aspects of a language, such as checking that a variable has been declared before use and that the use of the variable is consistent with its declaration. Attribute grammars can also be used to specify the operational semantics of a language by defining a translation into machine-specific lower-level code.

In this section, we will first introduce attribute grammars through examples, then provide a formal definition for an attribute grammar, and finally present attribute grammars for a couple of constructs in *j*--.

4.8.1 Examples

An attribute grammar may be informally defined as a context-free grammar that has been extended to provide context sensitivity using a set of attributes, assignment of attribute values, evaluation rules, and conditions. In a parse tree representing the input sentence (source program), attribute grammars can pass values from a node to its parent, using a *synthesized attribute*, or from a node to its child, using an *inherited attribute*. In addition to passing values up or down the tree, the attribute values may be assigned, modified, and checked at any node in the tree. We clarify these ideas using the following examples.

Let us try and write a grammar to recognize sentences of the form $a^n b^n c^n$. The sentences aabbcc and abc belong to this grammar, but the sentences abbcc and aac do not. Consider the following context-free grammar (from [Slonneger and Kurtz, 1995]:)

letterSequence ::= aSequence bSequence cSequence

aSequence ::= a | aSequence a

bSequence ::= b | aSequence b

cSequence ::= c | aSequence c

It is easily seen that the above grammar, in addition to generating the acceptable strings such as aabbcc, also generates the unacceptable strings such as abbcc. It is impossible to write a context-free grammar describing the language under consideration here. Attribute grammars come to our rescue. Augmenting our context-free grammar with an attribute describing the length of a letter sequence, we can use these values to make sure that the sequences of a, b, and c have the same length.

We associate a synthesized attribute *size* with the non-terminals aSequence, bSequence, and cSequence, and a condition at the root of the parse tree that the *size* attribute for each of the letter sequences has the same value. If the input consists of a single character, *size* is set to 1; if it consists of a sequence followed by a single character, *size* for the parent character is set to the *size* of the child character plus one. Added to the root is the condition that the sizes of the sequences be the same. Here is the augmented grammar:

letterSequence ::= aSequence bSequence cSequence

> condition : $size(\text{aSequence}) = size(\text{bSequence}) = size(\text{cSequence})$

aSequence ::= a
$$size(\text{aSequence}) \leftarrow 1$$
| aSequence a
$$size(\text{aSequence}) \leftarrow size(\text{aSequence}) + 1$$

bSequence ::= b
$$size(\text{bSequence}) \leftarrow 1$$
| bSequence b
$$size(\text{bSequence}) \leftarrow size(\text{bSequence}) + 1$$

cSequence ::= c
$$size(\text{cSequence}) \leftarrow 1$$
| cSequence c
$$size(\text{cSequence}) \leftarrow size(\text{cSequence}) + 1$$

The augmented attribute grammar successfully parses legal strings such as aabbcc, but does not parse illegal sequences such as abbcc; such illegal sequences satisfy the BNF part of the grammar, and do not satisfy the condition required of the attribute values.

Another approach is to pass information from one part of the parse tree to some node, and then have it inherited down into other parts of the tree. Let *size* be a synthesized attribute for the sequence of a's, and *iSize* be the inherited attribute for the sequences of b's, and c's. We synthesize the *size* of the sequence of a's to the root of the parse tree, and set the *iSize* attribute for the b sequence and the c sequence to this value and inherit it down the tree, decrementing the value by one every time we see another character in the sequence. When we reach the node where the sequence has a child consisting of a single character, we check if the inherited *iSize* attribute equals one. If so, the size of the sequence must be the same as the size of the a sequence; otherwise, the two sizes do not match and the parsing is unsuccessful. The following attribute grammar clarifies this:

letterSequence ::= aSequence bSequence cSequence

$$iSize(\text{bSequence}) \leftarrow size(\text{aSequence})$$
$$iSize(\text{cSequence}) \leftarrow size(\text{aSequence})$$

aSequence ::= a
$$size(\text{aSequence}) \leftarrow 1$$
| aSequence a

$$size(\text{aSequence}) \leftarrow size(\text{aSequence}) + 1$$

bSequence ::= b
  condition : $iSize(\text{bSequence}) = 1$
        | bSequence b
              $iSize(\text{bSequence}) \leftarrow iSize(\text{bSequence}) - 1$

cSequence ::= c
  condition : $iSize(\text{cSequence}) = 1$
        | cSequence c
              $iSize(\text{cSequence}) \leftarrow iSize(\text{cSequence}) - 1$

For the non-terminal a Sequence, *size* is a synthesized attribute; but for the non-terminals bSequence and cSequence, *iSize* is an inherited attribute passed from the parent to child. As before, the attribute grammar above cannot parse illegal sequences such as abbcc, as these sequences do not satisfy all the conditions associated with attribute values.

In this grammar, the sequence of a's determines the desired length against which the other sequences are checked. Consider the sequence abbcc. It could be argued that the sequence of a's is at fault and not the other two sequences. However, in a programming language with declarations, we use the declarations to determine the desired types against which the remainder of the program is checked. In other words, the declaration information is synthesized up and is inherited by the entire program for checking.

In our next example, we illustrate the use of attribute grammars in specifying semantics using an example from Donald Knuth's 1968 seminal paper on attribute grammars [Knuth, 1968]. The example grammar computes the values of binary numbers. The grammar uses both synthesized and inherited attributes.

Consider the following context-free grammar describing the structure of binary numbers. N stands for binary numbers, L stands for bit lists, and B stands for individual bits.

$N ::= L$

$N ::= L \cdot L$

$L ::= B$

$L ::= L \, B$

$B ::= 0$

$B ::= 1$

Examples of binary numbers that follow the above grammar are both integral numbers such as 0, 1, 10, 11, 100, and so on, and rational numbers with integral fractional part such as 1.1, 1.01, 10.01, and so on.

We want to compute the values of the binary numbers using an attribute grammar. For example, 0 has the value 0, 1 the value 1, 0.1 the value 0.5, 10.01 the value 2.25, and so on.

Knuth provides a couple of different attribute grammars to define the computation; the first one uses only synthesized attributes, and the second uses both inherited and synthesized attributes. We will illustrate the latter.

In this attribution, each bit has a synthesized attribute *value*, which is a rational number that takes the position of the bit into account. For example, the first bit in the binary number

10.01 will have the value 2 and the last bit will have the value 0.25. In order to define these attributes, each bit also has an inherited attribute *scale* used to compute the value of a 1-bit as $value = 2^{scale}$. So for the bits in 10.01, scale will be 1, 0, -1, -2, respectively.

If we have the values of the individual bits, these values can simply be summed up to the total value. Adding synthesized attributes *value* both for bit lists and for binary numbers does this. In order to compute the *scale* attribute for the individual bits, an inherited attribute *scale* is added also for bit lists (representing the scale of the rightmost bit in that list), and a synthesized attribute *length* for bit lists, holding the length of the list.

Knuth uses the abbreviations v, s, and l for the *value*, *scale*, and *length* attributes. Here is the resulting attribute grammar:

$N ::= L$
$$v(N) \leftarrow v(L)$$
$$s(L) \leftarrow 0$$

$N ::= L_1 \; . \; L_2$
$$v(N) \leftarrow v(L_2) + v(L_2)$$
$$s(L_1) \leftarrow 0$$
$$s(L_2) \leftarrow - \, l(L_2)$$

$L ::= B$
$$v(L) \leftarrow v(B)$$
$$s(B) \leftarrow s(L)$$
$$l(L) \leftarrow 1$$

$L_1 ::= L_2 \; B$
$$v(L_1) \leftarrow v(L_2) + v(B)$$
$$s(B) \leftarrow s(L_1)$$
$$s(L_2) \leftarrow s(L_1) + 1$$
$$l(L_1) \leftarrow l(L_2) + 1$$

$B ::= 0$
$$v(B) \leftarrow 0$$

$B ::= 1$
$$v(B) \leftarrow 2^{s(B)}$$

4.8.2 Formal Definition

An attribute grammar is a context-free grammar augmented with attributes, semantic rules, and conditions. Let $G = (N, T, S, P)$ be a context-free grammar. For each non-terminal $X \in N$ in the grammar, there are two finite disjoint sets $I(X)$ and $S(X)$ of inherited and synthesized attributes. For $X = S$, the start symbol $I(X) = \emptyset$.

Let $A(X) = I(A) \cup S(X)$ be the set of attributes of X. Each attribute $A \in A(X)$ takes a value from some semantic domain (such as integers, strings of characters, or structures of some type) associated with that attribute. These values are defined by semantic functions or semantic rules associated with the productions in P.

Consider a production $p \in P$ of the form $X_0 ::= X_1 X_2 \ldots X_n$. Each synthesized attribute $A \in S(X_0)$ has its value defined in terms of the attributes in $A(X_1) \cup A(X_2) \cup \cdots \cup A(X_n) \cup I(X_0)$. Each inherited attribute $A \in I(X_k)$ for $1 \leq k \leq n$ has its value defined in terms of the attributes in $A(X_0) \cup S(X_1) \cup S(X_2) \cup \cdots \cup S(X_n)$. Each production may also have a set of conditions on the values of the attributes in $A(X_0) \cup A(X_1) \cup A(X_2) \cup \cdots \cup A(X_n)$ that further constrain an application of the production. The derivation of a sentence in the attribute grammar is satisfied if and only if the context-free grammar is satisfied and all conditions are true.

4.8.3 *j--* Examples

We now look at how the semantics for some of the constructs in *j--* can be expressed using attribute grammars. Let us first consider examples involving just synthesized attributes. Semantics of literals involves just the synthesis of their type. This can be done using a synthesized attribute *type*:

literal ::= <int_literal>
        type(literal) ← int
  | <char_literal>
        type(literal) ← char
  | <string_literal>
        type(literal) ← String

Semantics of a multiplicative expression involves checking the types of its arguments making sure that they are numeric (integer), and synthesizing the type of the expression itself. Hence the following attribute grammar.

multiplicativeExpression ::= unaryExpression$_1$
              {* unaryExpression$_2$ }
  condition: *type*(unaryExpression$_1$) = int and *type*(unaryExpression$_2$) = int
        type(multiplicativeExpression) ← int

Now, consider a *j--* expression x * 5. The variable x must be of numeric (int) type. We cannot simply synthesize the type of x to int, but it must be synthesized elsewhere, that is, at the place where it is declared, and inherited where it is used. The language semantics also require that the variable x be declared before the occurrence of this expression. Thus, we must inherit the context (symbol table) that stores the names of variables and their associated types, so we can perform a lookup to see if the variable x is indeed declared. The attribute grammar below indicates how we synthesize the context for variables at the place where they are declared and how the context is inherited at the place the variables are used.

localVariableDeclarationStatement ::= type variableDeclarators ;
        foreach variable in variableDeclarators
           add(context, *defn*(name(variable), type))

variableDeclarator ::= <identifier> [= variableInitializer]
        name(variableDeclarator) ← <identifier>

primary ::= ...
    | variable
        defn ← *lookup*(context, *name*(variable))
        if defn = null
          error("Cannot find variable " + *name*(variable))
        else
          type(primary) ← *type*(defn)

Functions *add*() and *lookup*() are auxiliary functions defined for adding variable definitions into a context (symbol table) and for looking up a variable in a given context. In the above attribute grammar, the attribute *context* is synthesized in localVariableDeclarationStatement and is inherited in primary.

Attribute grammars might also be used for specifying the code to be produced for (or, at least the mathematical meaning of) the various syntactic forms in our source language.

For better or worse, attribute grammars have never found their way into the practice of compiler writing, but have rather been an academic tool for specifying the formal semantics of programming languages.

4.9 Further Readings

The *Java Language Specification* [Gosling et al., 2005] is the best source of information on the semantic rules for the Java language.

Attribute grammars were first introduced in [Knuth, 1968]. Chapter 3 of [Slonneger and Kurtz, 1995] offers an excellent treatment of attribute grammars; Chapter 7 discusses the application of attribute grammars in code generation. A general compilers text that makes extensive use of attribute grammars is [Waite and Goos, 1984]. Chapter 5 of [Aho et al., 2007] also treats attribute grammars.

[Gamma, 1995] describes the Visitor Pattern as well as many other design patterns. Another good text describing how design patterns may be used in Java programming is [Jia, 2003]. Yet another resource for design patterns is [Freeman et al., 2004].

4.10 Exercises

The *j--* compiler does not enforce all of the Java rules that it might. The following exercises allow the reader to rectify this.

Exercise 4.1. The compiler does not enforce the Java rule that only one type declaration (that is, class declaration in *j--*) be declared public. Repair this in one of the `analyze()` methods.

Exercise 4.2. The `analyze()` method in `JVariable` assigns the type `Type.ANY` to a `JVariable` (for example, x) in the AST that has not been declared in the symbol table. This prevents a cascading of multiple errors from the report of a single undeclared variable. But we could go further by declaring the variable (as `Type.ANY`) in the symbol table so as to suppress superfluous error reports in subsequent encounters of the variable. Make this improvement to analysis.

Exercise 4.3. Go through *The Java Language Specification* [Gosling et al., 2005], and for each node type in the *j--* compiler, compile a list of rules that Java imposes on the language but are not enforced in *j--*.

a. Describe how you might implement each of these.

b. Which of these rules cannot be enforced?

c. Implement the enforcement of those rules that have not been implemented in *j--* but that one can.

Exercise 4.4. Write an attribute grammar expressing the type rules for each of the statements in *j--*.

> statement ::= block
> | if parExpression statement [else statement]
> | while parExpression statement
> | return [expression] ;
> | ;
> | statementExpression ;

Exercise 4.5. Write an attribute grammar expressing the semantics of expressions in *j--*, that is, for productions from statementExpression through literal. See Appendix B for the *j--* syntax specification.

Implementation of analysis for new functionality is left for Chapter 5.

Chapter 5

JVM Code Generation

5.1 Introduction

Once the AST has been fully analyzed, all variables and expressions have been typed and any necessary tree rewriting has been done. Also a certain amount of setup needed for code generation has been accomplished. The compiler is now ready to traverse the AST one more time to generate the Java Virtual Machine (JVM) code, that is, build the class file for the program.

For example, consider the following very simple program:

```
public class Square {
    public int square(int x) {
        return x * x;
    }
}
```

Compiling this with our *j--* compiler,

```
> $j/j--/bin/j-- Square.java
```

produces a class file, `Square.class`. If we run the `javap` program on this, that is

```
> javap -verbose Square
```

we get the following symbolic representation of the class file:

```
public class Square extends java.lang.Object
  minor version: 0
  major version: 49
  Constant pool:
const #1 = Asciz      Square;
const #2 = class      #1; //  Square
const #3 = Asciz      java/lang/Object;
const #4 = class      #3; //  java/lang/Object
const #5 = Asciz      <init>;
const #6 = Asciz      ()V;
const #7 = NameAndType #5:#6;//  "<init>":()V
const #8 = Method     #4.#7; //  java/lang/Object."<init>":()V
const #9 = Asciz      Code;
const #10 = Asciz     square;
const #11 = Asciz     (I)I;

{
public Square();
  Code:
   Stack=1, Locals=1, Args_size=1
   0: aload_0
   1: invokespecial #8; //Method java/lang/Object."<init>":()V
   4: return
```

```
public int square(int);
  Code:
    Stack=2, Locals=2, Args_size=2
    0: iload_1
    1: iload_1
    2: imul
    3: ireturn

}
```

We cannot *run* this program because there is no `main()` method. But this is not our purpose; we simply wish to look at some code.

Now, the first line is the class header:

```
public class Square extends java.lang.Object
```

It tells us the name of the class is `Square` and its super class is `java.lang.Object`. The next two lines tell us something about the version of the JVM that we are using; we are not interested in this. Following this is a constant pool, which in this case defines eleven constants. The numerical tags #1,...,#11 are used in the code for referring to these constants. For example, #10 refers to the method name `square`; #11 refers to its descriptor (I)I[1]. The use of the constant pool may appear somewhat archaic, but it makes for a compact class file.

Notice there are two methods defined. The first is a *constructor*.

```
public Square();
  Code:
    Stack=1, Locals=1, Args_size=1
    0: aload_0
    1: invokespecial #8; //Method java/lang/Object."<init>":()V
    4: return
```

This constructor (generated as a method named `<init>`) is the implied empty constructor (having no arguments), which the compiler must generate if the user has not defined an explicit version. The first line of this code is not really code at all but rather a sequence of values that the run-time JVM will want to know when it allocates a stack frame:

```
Stack=1, Locals=1, Args_size=1
```

The `Stack=1` indicates that one memory location is required for partial results (in this case, for loading `this` onto the stack). The `Locals=1` says just one local variable is allocated in the stack frame, in this instance, the location at offset 0 for `this`. The `Args_size=1` says there is just one actual argument to this constructor (in this case, the implicit argument, `this`).

Then there is the code, proper:

```
    0: aload_0
    1: invokespecial #8; //Method java/lang/Object."<init>":()V
    4: return
```

At location 0 is a one-byte `aload_0` instruction that loads `this` onto the stack. Following that, at location 1, is a three-byte `invokespecial` instruction. The first byte holds the opcode for `invokespecial`, and the next two bytes hold a reference to the constant named #8 in the constant pool: Method #4.#7;. This constant in turn makes reference to additional constant entries #4 and #7. Taken as a whole, the `invokespecial` instruction invokes the

---

[1]Recall, a *descriptor* describes a method's signature. In this case, (I)I describes a method that takes one `int` as an argument and returns an `int` result.

constructor `<init>,<init>` with descriptor `()V`, of `Square`'s super class `java/lang/Object`, on the single argument `this`. After invoking the super constructor, the constructor simply returns, obeying the `return` instruction at location 4.

The second, explicit method `square()` is even simpler. The stack frame information,

```
Stack=2, Locals=2, Args_size=2
```

says we need two locations for computation (the two operands to the multiplication operation), two local variables allocated (for the implicit parameter `this`, and the explicit parameter `x`), and that there will be two actual arguments on the stack: `this` and the argument for `x`. The code consists of four one-byte instructions:

```
0:  iload_1
1:  iload_1
2:  imul
3:  ireturn
```

The instruction at location 0 loads the value for `x` onto the stack; the instruction at location 1 does the same. The instruction at location 2 pops those two values off the stack and performs an integer multiplication on them, leaving its result on the stack. The instruction at location 3 returns the integer (`x * x`) on the stack as the result of the method invocation.

Of course, to emit these instructions, we first create a `CLEmitter` instance, which is an abstraction of the class file we wish to build, and then call upon `CLEmitter`'s methods for generating the necessary headers and instructions.

For example, to generate the class header,

```
public class Square extends java.lang.Object
```

one would invoke the `addClass()` method on `output`, an instance of `CLEmitter`:

```
output.addClass(mods, "Square", "java/lang/Object", false);
```

where

- The `mods` denotes an `ArrayList` containing the single string "public",

- `Square` is the class name,

- `java/lang/Object` is the *internal form* for the fully qualified super class, and

- `false` indicates that the class is not synthetic.

As a simpler example, the one-byte, no-argument instruction `aload_1` may be generated by

```
output.addNoArgInstruction(ALOAD_1);
```

To fully understand `CLEmitter` and all of its methods for generating code, one should read through the `CLEmitter.java` source file, which is distributed as part of the *j--* compiler. One must sometimes be careful to pay attention to what the `CLEmitter` is expecting. For example, sometimes a method requires a fully qualified name in Java form such as `java.lang.Object`; other times an internal form is required, that is, `java/lang/Object`.

For another, more involved example of code generation, we revisit the `Factorial` class from Chapters 2 and 3. Recall the source code:

```
package pass;

import java.lang.System;
```

```
public class Factorial {
    // Two methods and a field

    public static int factorial(int n) {
        // position 1:
        if (n <= 0) {
            return 1;
        } else {
            return n * factorial(n - 1);
        }
    }

    public static void main(String[] args) {
        int x = n;

        // position 2:
        System.out.println(n + "! = " + factorial(x));
    }

    static int n = 5;
}
```

Running `javap` on the class produced for this by the *j--* compiler gives us

```
public class pass.Factorial extends java.lang.Object
  minor version: 0
  major version: 49
  Constant pool:
  ...
{
static int n;

public pass.Factorial();
  Code:
    Stack=1, Locals=1, Args_size=1
    0: aload_0
    1: invokespecial #8; //Method java/lang/Object."<init>":()V
    4: return

public static int factorial(int);
  Code:
    Stack=3, Locals=1, Args_size=1
    0: iload_0
    1: iconst_0
    2: if_icmpgt 10
    5: iconst_1
    6: ireturn
    7: goto  19
    10: iload_0
    11: iload_0
    12: iconst_1
    13: isub
    14: invokestatic #13; //Method factorial:(I)I
    17: imul
    18: ireturn
    19: nop

public static void main(java.lang.String[]);
  Code:
    Stack=3, Locals=2, Args_size=1
    0: getstatic #19; //Field n:I
    3: istore_1
    4: getstatic #25; //Field java/lang/System.out:Ljava/io/PrintStream;
    7: new   #27; //class java/lang/StringBuilder
```

```
   10: dup
   11: invokespecial    #28; //Method java/lang/StringBuilder."<init>":()V
   14: getstatic    #19; //Field n:I
   17: invokevirtual    #32; //Method java/lang/StringBuilder.append:
                                   (I)Ljava/lang/StringBuilder;
   20: ldc #34; //String ! =
   22: invokevirtual    #37; //Method java/lang/StringBuilder.append:
                                   (Ljava/lang/String;)Ljava/lang/StringBuilder;
   25: iload_1
   26: invokestatic #13; //Method factorial:(I)I
   29: invokevirtual    #32; //Method java/lang/StringBuilder.append:
                                   (I)Ljava/lang/StringBuilder;
   32: invokevirtual    #41; //Method java/lang/StringBuilder.toString:
                                   ()Ljava/lang/String;
   35: invokevirtual    #47; //Method java/io/PrintStream.println:
                                   (Ljava/lang/String;)V
   38: return

public static {};
  Code:
   Stack=2, Locals=0, Args_size=0
   0: iconst_5
   1: putstatic #19; //Field n:I
   4: return
}
```

We have removed the constant pool, but the comments contain the constants that the program refers to. Notice the last method, `static`. That implements what is known as the *static block*, where class initializations can go. Here, the static field n is initialized to 5. The method that does this in the JVM code is `<clinit>` (for class initialization).

The following sections address the task of generating code for various *j--* constructs.

5.2 Generating Code for Classes and Their Members

JCompilationUnit's `codegen()` drives the generation of code for classes. For each type (that is, class) declaration, it

- Invokes `codegen()` on the JClassDeclaration for generating the code for that class,

- Writes out the class to a class file in the destination directory, and

- Adds the in-memory representation of the class to a list that stores such representations for all the classes within a compilation unit; this list is used in translating JVM byte code to native (SPIM) code, in Chapter 6.

```
public void codegen(CLEmitter output) {
    for (JAST typeDeclaration : typeDeclarations) {
        typeDeclaration.codegen(output);
        output.write();
        clFiles.add(output.clFile());
    }
}
```

5.2.1 Class Declarations

The `codegen()` method for `JClassDeclaration` does the following:

- It computes the fully- qualified name for the class, taking any package name into account.

- It invokes an `addClass()` on the `CLEmitter` for adding the class header to the start of the class file.

- If there is no explicit constructor with no arguments defined for the class, it invokes the private method `codegenImplicitConstructor()` to generate code for the implicit constructor as required by the language.

- It generates code for its members by sending the `codegen()` message to each of them.

- If there are any static field initializations in the class declaration, then it invokes the private method `codegenClassInit()` to generate the code necessary for defining a static block, a block of code that is executed after a class is loaded.

In the case of our `Factorial` example, there is no explicit constructor, so one is generated. The method, `codegenImplicitConstructor()` is invoked to generate the following JVM code[2]:

```
public <init>();
  Code:
    Stack=1, Locals=1, Args_size=1
  0: aload_0
  1: invokespecial #8; //Method java/lang/Object."<init>":()V
  4: return
```

The constructor is simply invoking the superclass constructor, that is, `Object()`. In Java, such a constructor would look like

```
public Factorial() {
    this.super();
}
```

The `Factorial` example has one static field with an initialization:

```
static int n = 5;
```

During analysis, the initialization was transformed into an explicit assignment, which must be executed after the `Factorial` class is loaded. Seasoned Java programmers will recognize this to be in a static block; in Java, the static block would look like

```
static {
    n = 5;
}
```

and would occur as a member of `Factorial`. Of course, *j*-- does not have static blocks[3]; they may be represented in the JVM as, for example[4],

---

[2]Actually, `javap` shows the name as `Factorial()` but when invoking the **addMethod()** method on the **CLEmitter**, one passes the argument `"<init>"` for the constructor's name; a constructor is simply an initialization method. The JVM expects this internal name.

[3]Its implementation is left as an exercise.

[4]Again, `javap` represents `<clinit>` as `static`, but in the argument to the **addMethod()**, method is `"<clinit>"`; `<clinit>` stands for *class initialization*. Again, the JVM expects this internal name.

```
public <clinit> {};
  Code:
    Stack=2, Locals=0, Args_size=0
    0: iconst_5
    1: putstatic #19; //Field n:I
    4: return
}
```

5.2.2 Method Declarations

The code generated for a `JMethodDeclaration` is pretty simple:

```
public void codegen(CLEmitter output) {
    output.addMethod(mods, name, descriptor, null, false);
    if (body != null) {
        body.codegen(output);
    }

    // Add implicit RETURN
    if (returnType == Type.VOID) {
        output.addNoArgInstruction(RETURN);
    }
}
```

It generates code for the method header, the body and then, for void methods, an implicit return-statement. Of course, the return may be superfluous if one already exits in the source code, but any good optimizer would remove the extra one.

5.2.3 Constructor Declarations

A constructor declaration is handled very much like a method declaration, but requires two additional tasks:

1. After the method header `<init>` has been emitted, `JConstructorDeclaration`'s `codegen()` looks to see if a super class constructor has been explicitly invoked:

   ```
   if (!invokesConstructor) {
       output.addNoArgInstruction(ALOAD_0);
       output.addMemberAccessInstruction(INVOKESPECIAL,
           ((JTypeDecl) context.classContext().definition())
               .superType().jvmName(), "<init>", "()V");
   }
   ```

   The flag `invokesConstructor` is set to `false` by default and is set to `true` during analysis of the body if the first statement in that body is an invocation of the super class constructor.

2. Any instance field initializations (after analysis, represented as assignments) are generated:

   ```
   for (JFieldDeclaration field : definingClass
           .instanceFieldInitializations()) {
       field.codegenInitializations(output);
   }
   ```

Notice that `JConstructorDeclaration` extends `JMethodDeclaration`.

5.2.4   Field Declarations

Because the analysis phase has moved initializations, `codegen()` for `JFieldDeclaration` need only generate code for the field declaration itself:

```
public void codegen(CLEmitter output) {
    for (JVariableDeclarator decl : decls) {
        // Add field to class
        output.addField(mods, decl.name(), decl.type()
            .toDescriptor(), false);
    }
}
```

5.3   Generating Code for Control and Logical Expressions

5.3.1   Branching on Condition

Almost all control statements in *j--* are controlled by some Boolean expression. Indeed, control and Boolean expressions are intertwined in *j--*. For example, consider the if-then-else statement below:

```
if (a > b) {
    c = a;
} else {
    c = b;
}
```

The code produced for this is as follows:

```
 0: iload_1
 1: iload_2
 2: if_icmple  10
 5: iload_1
 6: istore_3
 7: goto  12
10: iload_2
11: istore_3
12: ...
```

Notice a couple of things about this code.

1. Rather than compute a Boolean value (`true` or `false`) onto the stack depending on the condition and then branching on the value of *that*, the code branches directly on the condition itself. This is faster, makes use of the underlying JVM instruction set, and makes for more compact code.

2. A branch is made over the code for the then-part to the else-part if the condition's *complement* is `true`, that is, if the condition is `false`; in our example, an `if_icmple` instruction is used.

   ```
   branch to elseLabel if <condition> is false
     <code for thenPart>
     branch to endLabel
   elseLabel:
     <code for elsePart>
   endLabel:
   ```

Consider the case where we were implementing the Java do-while statement; for example,

```
do {
    a++;
}
while (a < b);
```

The do-while statement is not in *j--* but we could add it[5]. The code we generate might have the form

```
topLabel:
<code for body>
branch to topLabel if <condition> is true
```

While we branch on the condition `false` for the if-then-else statement, we would branch on the condition `true` for the do-while statement.

In generating code for a condition, one needs a method of specifying three arguments:

1. The `CLEmitter`,

2. The target label for the branch, and

3. A `boolean` flag, `onTrue`. If `onTrue` is `true`, then the branch should be made on the condition; if `false`, the branch should be made on the condition's complement.

Thus, every boolean expression must support a version of `codegen()` with these three arguments. For example, consider that for `JGreaterThanOp` is:

```
public void codegen(CLEmitter output, String targetLabel,
    boolean onTrue) {
    lhs.codegen(output);
    rhs.codegen(output);
    output.addBranchInstruction(onTrue ? IF_ICMPGT : IF_ICMPLE,
        targetLabel);
}
```

A method of this sort is invoked on the condition controlling execution; for example, the following `codegen()` for `JIfStatement` makes use of such a method in producing code for the if-then-else statement.

```
public void codegen(CLEmitter output) {
    String elseLabel = output.createLabel();
    String endLabel = output.createLabel();
    condition.codegen(output, elseLabel, false);
    thenPart.codegen(output);
    if (elsePart != null) {
        output.addBranchInstruction(GOTO, endLabel);
    }
    output.addLabel(elseLabel);
    if (elsePart != null) {
        elsePart.codegen(output);
        output.addLabel(endLabel);
    }
}
```

Notice that method `createLabel()` creates a unique label each time it is invoked, and the `addLabel()` and `addBranchInstruction()` methods compute the necessary offsets for the actual JVM branches.

---

[5]Implementing the do-while statement is left as an exercise.

5.3.2   Short-Circuited &&

Consider the logical && operator. In a logical expression like

$$arg_1 \ \&\& \ arg_2$$

The semantics of Java, and so of *j--*, require that the evaluation of such an expression be short-circuited. That is, if arg_1 is false, arg_2 is not evaluated; false is returned. If arg_1 is true, the value of the entire expression depends on arg_2. This can be expressed using the Java conditional expression, as arg_1 ? arg_2 : false. How do we generate code for this operator? The code to be generated depends of whether the branch for the entire expression is to be made on true, or on false:

```
Branch to target when            Branch to target when
    arg1 && arg2 is true:            arg1 && arg2 is false:

    branch to skip if                branch to target if
        arg1 is false                    arg1 is false
    branch to target when            branch to target if
        arg2 is true                     arg2 is false
skip: ...
```

For example, the code generated for

```
if (a > b && b > c) {
    c = a;
}
else {
    c = b;
}
```

would be

```
0: iload_1
1: iload_2
2: if_icmple 15
5: iload_2
6: iload_3
7: if_icmple 15
10: iload_1
11: istore_3
12: goto 17
15: iload_2
16: istore_3
17: ...
```

The codegen() method in JLogicalAndOp generates the code for the && operator:

```
public void codegen(CLEmitter output, String targetLabel,
    boolean onTrue) {
    if (onTrue) {
        String falseLabel = output.createLabel();
        lhs.codegen(output, falseLabel, false);
        rhs.codegen(output, targetLabel, true);
        output.addLabel(falseLabel);
    } else {
        lhs.codegen(output, targetLabel, false);
        rhs.codegen(output, targetLabel, false);
    }
}
```

Notice that our method prevents unnecessary branches to branches. For example, consider the slightly more complicated condition in

```
if (a > b && b > c && c > 5) {
    c = a;
}
else {
    c = b;
}
```

The JVM code produced for this targets the same exit on `false`, for each of the `&&` operations:

```
 0:    iload_1
 1:    iload_2
 2:    if_icmple      18
 5:    iload_2
 6:    iload_3
 7:    if_icmple      18
10:    iload_3
11:    iconst_5
12:    if_icmple      18
15:    iinc     1, -1
18:    ...
```

Branches to branches are avoided even when generating code for a deeply nested and complex Boolean expression because we pass the target through nested recursive calls to this special version of `codegen()`.

The implementation of a `||` operator for a short-circuited logical or operation (not yet in the *j--* language) is left as an exercise.

5.3.3   Logical Not !

The code generation for the `JLogicalNot` operator `!` is trivial: a branch on `true` becomes a branch on `false`, and a branch on `false` becomes a branch on `true`:

```
public void codegen(CLEmitter output, String targetLabel,
    boolean onTrue) {
    arg.codegen(output, targetLabel, !onTrue);
}
```

5.4   Generating Code for Message Expressions, Field Selection, and Array Access Expressions

5.4.1   Message Expressions

Message expressions are what distinguish object-oriented programming languages from the more traditional programming languages such as C. *j--*, like Java, is object-oriented.

Much of the work in decoding a message expression is done in analysis. By the time the compiler has entered the code generation phase, any ambiguous targets have been reclassified, instance messages have been distinguished from class (static) messages, implicit targets have been made explicit, and the message's signature has been determined.

In fact, by examining the target's type (and if necessary, its sub-types) analysis can determine whether or not a viable message matching the signature of the message exists. Of course, for instance messages, the actual message that is invoked cannot be determined until run-time. That depends on the actual run-time object that is the target of the expression and so is determined by the JVM. But the code generated for a message expression does state everything that the compiler can determine.

JMessageExpression's `codegen()` proceeds as follows:

1. If the message expression involves an instance message, `codegen()` generates code for the target.

2. The message invocation instruction is determined: `invokevirtual` for instance messages and `invokestatic` for static messages.

3. The `addMemberAccessInstruction()` method is invoked to generate the message invocation instruction; this method takes the following arguments:

    (a) The instruction (`invokevirtual` or `invokestatic`).

    (b) The JVM name for the target's type.

    (c) The message name.

    (d) The descriptor of the invoked method, which was determined in analysis.

4. If the message expression is being used as a statement expression and the return type of the method referred to in (d), and so the type of the message expression itself is non-void, then method `addNoArgInstruction()` is invoked for generating a `pop` instruction. This is necessary because executing the message expression will produce a result on top of the stack, and this result is to be thrown away.

For example, the code generated for the message expression

```
... = s.square(6);
```

where `square` is an instance method, would be

```
aload s'
bipush  6
invokevirtual    #6; //Method square:(I)I
```

Notice this leaves an integer result 36 on the stack. But if the message expression were used as a statement, as in

```
s.square(6);
```

our compiler generates a `pop` instruction to dispose of the result[6]:

```
aload s'
bipush  6
invokevirtual    #6; //Method square:(I)I
pop
```

We invoke static methods using `invokestatic` instructions. For example, say we wanted to invoke the static method `csquare`; the message expression

```
.. = Square1.csquare(5);
```

would be translated as

---

[6]`s'` indicates the stack frame offset for the local variable `s`.

```
iconst_5
invokestatic      #5;  //Method csquare:(I)I
```

If we were to implement interfaces (this is left as an exercise), we would want to generate the `invokeinterface` instruction.

5.4.2   Field Selection

When we talk about generating code for a field selection, we are interested in the case where we want the *value*[7] for the field selection expression. The case where the field selection occurs as the target of an assignment statement is considered in the next section.

As for message expressions, much of the work in decoding a field selection occurs during analysis. By the time the code generation phase begins, the compiler has determined all of what it needs to know to generate code. `JFieldSelection`'s `codegen()` works as follows:

1. It generates code for its target. (If the target is a class, no code is generated.)

2. The compiler must again treat the special case `a.length` where `a` is an array. As was noted in the discussion of analysis, this is a language hack. The code generated is also a hack, making use of the special instruction `arraylength`.

3. Otherwise, it is treated as a proper field selection. The field selection instruction is determined: `getfield` for instance fields and `getstatic` for static fields.

4. The `addMemberAccessInstruction()` method is invoked with the following arguments:

   (a) The instruction (`getfield` or `getstatic`)
   (b) The JVM name for the target's type
   (c) The field name
   (d) The JVM descriptor for the type of the field, and so the type of the result

For example, given that `instanceField` is indeed an instance field, the field selection

```
s.instanceField
```

would be translated as

```
aload s'
getfield instanceField:I
```

On the other hand, for a static field selection such as

```
Square1.staticField
```

our compiler would generate

```
getstatic staticField:I
```

---

[7]Actually, as we shall see in the next section, the field expression's *r-value*.

5.4.3   Array Access Expressions

Array access expressions in Java and so in *j--* are pretty straightforward. For example, if the variable a references an array object, and i is an integer, then

```
a[i]
```

is an array access expression. The variable a may be replaced by any primary expression that evaluates to an array object, and i may be replaced by any expression evaluating to an integer. The JVM code is just as straightforward. For example, if a is an integer array, then the code produced for the array access expression would be

```
aload a'
iload i'
iaload
```

Instruction `iaload` pops an index i and an (integer) array a off the stack and replaces those two items by the value for `a[i]`.

In *j--*, as in Java, multiply dimensioned arrays are implemented as arrays of arrays[8]. That is, *j--* arrays are vectors.

5.5   Generating Code for Assignment and Similar Operations

5.5.1   Issues in Compiling Assignment

l-values and r-values

Consider a simple assignment statement;

```
x = y;
```

It does not seem to say very much; it asks that the value of variable y be stored (as the new value) in variable x. Notice that we want different values for x and y when interpreting such an assignment. We often say we want the l-value for x, and the r-value for y. We may think of a variable's l-value as its address or location (or in the case of local variables, its stack frame offset), and we may think of a variable's r-value as its content or the value stored in (or, as the value of) the variable. This relationship is illustrated in Figure 5.1.

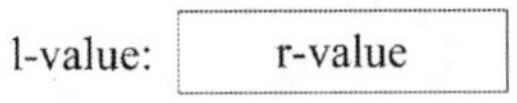

l-value: | r-value |

FIGURE 5.1 A variable's l-value and r-value.

The names l-value and r-value come from the corresponding positions in the assignment expression: the left-hand side of the = and the right-hand side of the =. Think of

l-value = r-value

Notice that while all expressions have r-values, many have no l-values. For example, if a is an array of ten integers and o is an object with field f, C is a class with static field sf, and x is a local variable, then all of

---

[8]Such arrays are known as *Illife vectors*, after their inventor, John K. Illife.

```
a[3]
o.f
C.sf
x
```

have both l-values and r-values. That is, any of these may appear on the left-hand side or on the right-hand side of an assignment operator. On the other hand, while all of the following have r-values, none has an l-value:

```
5
x+5
Factorial.factorial(5)
```

None of these may meaningfully appear on the left-hand side of an assignment statement.

In compiling an assignment, compiling the right-hand side expression to produce code for computing its r-value and leaving it on the stack is straightforward. But what code must be generated for the left-hand side? In some cases, no code need be generated; for example, assuming x and y are local integer variables, then compiling

```
x = y;
```

produces

```
iload y'
istore x'
```

where x' and y' are the stack offsets for x and y, respectively. On the other hand, compiling

```
a[x] = y;
```

produces

```
aload a'
iload x'
iload y'
iastore
```

This code loads (a reference to) the array a and an index x onto the stack. It then loads the r-value for y onto the stack. The `iastore` (integer array store) instruction pops the value for y, the index x, and the array a from the stack, and it stores y in the element indexed by x in array a.

Assignment Expressions versus Assignment Statements

Another issue is that an assignment may act as an expression producing a result (on the stack) or as a statement (syntactically, as a statement expression). That is, we can have

```
x = y;
```

or

```
z = x = y;
```

In the first case, the assignment x = y; is a statement; no value is left on the stack. But in the second case, the x = y must assign the value of y to x but also leave a value (the r-value for y) on the stack so that it may be popped off and assigned to z. That is, the code might look something like

```
iload y'
dup
istore x'
istore z'
```

The dup duplicates the r-value for y on top of the stack so there are two copies. Each of the istore operations pops a value and stores it in the appropriate variable.

In parsing, when an expression is used as a statement, Parser's statementExpression () method sets a flag isStatementExpression in the expression node to true. The code generation phase can make use of this flag in deciding when code must be produced for duplicating r-values on the run-time stack.

Additional Assignment-Like Operations

The most important property of the assignment is its *side effect*. One uses the assignment operation for its side effect, overwriting a variable's r-value with another. There are other Java (and *j--*) operations having the same sort of side effect. For example,

```
x--
++x
x += 6
```

are all expressions that have side effects. But they also denote values. The context in which they are used determines whether or not we want the value left on the stack, or we simply want the side effect.

5.5.2   Comparing Left-Hand Sides and Operations

The table below compares the various operations (labeled down the left) with an assortment of left-hand sides (labeled across the top). We do not deal with string concatenation here, but leave that to a later section.

	x	a[i]	o.f	C.sf
lhs = y	iload y'	aload a'	aload o'	iload y'
	[dup]	iload i'	iload y	[dup]
	istore x'	iload y'	[dup_x1]	putstatic sf
	[dup_x2]	putfield f		
	iastore			
lhs += y	iload x'	aload a'	aload o'	getstatic sf
	iload y'	iload i'	dup	iload y'
	iadd	dup2	getfield f	iadd
	[dup]	iaload	iload y'	[dup]
	istore x'	iload y'	iadd	putstatic sf
	iadd	[dup_x1]		
	[dup_x2]	putfield f		
	iastore			
++lhs	iinc x',1	aload a'	aload o'	getstatic sf
	[iload x']	iload i'	dup	iconst_1
	dup2	getfield f	iadd	
	iaload	iconst_1	[dup]	
	iconst_1	iadd	putstatic sf	
	iadd	[dup_x1]		
	[dup_x2]	putfield f		
	iastore			
lhs--	[iload x']	aload a'	aload o'	getstatic sf
	iinc x',1	iload i'	dup	[dup]
	dup2	getfield f	iconst_1	
	iaload	[dup_x1]	isub	
	[dup_x2]	iconst_1	putstatic sf	
	iconst_1	isub		
	isub	putfield f		
	iastore			

The instructions in brackets [...] must be generated if and only if the operation is a sub-expression of some other expression, that is, if the operation is not a statement expression.

Notice that there are all sorts of stack instructions for duplicating values and moving them into various locations on the stack:

dup duplicates the top value on the stack (pushing a copy on top)

dup2 duplicates the top pair of items on the stack (pushing a second pair on top)

dup_x2 duplicates the top value, but copies it below the two items that are below the original top value

dup_x1 duplicates the top value, but copies it below the single item that is below the original top value

Figure 5.2 illustrates the effect of each of these. As one might guess, these instructions were included in the JVM instruction set for just these sorts of operations.

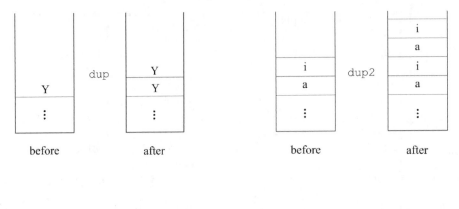

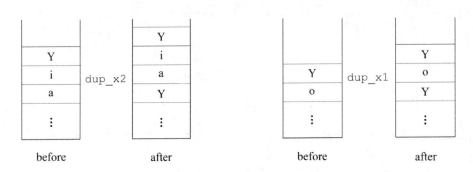

FIGURE 5.2 The effect of various duplication instructions.

The table above suggests that all of this is very complicated, but there is a pattern here, which we may take advantage of in clarifying what code to generate.

5.5.3　Factoring Assignment-Like Operations

The table above suggests four sub-operations common to most of the assignment-like operations in *j--*. They may also be useful if and when we want to extend *j--* and when adding additional operations. These four sub-operations are

1. `codegenLoadLhsLvalue()` – this generates code to load any up-front data for the left-hand side of an assignment needed for an eventual store, that is, its l-value.

2. `codegenLoadLhsRvalue()` – this generates code to load the r-value of the left-hand side, needed for implementing, for example, the `+=` operator.

3. `codegenDuplicateRvalue()` – this generates code to duplicate an r-value on the stack and put it in a place where it will be on top of the stack once the store is executed.

4. `codegenStore()` – this generates the code necessary to perform the actual store.

The code needed for each of these differs for each potential left-hand side of an assignment: a simple local variable `x`, an indexed array element `a[i]`, an instance field `o.f`, and a static field `C.sf`. The code necessary for each of the four operations, and for each left-hand side form, is illustrated in the table below.

	x	a[i]	o.f	C.sf
codegenLoadLhsLvalue()	[none]	aload a' aload i'	aload o'	[none]
codegenLoadLhsRvalue()	iload x' iaload	dup2 getfield f	dup	getstatic sf
codegenDuplicateRvalue()	dup	dup_x2	dup_x1	dup
codegenStore()	istore x'	iastore	putfield f	putstatic sf

Our compiler defines an interface JLhs, which declares four abstract methods for these four sub-operations. Each of JVariable, JArrayExpression, and JFieldSelection implements JLhs. Of course, one must also be able to generate code for the right-hand side expression. But codegen() is sufficient for that—indeed, that is its purpose.

For example, JPlusAssignOp's codegen() makes use of all of these operations:

```
public void codegen(CLEmitter output) {
    ((JLhs) lhs).codegenLoadLhsLvalue(output);
    if (lhs.type().equals(Type.STRING)) {
        rhs.codegen(output);
    } else {
        ((JLhs) lhs).codegenLoadLhsRvalue(output);
        rhs.codegen(output);
        output.addNoArgInstruction(IADD);
    }
    if (!isStatementExpression) {
        // Generate code to leave the r-value atop stack
        ((JLhs) lhs).codegenDuplicateRvalue(output);
    }
    ((JLhs) lhs).codegenStore(output);
}
```

5.6 Generating Code for String Concatenation

The implementation of most unary and binary operators is straightforward; there is a JVM instruction for each j-- operation. A case for which this does not apply is string concatenation.

In j--, as in Java, the binary + operator is overloaded. If both of its operands are integers, it denotes addition. But if either operand is a string then the operator denotes *string* concatenation and the result is a string. String concatenation is the only j-- operation where the operand types don't have to match[9].

The compiler's analysis phase determines whether or not string concatenation is implied. When it is, the concatenation is made explicit; that is, the operation's AST is rewritten, replacing JAddOp with a JStringConcatenationOp. Also, when x is a string, analysis replaces

```
x += <expression>
```

by

```
x = x + <expression>
```

[9]We leave the implementation of other implicit type conversions as an exercise.

So, code generation is left with generating code for only the explicit string concatenation operation. To implement string concatenation, the compiler generates code to do the following:

1. Create an empty string buffer, that is, a `StringBuffer` object, and initialize it.

2. Append any operands to that buffer. That `StringBuffer`'s `append()` method is overloaded to deal with any type makes handling operands of mixed types easy.

3. Invoke the `toString()` method on the string buffer to produce a `String`.

`JStringConcatenationOp`'s `codegen()` makes use of a helper method `nestedCodegen()` for performing only step 2 for any nested string concatenation operations. This eliminates the instantiation of unnecessary string buffers. For example, given the *j*-- expression

```
x + true + "cat" + 0
```

the compiler generates the following JVM code:

```
new java/lang/StringBuilder
dup
invokespecial StringBuilder."<init>":()V
aload x'
invokevirtual append:(Ljava/lang/String;)StringBuilder;
iconst_1
invokevirtual append:(Z)Ljava/lang/StringBuilder;
ldc "cat"
invokevirtual append:(Ljava/lang/String;)Ljava/lang/StringBuilder;
iconst_0
invokevirtual append:(I)Ljava/lang/StringBuilder;
invokevirtual StringBuilder.toString:()Ljava/lang/String;
```

5.7 Generating Code for Casts

Analysis determines both the validity of a cast and the necessary `Converter`, which encapsulates the code generated for the particular cast. Each `Converter` implements a method `codegen()`, which generates any code necessary to the cast. For example, consider the converter for casting a reference type to one of its sub-types; such a cast is called *narrowing* and requires that a `checkcast` instruction be generated:

```
class NarrowReference implements Converter {

    private Type target;

    public NarrowReference(Type target) {
        this.target = target;
    }

    public void codegen(CLEmitter output) {
        output.addReferenceInstruction(CHECKCAST,
            target.jvmName());
    }
}
```

On the other hand, when any type is cast to itself (the identity cast), or when a reference type is cast to one of its super types (called *widening*), no code need be generated.

Casting an `int` to an `Integer` is called *boxing* and requires an invocation of the `Integer` `.valueOf()` method:

```
invokestatic java/lang/Integer.valueOf:(I)Ljava/lang/Integer;
```

Casting an `Integer` to an `int` is called *unboxing* and requires an invocation of the `Integer.intValue()` method:

```
invokevirtual java/lang/Integer.intValue:()I
```

Certain casts, from one primitive type to another, require that a special instruction be executed. For example, the `i2c` instruction converts an `int` to a `char`. There is a `Converter` defined for each valid conversion in *j--*.

5.8 Further Readings

[Lindholm and Yellin, 1999] is the authoritative reference for the JVM and its instruction set; Chapter 7 in that text discusses some of the issues in compiling a high-level language such as Java to the JVM.

An excellent introduction to compiling Microsoft's® .NET Common Language Runtime (CLR) is [Gough and Gough, 2001].

[Strachey, 2000] introduces the notion of l-values and r-values.

A good introduction to the behavior of Java generics—one of our exercises below— is [Bracha, 2004].

5.9 Exercises

Exercise 5.1. Add interfaces (as a type declaration) to *j--*. Make any changes that are necessary to classes and class declarations for accommodating interfaces. Test your implementation thoroughly.

An interface defines a type that obeys a contract. Declare a variable to be of an interface type and you can be sure that it will refer to objects that support certain behaviors. Any class that implements the interface must support those same behaviors. Think of an interface as a purely abstract class.

From the point of view of analysis, interfaces have all sorts of rules that must be enforced. Method `analyze()`, in both `JInterfaceDeclaration`, and `JClassDeclaration` is where most of these rules may be enforced. These rules include

1. Although an interface may have `public`, `protected`, or `private` modifiers, only `public` is meaningful for interfaces not nested within other classes.

2. The compiler must also allow the modifiers `abstract`, `static`, and `strictfp`. The `abstract` modifier is redundant (as all interfaces are abstract) but the compiler must allow it. The `static` modifier has meaning only for member (nested) interfaces. The `strictfp` modifier has meaning if and only if one is supporting (`float` or `double`) floating-point arithmetic.

3. An interface may extend any number of (super-) interfaces.

4. A class may implement as many interfaces as it likes. Notice that this will require our changing the parser's `classDeclaration()` method so that it parses an (optional) implements-clause and stores a (possibly empty) list of interfaces.

5. Every field declaration in an interface is implicitly `public`, `static`, and `final`, but any of these three modifiers may be present even if unnecessary. Every field must be initialized. The initializing expression need not be constant and may reference other fields appearing textually before the reference.

6. The initializing expression in an interface's field declaration must involve only constants or other static fields.

7. There are rules about the inheritance of fields; see the language specification.

8. Every method declaration in an interface is implicitly `public` and `abstract` but either of these modifiers may be present in the declaration. Because the methods are abstract, its body is always a semicolon.

9. An interface's method can never be declared `static` or `final`, but class methods implementing them can be `final`.

10. There are additional rules governing method inheritance, overriding, and overloading abstract methods in interfaces; see the language specification.

11. Any class implementing an interface (or a list of interfaces) must implement all of the methods of the interface(s).

In code generation, there are several facts to consider:

1. In the JVM, an interface is a kind of class. So, just as for a class, one uses `CLEmitter`'s `addClass()` method for introducing an interface.

2. The list of modifiers for the `addClass()`'s formal parameter `accessFlags` must include both `interface` and `abstract`. The `superClass` argument should be the JVM name for `java.lang.Object`. Of course, the argument passed for `superInterfaces` should be the (possibly empty) list of the super-interfaces (from the extends-clause).

3. The list of modifiers of any field defined by `addField()` must include `public`, `static`, and `final`.

4. The list of modifiers of any method defined by `addMethod()` must include `public` and `abstract`.

5. In invoking a method on an interface, that is, an object declared to be of a type defined by an interface, one uses the `invokeinterface` instruction. Although less efficient than `invokevirtual`, `invokeinterface` deals with the possibility of interfaces that inherited multiple interfaces, and multiple classes implementing the same interface.

Otherwise, use `JClassDeclaration` as a guide to, and so a starting point for, implementing `JInterfaceDeclaration`.

Exercise 5.2. Study the *Java Language Specification* [Gosling et al., 2005] and determine what must be done to incorporate the final modifier into *j--*. Add the final modifier to *j--*, adding it to your compiler and testing it thoroughly.

Exercise 5.3. Add static initializers to *j--*, adding it to your compiler and testing it thoroughly.

A *static initializer* is a class (or interface) member. It is a block of initializing code that is executed before the class enclosing it is referenced. It is identified by the `static` keyword as in

```
static {
    ...
}
```

The code in the static initializer is executed before the enclosing class is referenced. As such, the code goes into the `<clinit>` method the same place that class field initializations go. To learn more about static initializers, see Section 8.7 of the *Java Language Specification* [Gosling et al., 2005]. The order that fields and static initializers are executed (and so the order they appear in `<clinit>`) is the order that the members appear in the class; the only exception to this is "that `final` class variables and fields of interfaces whose values are compile-time constants are initialized first." See Section 12.4.2 of [Gosling et al., 2005].

Rules that static initializers must follow include

1. It is a compile-time error to `throw` an exception or to `return` in the initializing code.

2. Static initializers may not refer to `this` or `super`.

3. There are restrictions in referring to class variables that are defined textually further in the code.

Exercise 5.4. Add instance initializers to *j--*, adding it to your compiler and testing it thoroughly.

An *instance initializer* is a block of initializing code that is executed each time an instance of the enclosing class is instantiated. Therefore, its code goes into the constructors, that is `<init>`. Make sure this code is not executed twice, that is, in a constructor whose code starts off with `this()`.

Rules that instance initializers must follow include

1. It is a compile-time error to `throw` an exception or to `return` in the initializing code.

2. Instance initializers may refer to `this`.

Exercise 5.5. Java permits the declaration of methods of *variable arity*. For example, one might declare a method

```
public void foo(int x, String y, int... z) { <body> }
```

Invocations of `foo` require an `int` value for x, a `String` value for y, and then any number (including zero) of `int`s, all of which are gathered into an implicitly declared integer array z. Implementing variable arity methods requires declaring the implicit array at method declaration time, and then gathering the (extra) actual arguments into an array at method invocation time. Add variable arity methods to *j--*, adding them to your compiler and testing them thoroughly.

Exercise 5.6. Add the do-while statement to *j--*, adding it to your compiler and testing it thoroughly.

The do-while statement takes the form

```
do {
    Statement
}
while (Test);
```

The Statement is always executed at least once, and the Test is at the bottom of the loop. As an example, consider

```
do {
    sum = sum + i;
    i = i      1;
} while (i > 0);
```

The Oracle Java compiler, `javac` would produce for this:

```
top:      iload sum'
      iload i'
      iadd
      istore sum'
      iload i'
      iconst_1
      isub
      istore i'
      iload i'
      ifgt    top
```

Compiling the do-while statement is very much like compiling the while-statement.

Exercise 5.7. Add the classic for-statement to *j--*, adding it to your compiler and testing it thoroughly. The classic for-statement may be reformulated as a while-statement, even in Java. For example, the template

```
for (Initialization; Test; Increment)
    Statement
```

may be expressed as

```
Initialization
while (Test) {
    Statement;
    Increment;
}
```

We must take into account the fact that either Initialization, Test, or Increment may be empty, and translate appropriately. This means that the for-statement may be translated in either of two ways:

1. We can generate JVM code directly, following the patterns illustrated in this chapter.

2. We can first rewrite the AST, replacing the `JForStatement` node with the block above, and then apply `codegen()` to that sub-tree. Notice how the enclosing block captures the limited scope of any variables declared in Initialization.

Which way is better?

Exercise 5.8. Add the enhanced for-statement to *j--*, adding it to your compiler and testing it thoroughly. The Java enhanced for-statement is used to iterate over collections. Syntactically, it looks something like

```
for (Type Identifier : Expression)
    Statement
```

How this can be interpreted depends on the type of Expression. If Expression's type is a sub-type of `Iterable`, let I be the type of `Expression.iterator()`. Then our enhanced for-statement may be expressed as the following classic for-statement:

```
for (I i' = Expression.iterator(); i'.hasNext() ;) {
    Type Identifier = i'.next();
    Statement
}
```

The variable i' is compiler generated in such a way as not to conflict with any other variables. Otherwise, Expression must have an array type T[]. In this case, our enhanced for- statement may be expressed as

```
T[] a' = Expression;
for (int i' = 0; i' < a'.length; i'++) {
    Type Identifier= a'[i'];
    Statement
}
```

The variables a' and i' are compiler generated in such a way as not to conflict with any other variables. This can be compiled similarly to the classic for- statement.

Exercise 5.9. Study the *Java Language Specification* [Gosling et al., 2005] to determine what it would take to implement the continue-statement in your compiler. Then add the continue-statement to *j--*, adding them to your compiler and testing it thoroughly.

Exercise 5.10. Study the *Java Language Specification* [Gosling et al., 2005] to determine what it would take to implement the break-statement in your compiler. Then add the break-statement to *j--*, adding it to your compiler and testing it thoroughly.

Exercise 5.11. Add conditional expressions to *j--*, adding them to your compiler and testing them thoroughly. Conditional expressions are compiled in a manner identical to if-else statements. The only difference is that in this case, both the consequent and the alternative are expressions. For example, consider the assignment

```
z = x > y ? x - y : y - x;
```

The `javac` compiler produces the following code for this:

```
43: iload_1
44: iload_2
45: if_icmple      54
48: iload_1
49: iload_2
50: isub
51: goto 57
54: iload_2
55: iload_1
56: isub
57: istore_3
```

As for the if-then statement, we compile the Boolean test expression using the 3-argument version of `codegen()`.

Exercise 5.12. Add the conditional or operator || to *j--*, adding it to your compiler and testing it thoroughly. Make sure to avoid unnecessary branches to branches. The conditional ||, like the conditional &&, is short-circuited. That is, in

$$e_1 \; || \; e_2$$

if e_1 evaluates to true, then e_2 is not evaluated and the whole expression is true. For example, `javac` compiles

```
if (x < 0 || y > 0) {
    x = y;
}
```

to produce

```
58:  iload_1
59:  iflt 66
62:  iload_2
63:  ifle 68
66:  iload_2
67:  istore_1
68:  ...
```

Exercise 5.13. Write tests that involve conditional branches on nested Boolean expressions that involve the, !, && and (if you have implemented it) || operator to insure that the compiler is not generating branches to branches.

Exercise 5.14. Add the switch-statement to *j--*, adding it to your compiler and testing it thoroughly.

The switch-statement is more involved than other control constructs. Fortunately, the JVM provides special support for it. Code generation for the switch-statement is discussed in Section 6.10 of the JVM specification [Lindholm and Yellin, 1999].

As an example, let us consider a method that makes use of a simple switch-statement.

```
int digitTight(char c) {
switch (c) {
    case '1': return 1;
    case '0': return 0;
    case '2': return 2;
    default: return -1;
    }
}
```

This method converts digit characters '0', '1', and '2' to the integers they denote; any other character is converted to -1. This is not the best way to do this conversion but it makes for a simple example of the switch-statement. Let us take a look at the code that javac produces, using the `javap` tool:

```
int digitTight(char);
  Code:
    Stack=1, Locals=2, Args_size=2
    0: iload_1
    1: tableswitch{ //48 to 50
            48: 30;
            49: 28;
            50: 32;
            default: 34 }
    28: iconst_1
    29: ireturn
    30: iconst_0
    31: ireturn
    32: iconst_2
    33: ireturn
    34: iconst_m1
    35: ireturn
```

Notice a few things here. First, the JVM uses a table to map the Unicode character representations to the code locations for each of the corresponding cases. For example, 48 (the Unicode representation for the character '0') maps to location 30, the location of the JVM code for the `return 0;`. Notice also that the characters have been sorted on their Unicode values: 48, 49, and 50. Finally, because the characters are consecutive, the `tableswitch` instruction can simply calculate an offset in the table to obtain the location to which it wants to branch.

Now consider a second example, where the range of the characters we are switching on is a little more sparse; that is, the characters are not consecutive, even when sorted.

```
int digitSparse(char c) {
    switch (c) {
    case '4': return 4;
    case '0': return 0;
    case '8': return 8;
    default: return -1;
    }
}
```

The code that `javac` produces for this takes into account this sparseness.

```
int digitSparse(char);
  Code:
    Stack=1, Locals=2, Args_size=2
    0: iload_1
    1: lookupswitch{ //3
                48: 38;
                52: 36;
                56: 40;
                default: 43 }
    36: iconst_4
    37: ireturn
    38: iconst_0
    39: ireturn
    40: bipush  8
    42: ireturn
    43: iconst_m1
    44: ireturn
```

Again, a table is used, but this time a `lookupswitch` instruction is used in place of the previous `tableswitch`. The table here represents a list of pairs, each mapping a Unicode representation for a character to a location. Again, the table is sorted on the Unicode representations but the sequence of characters is too sparse to use the offset technique for finding the appropriate code location. Rather, `lookupswitch` searches the table for the character it is switching on and then branches to the corresponding location. Because the table is sorted on the characters, a binary search strategy may be used for searching large tables.

Consider one more example, where the sorted characters are almost but not quite sequential:

```
int digitClose(char c) {
    switch (c) {
    case '1': return 1;
    case '0': return 0;
    case '3': return 3;
    default: return -1;
    }
}
```

That the range of characters we are switching on is not too sparse suggests our using the `tableswitch` instruction, which uses an offset to choose the proper location to branch to. The `javac` compiler does just this, as illustrated by the following output provided by the `javap` tool.

```
int digitClose(char);
  Code:
    Stack=1, Locals=2, Args_size=2
    0: iload_1
    1: tableswitch{ //48 to 51
                48: 34;
                49: 32;
                50: 38;
                51: 36;
                default: 38 }
    32: iconst_1
    33: ireturn
    34: iconst_0
    35: ireturn
    36: iconst_3
    37: ireturn
    38: iconst_m1
    39: ireturn
```

Notice a sequential table is produced. Of course, the entry for character '2' (Unicode 50) maps to location 38 (the default case) because it is not one of the explicit cases.

So our compiler must construct a list of value-label pairs, mapping a case value to a label we will emit to mark the location of code to branch to. We then sort that list on the case values and decide, based on sparseness, whether to use a `tableswitch` instruction (not sparse) or a `lookupswitch` (sparse) instruction. `CLEmitter` provides a method for each choice.

`CLEmitter`'s `addTABLESWITCHInstruction()` method provides for several arguments: a `default` label, a lower bound on the case values, an upper bound on the case values, and a sequential list of labels.

`CLEmitter`'s `addLOOKUPSWITCHInstruction()` method provides for a different set of arguments: a `default` label, a count of the number of value-label pairs in the table, and a `TreeMap` that maps case values to labels.

Of course, our compiler must decide which of the two instructions to use.

The next three exercises deal with exception handling. Exception handling in Java is captured by the try-catch-finally and throws-statement. Additionally, there is the throws-clause, which serves to help the compiler ensure that there is a `catch` for every exception thrown.

To illustrate the JVM code that is generated for exceptions, consider the following class declaration:

```
import java.io.FileReader;
import java.io.FileWriter;
import java.lang.IndexOutOfBoundsException;
import java.lang.System;
import java.io.FileNotFoundException;
import java.io.IOException;

public class Copy {

    private static final int EOF = -1;  // end of file character rep.

    public static void main(String[] args) throws IOException {
        FileReader inStream  = null;
        FileWriter outStream = null;
        int ch;

        try {
            // open the files
```

```
            inStream  = new FileReader(args[0]);
            outStream = new FileWriter(args[1]);

            // copy
            while ((ch = inStream.read()) != EOF) {
                outStream.write(ch);
            }
        }
        catch (IndexOutOfBoundsException e) {
            System.err.println(
                "usage: java Copy1 sourcefile targetfile");
        }
        catch (FileNotFoundException e) {
            System.err.println(e);  // rely on e's toString()
        }
        catch (IOException e) {
            System.err.println(e);
        }
        finally { // close the files
        inStream.close();
        outStream.close();
        }
    }

    void foo()
        throws IOException {
    throw new IOException();
    }
}
```

Notice that the closing of files in the finally-clause might cause an exception to be thrown. We could either nest this closing of files within its own try-catch block or we could declare that the method main might possibly throw an IOException. We do the latter here.

Exercise 5.15. Add the throw-statement to *j--*, adding it to your compiler and testing it thoroughly. The throw-statement is straightforwardly compiled to JVM code using the `athrow` instruction; for example,

```
throw new IOException();
```

produces the following JVM code, which constructs, initializes, and throws the exception object:

```
0: new     #16; //class java/io/IOException
3: dup
4: invokespecial #17; //method java/io/IOException."<init>":()V
7: athrow
```

Exercise 5.16. Add the try-catch-finally-statement to *j--*, adding it to your compiler and testing it thoroughly.

Let us look at the code generated by `javac` for method `main()` (above):

```
public static void main(java.lang.String[])     throws java.io.IOException;
  Code:
  Stack=4, Locals=6, Args_size=1
  0: aconst_null
  1: astore_1
  2: aconst_null
  3: astore_2
  4: new     #2; //class java/io/FileReader
  7: dup
  8: aload_0
```

```
 9: iconst_0
10: aaload
11: invokespecial    #3; //Method java/io/FileReader."<init>"
                            :(Ljava/lang/String;)V
14: astore_1
15: new  #4; //class java/io/FileWriter
18: dup
19: aload_0
20: iconst_1
21: aaload
22: invokespecial    #5; //Method java/io/FileWriter."<init>"
                            :(Ljava/lang/String;)V
25: astore_2
26: aload_1
27: invokevirtual    #6; //Method java/io/FileReader.read:()I
30: dup
31: istore_3
32: iconst_m1
33: if_icmpeq    44
36: aload_2
37: iload_3
38: invokevirtual    #7; //Method java/io/FileWriter.write:(I)V
41: goto 26
44: aload_1
45: invokevirtual    #8; //Method java/io/FileReader.close:()V
48: aload_2
49: invokevirtual    #9; //Method java/io/FileWriter.close:()V
52: goto 131
55: astore   4
57: getstatic    #11; //Field java/lang/System.err:Ljava/io/PrintStream;
60: ldc #12; //String usage: java Copy1 sourcefile targetfile
62: invokevirtual    #13; //Method java/io/PrintStream.println:
                            (Ljava/lang/String;)V
65: aload_1
66: invokevirtual    #8; //Method java/io/FileReader.close:()V
69: aload_2
70: invokevirtual    #9; //Method java/io/FileWriter.close:()V
73: goto 131
76: astore   4
78: getstatic    #11; //Field java/lang/System.err:Ljava/io/PrintStream;
81: aload   4
83: invokevirtual    #15; //Method java/io/PrintStream.println:
                            (Ljava/lang/Object;)V
86: aload_1
87: invokevirtual    #8; //Method java/io/FileReader.close:()V
90: aload_2
91: invokevirtual    #9; //Method java/io/FileWriter.close:()V
94: goto 131
97: astore   4
99: getstatic    #11; //Field java/lang/System.err:Ljava/io/PrintStream;
102: aload   4
104: invokevirtual    #15; //Method java/io/PrintStream.println:
                            (Ljava/lang/Object;)V
107: aload_1
108: invokevirtual    #8; //Method java/io/FileReader.close:()V
111: aload_2
112: invokevirtual    #9; //Method java/io/FileWriter.close:()V
115: goto    131
118: astore 5
120: aload_1
121: invokevirtual    #8; //Method java/io/FileReader.close:()V
124: aload_2
125: invokevirtual    #9; //Method java/io/FileWriter.close:()V
128: aload   5
```

```
130: athrow
131: return
Exception table:
  from    to  target type
     4    44      55  Class java/lang/IndexOutOfBoundsException
     4    44      76  Class java/io/FileNotFoundException
     4    44      97  Class java/io/IOException
     4    44     118  any
    55    65     118  any
    76    86     118  any
    97   107     118  any
   118   120     118  any
```

The code for the try block ranges from location 4 up to (but not including location) 44.

```
 4: new     #2; //class java/io/FileReader
 7: dup
 8: aload_0
 9: iconst_0
10: aaload
11: invokespecial    #3; //Method java/io/FileReader."<init>":
                          (Ljava/lang/String;)V
14: astore_1
15: new     #4; //class java/io/FileWriter
18: dup
19: aload_0
20: iconst_1
21: aaload
22: invokespecial    #5; //Method java/io/FileWriter."<init>":
                          (Ljava/lang/String;)V
25: astore_2
26: aload_1
27: invokevirtual    #6; //Method java/io/FileReader.read:()I
30: dup
31: istore_3
32: iconst_m1
33: if_icmpeq     44
36: aload_2
37: iload_3
38: invokevirtual    #7; //Method java/io/FileWriter.write:(I)V
41: goto 26
```

Much of the exception handling control is specified in the exception table:

```
  from    to  target type
     4    44      55  Class java/lang/IndexOutOfBoundsException
     4    44      76  Class java/io/FileNotFoundException
     4    44      97  Class java/io/IOException
     4    44     118  any
    55    65     118  any
    76    86     118  any
    97   107     118  any
   118   120     118  any
```

The first entry,

```
     4    44      55  Class java/lang/IndexOutOfBoundsException
```

says that the exception handling applies to locations 4 up to (but not including) 44; control is transferred to location 55 if and when an exception of the given type (`java.lang. IndexOutOfBoundsException` here) is raised. This entry captures the control for the first catch-clause. Similarly, the next two table entries,

```
     4    44      76  Class java/io/FileNotFoundException
```

```
    4    44    97    Class java/io/IOException
```

capture the control for the next two catch-clauses in the source code. The next five entries in the table

```
    4    44   118    any
   55    65   118    any
   76    86   118    any
   97   107   118    any
  118   120   118    any
```

apply to the try block, the three catch-clauses, and the finally-clause itself. These insure that if an exception is raised in any of these, the code for the finally-clause is executed:

```
118: astore  5
120: aload_1
121: invokevirtual   #8; //Method java/io/FileReader.close:()V
124: aload_2
125: invokevirtual   #9; //Method java/io/FileWriter.close:()V
128: aload   5
130: athrow
```

Notice copies of the finally-clause code follow the try block and each of the catch-clauses, insuring that the finally code is always executed[10].

Invoke `CLEmitter`'s `addExceptionHandler()` method for emitting each entry in the table. This method uses labels for identifying locations and is discussed in Appendix D (D.3.2).

Exercise 5.17. Add the throws-clause to *j--*, adding it to your compiler and testing it thoroughly. Make sure all the rules of exception handling are covered.

Much of the work here is ensuring that any exception that might be possibly thrown in the invocation of the method is either caught by a catch-clause in a surrounding try-catch block, or is covered by one of the exceptions in the throws declaration. See Section 11.2 of the *Java Language Specification* [Gosling et al., 2005].

As far as the generated code goes, one needs to ensure that the invocation of the method for adding a new method,

```
public void addMethod(ArrayList<String> accessFlags,
                      String name,
                      String descriptor,
                      ArrayList<String> exceptions,
                      boolean isSynthetic)
```

includes a list of the exceptions that the method throws, in internal form.

Exercise 5.18. Modify the `Copy` class to catch possible exceptions in the finally-clause:

```
finally { // close the files
    try {
        if (inStream != null) {
            inStream.close();
        }
    }
    catch (Exception e) {
        System.err.println("Unable to close input stream.");
    }
    try {
```

---

[10]Another way to handle the finally-clause is to treat it as a subroutine, making use of the JVM's `jsr` and `ret` instructions. Consult Chapter 7 of [Lindholm and Yellin, 1999] for hints on using these instructions for exception handling.

```
        if (outStream != null) {
            outStream.close();
        }
    }
    catch (Exception e) {
        System.err.println("Unable to close output stream.");
    }
}
```

Execute `javap` on the class files produced by both `javap` and by your extensions to *j--*. Make sure you are generating the proper JVM code.

Exercise 5.19. Study the *Java Language Specification* [Gosling et al., 2005] to determine what it would take to implement the assert-statement in your compiler. Then add the assert-statement to *j--*, adding it to your compiler and testing it thoroughly.

The next three exercises involve adding primitive types to our compiler. (There is no reason why we cannot add built-in reference types such as matrices for which we can overload the arithmetic operators; but doing so would go against the spirit of Java.) And, because the primitive types in Java that are not already in *j--* are numeric types, the new types must work in concert with the current types.

Exercise 5.20. Add the primitive type `double` to *j--*, adding it to your compiler and testing it thoroughly. Adding double-precision floating-point numbers to *j--* introduces several wrinkles:

1. Arithmetic requires new JVM instructions, for example, `dadd`.

2. Each double-precision floating-point number occupies two words. This requires that the offset be incremented by 2 each time a `double` variable is declared.

3. Mixed arithmetic introduces the need for implicit conversion.

For example, consider the following contrived method, where d and e are double-precision variables and i is an integer variable:

```
void foo() {
    int i = 4;
    double d = 3.0;
    double e = 5.0;

    e = d * i;
}
```

The `javac` compiler produces the following JVM code:

```
void foo();
  Code:
   Stack=4, Locals=6, Args_size=1
   0:   iconst_4
   1:   istore_1
   2:   ldc2_w    #2; //double 3.0d
   5:   dstore_2
   6:   ldc2_w    #4; //double 5.0d
   9:   dstore    4
   11:  dload_2
   12:  iload_1
   13:  i2d
   14:  dmul
   15:  dstore    4
   17:  return
```

Notice several things here. First, notice that the stack offsets take account of i being a one-word int, and d and e being two-word doubles. Second, notice the mix of types an iload instruction is used to load the integer and a dload instruction is used to load the double-precision value. The arithmetic is done in double precision, using the dmul instruction. Therefore, the integer i must be promoted (converted to a more precise representation) to a double-precision value using the i2d instruction.

In the assignment and arithmetic operations, analyze() must be modified for dealing with this mixed arithmetic. In the case of a mix of int and double, the rule is pretty simple: convert the int to a double. One will want to consult the *Java Language Specification* [Gosling et al., 2005], especially Chapters 5 and 15. The rules are pretty easy to ferret out for just ints and doubles.

Exercise 5.21. Add the primitive type long to *j--*, adding it to your compiler and testing it thoroughly. When adding additional primitive types (to int and double), the rules become more complicated. One must worry about when to do conversions and how to do them. One must follow the *Java Language Specification* [Gosling et al., 2005].

Once one has more than two numeric types to deal with, it is probably best to re-think the conversion scheme and build a simple framework for deciding how to promote one type to another. There are fifteen primitive type conversion instructions available to us in the JVM, and we will want to make effective use of them.

Exercise 5.22. Add the primitive type float to *j--*, adding it to your compiler and testing it thoroughly.

Exercise 5.23. The JVM provides instructions making it relatively straightforward to add synchronized-statement. Study the *Java Language Specification* [Gosling et al., 2005] to determine what it would take to implement the synchronized-statement in your compiler. Then add the synchronized-statement to *j--*, adding it to your compiler and testing it thoroughly.

Exercise 5.24. (Small project) Study the *Java Language Specification* [Gosling et al., 2005] to determine what it would take to implement nested classes in your compiler. Then add nested classes to *j--*, adding it to your compiler and testing them thoroughly.

In implementing nested classes, one must take into account and enforce both the scope rules and access rules. The *j--* compiler includes a little helpful machinery as it stands, in the form of a class context entry in the symbol table where nested classes might be stored.

Exercise 5.25. (Small project) Study the *Java Language Specification* [Gosling et al., 2005] to determine what it would take to implement enum types in your compiler. Then add enum types to *j--*, adding it to your compiler and testing them thoroughly.

The implementation of enum types simply requires enumerating the values supplied in the type declaration and storing these values along with their enumeration; of course, one must take into account explicit assignation in the declaration.

Exercise 5.26. (Large project) Study the *Java Language Specification* [Gosling et al., 2005] and other articles to determine what it would take to implement generic types in your compiler. Then add generic types to *j--*, adding it to your compiler and testing it thoroughly. Implementing generic types is a major project. Generics are described in [Bracha, 2004].

Exercise 5.27. (Large project) Add implicit type conversions to your implementation of *j--*, implementing them in your compiler and testing them thoroughly.

What makes this difficult is that it leads to complexity in the identification and selection from applicable methods in resolving method invocation. There are hints to an implementation in the steps outlined in [Gosling et al., 2005].

Chapter 6

Translating JVM Code to MIPS Code

6.1 Introduction

6.1.1 What Happens to JVM Code?

Compilation is not necessarily over with after the class file is constructed. At "execution," the class is loaded into the JVM and then interpreted. In the Oracle HotSpot^{TM}VM, once a method has been executed several times, it is compiled to native code—code that can be directly executed by the underlying computer. Once these *hotspots* in the code are compiled, the native code is cached so that it may be re-used in subsequent invocations of the method. So at run-time, control shifts back and forth between JVM code and native code. Of course, the native code runs much faster than the interpreted JVM code. This regimen takes advantage of the fact that nearly all programs spend most of their time executing small regions of code.

This scenario is further complicated by the existence of at least two run-time environments. The VM that runs on a server performs many optimizations on the native code. While these optimizations are expensive, they produce very fast code that, on a server, runs over and over again. On the other hand, the VM that runs on a client computer, such as a user's personal workstation, performs fewer optimizations and thus minimizes the one-time cost of compiling the method to native code.

Compiling JVM code to native code involves the following:

Register allocation. In the JVM code, all local variables, transients and instruction operands are stored on the run-time stack. But the computer architectures we are compiling to have a fixed number of (often eight or thirty-two) high-speed registers. The native instructions often require that operands be in registers and permanently assigning as many local variables and transients to registers makes for faster running programs. When all registers are in use and the computation requires another, one register must be spilled to memory. Register spilling, in turn, degrades performance, so we want to minimize that.

Optimization. The code produced can be improved so as to run faster. In some instances, invocations of methods having short bodies can be in lined, that is, replaced by the bodies' code.

Instruction selection. We must choose and generate the native code instructions sufficient to perform the computations.

Run-time support. A certain amount of functional support is required at run-time. For example, we need support for creating new objects on the heap. We also implement a class that gives us basic input and output functionality for testing our programs.

One might reasonably ask, "Why don't we simply compile the entire JVM program to native code at the very start?" One reason is Java's dynamic quality means we are not exactly sure which code will be needed at compile time. Further, new classes may be loaded while a Java program is executing. That HotSpot compilation can compete with static compilation in performance is counter-intuitive to programmers. But Oracle has benchmarks that demonstrate that the HotSpot VM often out-performs programs that have been statically compiled to native code.

6.1.2   What We Will Do Here, and Why

Here we shall translate a small subset of JVM instructions to the native code for the MIPS architecture. MIPS is a relatively modern reduced instruction set computer (RISC), which has a set of simple but fast instructions that operate on values in registers; for this reason it is often referred to as a register-based architecture—it relies on loads and stores for moving values between memory and its thirty-two general-purpose registers. The RISC architecture differs from the traditional complex instruction set computer (CISC), which has fewer but more complex (and so slower) instructions with a wider range of operand addressing capabilities; CISC operands can be expressed as registers, memory locations, or a combination of both (allowing indexing), and so CISC architectures normally have fewer (for example, eight) general-purpose registers. A popular CISC architecture is Intel's family of i86x computers.

More accurately, we will target the MIPS *assembly language*, which is directly interpreted by James Larus's SPIM simulator [Larus, 2010], and is readily available for many environments. We call this assembly language SPIM code. (SPIM is MIPS spelled backward.)

Assembly code is a symbolic language for expressing a native code program. It captures the native code program nicely because there is a one-to-one correspondence between it and the bit pattern representing each individual instruction but it is more meaningful to the reader.

Normally, a compiler will produce this assembly code and then use an *assembler*, i.e., a simple translator for translating the assembly code to the bit representation of native code.

But the SPIM interpreter interprets the MIPS assembly code directly. This serves our purpose just fine; we can produce sequences of MIPS instructions that are both readable and can be executed directly for testing the code we generate.

Our goal here is illustrated in Figure 6.1. The work that we have already done is shown using the dashed line. The work we intend to do here in this chapter and the next is shown using the solid line.

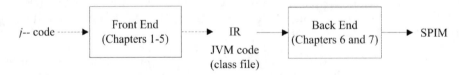

FIGURE 6.1 Our *j*-- to SPIM compiler.

Here we can re-define what constitutes the IR, the front end, and the back end:

- JVM code is our new IR.

- The *j*-- to JVM translator (discussed in Chapters 1–5) is our new front end.

- The JVM-to-SPIM translator (discussed in this chapter and the next) is our new back end.

6.1.3   Scope of Our Work

We translate a sufficient subset of the JVM to SPIM code to give the student a taste of native code generation, some of the possibilities for optimization, and register allocation. More precisely, we translate enough JVM code to SPIM code to handle the following *j--* program[1]. The class `spim.SPIM` is a class that we have defined for accessing special SPIM system routines for doing simple IO.

```java
import spim.SPIM;

// Prints factorial of a number computed using recursive and iterative
// algorithms.
public class Factorial {
    // Return the factorial of the given number computed recursively.
    public static int computeRec(int n) {
        if (n <= 0) {
            return 1;
        } else {
            return n * computeRec(n - 1);
        }
    }

    // Return the factorial of the given number computed iteratively.
    public static int computeIter(int n) {
        int result = 1;
        while ( n > 0 ) {
            result = result * n--;
        }
        return result;
    }

    // Entry point; print factorial of a number computed using
    // recursive and iterative algorithms.
    public static void main(String[] args) {
        int n = 7;
        SPIM.printInt(Factorial.computeRec(n));
        SPIM.printChar('\n');
        SPIM.printInt(Factorial.computeIter(n));
        SPIM.printChar('\n');
    }
}
```

We handle static methods, conditional statements, while loops, recursive method invocations, and enough arithmetic to do a few computations. We must deal with some objects, for example, constant strings. Although the program above refers to an array, it does not really do anything with it so we do not implement array objects. Our run-time support is minimal.

In order to determine just what JVM instructions must be handled, it is worth looking at the output from running `javap` on the class file produced for this program.

```
public class Factorial extends java.lang.Object
  minor version: 0
  major version: 49
  Constant pool:
... <the constant pool is elided here> ...

{
public Factorial();
  Code:
```

---

[1] We also compile a few more programs, which are included in the code tree but are not discussed further here: **Fibonacci**, **GCD**, **Formals**, and **HelloWorld**.

```
       Stack=1, Locals=1, Args_size=1
       0:  aload_0
       1:  invokespecial #8; //Method java/lang/Object."<init>":()V
       4:  return

public static int computeRec(int);
  Code:
       Stack=3, Locals=1, Args_size=1
       0:  iload_0
       1:  iconst_0
       2:  if_icmpgt 10
       5:  iconst_1
       6:  ireturn
       7:  goto  19
       10: iload_0
       11: iload_0
       12: iconst_1
       13: isub
       14: invokestatic #13; //Method computeRec:(I)I
       17: imul
       18: ireturn
       19: nop

public static int computeIter(int);
  Code:
       Stack=2, Locals=2, Args_size=1
       0:  iconst_1
       1:  istore_1
       2:  iload_0
       3:  iconst_0
       4:  if_icmple 17
       7:  iload_1
       8:  iload_0
       9:  iinc  0, -1
       12: imul
       13: istore_1
       14: goto 2
       17: iload_1
       18: ireturn

public static void main(java.lang.String[]);
  Code:
       Stack=1, Locals=2, Args_size=1
       0:  bipush     7
       2:  istore_1
       3:  iload_1
       4:  invokestatic  #13; //Method computeRec:(I)I
       7:  invokestatic  #22; //Method spim/SPIM.printInt:(I)V
       10: bipush    10
       12: invokestatic #26; //Method spim/SPIM.printChar:(C)V
       15: iload_1
       16: invokestatic #28; //Method computeIter:(I)I
       19: invokestatic #22; //Method spim/SPIM.printInt:(I)V
       22: bipush    10
       24: invokestatic #26; //Method spim/SPIM.printChar:(C)V
       27: return
}
```

All methods (other than the default constructor) are static; thus, the plethora of invokestatic instructions. Otherwise, we see lots of loads, stores, a few arithmetic operations and both conditional and unconditional branches. Translating additional instructions is left as a set of exercises.

6.2 SPIM and the MIPS Architecture

SPIM and the MIPS computer it simulates are both nicely described by SPIM's author in [Larus, 2009]. Our brief description here is based on that.

6.2.1 MIPS Organization

The MIPS computer organization is sketched out in Figure 6.2.

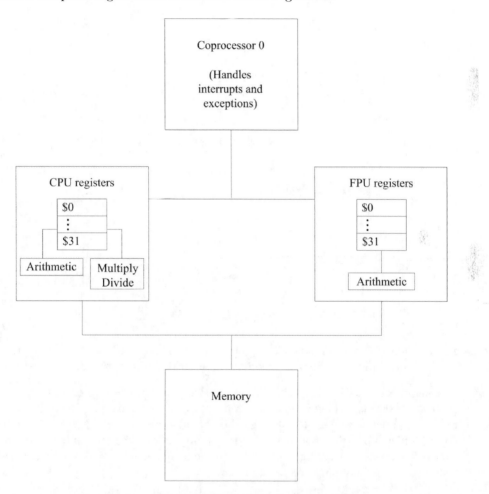

FIGURE 6.2 The MIPS computer organization.

It has an integer central processing unit (CPU), which operates on thirty-two general-purpose registers (numbered $0–$31); a separate floating- point coprocessor 1 (FPU), with its own 32 registers ($f0–$f31) for doing single- precision (32-bit) and double- precision (64-bit) floating point arithmetic, and coprocessor 0 for handling exceptions and interrupts, also with its own set of registers, including a status register. There are instructions for moving values from one register set to another. Our translation will focus on the integer-processing unit; exercises may involve the reader's dealing with the other coprocessors.

Programming the raw MIPS computer can be quite complicated, given its (time) delayed branches, delayed loads, caching, and memory latency. Fortunately, the MIPS assembler models a virtual machine that both hides these timing issues and uses pseudo-instructions to provide a slightly richer instruction set. By default, SPIM simulates this virtual machine, although it can be configured to model the more complicated raw MIPS computer. We will make use of the simpler, default assembly language.

6.2.2   Memory Organization

Memory is, by convention, divided into four segments, as illustrated in Figure 6.3 and derived from [Larus, 2009].

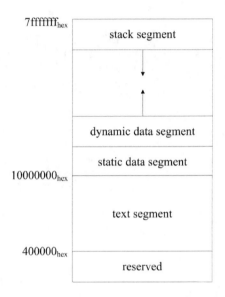

FIGURE 6.3 SPIM memory organization.

Text segment. The program's instructions go here, near the bottom end of memory, starting at 400000_{hex}. (The memory location below 400000_{hex} are reserved.)

Static data segment. Static data, which exist for the duration of the program, go here, starting at 1000000_{hex}. This would be a good place to put constants (including constant strings) and static fields. To access these values conveniently, MIPS designates one of its thirty-two registers as \$gp (register number 28), which points to the middle of a 64K block data in this segment. Individual memory locations may be addressed at fixed (positive or negative) offsets from \$gp.

Dynamic data segment. Often called the heap, this is where objects and arrays are dynamically allocated during execution of the program. Many systems employ a garbage collection to reclaim the space allocated to objects that are no longer of use. This segment starts just above the static data segment and grows upward, toward the run-time stack.

Stack segment. The stack is like that for the JVM. Every time a routine is called, a new stack frame is pushed onto the stack; every time a return is executed, a frame is popped off. The stack starts at the top of memory and grows downward toward the heap. That the dynamic data segment and stack segment start as far apart as possible and grow toward each other leads to an effective use dynamic memory. We use the

$fp register (30) to point to the current stack frame and the $sp register (29) to point to the last element on the stack.

MIPS and SPIM require that quantities be aligned on byte addresses that are multiples of their size; for example, 32-bit words must be at addresses that are multiples of four, and 64-bit doubles must be at addresses that are multiples of eight. Fortunately, the MIPS assembler (and thus SPIM) provides program directives for correctly aligning data quantities. For example, in

```
b:   .byte   10
w:   .word   324
```

the byte 10 is aligned on a byte boundary and may be addressed using the label b, and the word 324 is aligned on a 4-byte boundary and may be addressed as w; w is guaranteed to address a word whose address is evenly divisible by 4. There are assembler directives for denoting and addressing data items of several types in the static data segment. See Appendix E and [Larus, 2009].

Interestingly, SPIM does not define a byte order for quantities of several bytes; rather, it adopts the byte order of the machine on which it is running. For example, on an Intel x86 machine (Windows or Mac OS X on an Intel x86), the order is said to be *little-endian*, meaning that the bytes in a word are ordered from right to left. On Sparc or a Power PC (Solaris or Mac OS X on a Sparc), the order is said to be *big-endian*, meaning that the bytes in a word are ordered from left to right. For example, given the directive

```
w:   .byte 0, 1, 2, 3
```

if we address w as a word, the 0 will appear in the lowest-order byte on a little-endian machine (as in Figure 6.4(a)) and in the highest order byte on a big-endian machine (as in Figure 6.4(b)).

little-endian
(a)

big-endian
(b)

FIGURE 6.4 Little-endian versus big-endian.

For most of our work, the endian-ness of our underlying machine on which SPIM is running should not matter.

6.2.3 Registers

Many of the thirty-two 32-bit general-purpose registers, by convention, are designated for special uses and have alternative names that remind us of these uses:

$zero (register number 0) always holds the constant 0.

$at (1) is reserved for use by the assembler and should not be used by programmers or in code generated by compilers.

$v0 and $v1 (2 and 3) are used for expression evaluation and as the results of a function.

$a0 – $a3 (4–7) are used for passing the first four arguments to routines; any additional arguments are passed on the stack.

$t0 – $t7 (8–15) are meant to hold temporary values that need not be preserved across routine calls. If they must be preserved, it is up to the caller (the routine making the call) to save them; hence they are called *caller-saved registers*.

$s0 – $s7 (16–23) are meant to hold values that must be preserved across routine calls. It is up to the callee (the routine being called) to save these registers; hence they are called *callee-saved registers*.

$t8 and $t9 (24 and 25) are caller-saved temporaries.

$k0 and $k1 (26 and 27) are reserved for use by the operating system kernel and so should not be used by programmers or in code generated by compilers.

$gp (28) is a global pointer; it points to the middle of a 64K block of memory in the static data segment.

$sp (29) is the stack pointer, pointing to the last location on the stack.

$fp (30) is the stack frame pointer, pointing to the latest frame on the stack.

$ra (31) is the return address register, holding the address to which execution should continue upon return from the latest routine.

It is a good idea to follow these conventions when generating code; this allows our code to cooperate with code produced elsewhere.

6.2.4 Routine Call and Return Convention

SPIM assumes we follow a particular protocol in implementing routine calls when one routine (the caller) invokes another routine (the callee).

Most bookkeeping for routine invocation is recorded in a stack frame on the run-time stack segment, much as like is done in the JVM; but here we must also deal with registers. The stack frame for an invoked (callee) routine is illustrated in Figure 6.5.

SPIM has pretty well defined protocols for preparing for a call in the calling code, making a call, saving registers in the frame at the start of a method, restoring the saved registers from the frame at the end of a method, and executing a return to the calling code. We follow these protocols closely in our run-time stack management and they are described where we discuss our run-time stack in Section 6.3.5.

6.2.5 Input and Output

The MIPS computer is said to have memory-mapped IO, meaning the input and output device registers are referred to as special, reserved addresses in memory. Although SPIM does simulate this for a simple console, it also provides a set of system calls for accessing simple input and output functions. We shall use these system calls for our work.

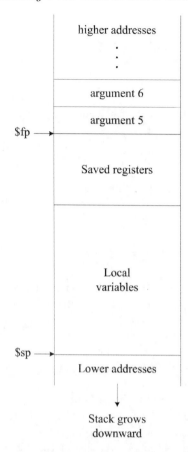

FIGURE 6.5 A stack frame.

6.3   Our Translator

6.3.1   Organization of Our Translator

Our JVM-to-SPIM translator is based on those described by Christian Wimmer [Wimmer, 2004] and Hanspeter Mössenböck [Mössenböck, 2000], which are in turn versions of the Sun (now Oracle) HotSpot Client Compiler. Our translator differs in several ways from those described by Wimmer and Mössenböck, and while theirs produce native code for the Intel 86x, ours targets SPIM, a simulator for the MIPS32 architecture.

Roughly, the phases proceed as illustrated in Figure 6.6.

In the first phase, the JVM code is parsed and translated to a control-flow graph, composed of basic blocks. A basic block consists of a linear sequence of instructions with just one entry point at the start of the block, and one exit point at the end; otherwise, there are no branches into or out of the block. The instructions are of a high level and preserve the tree structure of expressions; following Wimmer, we call this a high-level intermediate representation (HIR). Organizing the code in this way makes it more amenable to analysis.

In the second (optional) phase, various optimizations may be applied to improve the code, making it faster and often smaller.

In the third phase, a lower-level representation of the code is constructed, assigning ex-

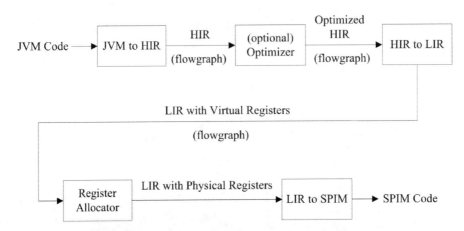

FIGURE 6.6 Phases of the JVM-to-SPIM translator.

plicit virtual registers to the operands of the instructions; there is no limit to the number of virtual registers used. Following Wimmer, we call this a low-level intermediate representation (LIR). The instructions are very close to the instructions in our target instruction set, SPIM.

In the fourth phase, we perform register allocation. The thirty-two physical registers in the MIPS architecture (and thus in SPIM) are assigned to take the place of virtual registers. Register allocation is discussed in Chapter 7.

In the fifth and final phase, we generate SPIM code—our goal.

The names of all classes participating in the translation from JVM code to SPIM code begin with the letter N (the N stands for Native). The translation is directed by the driver, NEmitter; most of the steps are directed by its constructor NEmitter().

NEmitter() iterates through the classes and methods for each class, constructing the control-flow graph of HIR instructions for each method, doing any optimizations, rewriting the HIR as LIR and performing register allocation.

SPIM code is emitted by the driver (Main or JavaCCMain) using NEmitter's write() method.

6.3.2 HIR Control-Flow Graph

The Control-Flow Graph

The first step is to scan through the JVM instructions and construct a flow graph of basic blocks. A *basic block* is a sequence of instructions with just one entry point at the start and one exit point at the end; otherwise, there are no branches into or out of the instruction sequence.

To see what happens here, consider the JVM code for the method computeIter() from our Factorial example:

```
public static int computeIter(int);
  Code:
   Stack=2, Locals=2, Args_size=1

   0: const_1
   1: istore_1

   2: iload_0
   3: iconst_0
```

```
 4: if_icmple 17

 7: iload_1
 8: iload_0
 9: iinc  0, -1
12: imul
13: istore_1
14: goto 2

17: iload_1
18: ireturn
```

We have inserted line breaks to delineate basic blocks. The entry point is the `iconst_1` instruction at location 0 so that begins a basic block; let us call it B1. The `iload_0` at location 2 is the target of the `goto` instruction (at 14) so that also must start a basic block; we call it B2. B2 extends through the `if_icmple` instruction; the block must end there because it is a branch. The next basic block (B3) begins at location 7 and extends to the `goto` at location 14. Finally, the `iload_1` at location 17 is the target of the `if_icmple` branch at location 4 so it starts basic block B4. B4 extends to the end of the method and is terminated by the `ireturn` instruction.

The control-flow graph, expressed as a graph constructed from these basic blocks is illustrated in Figure 6.7; we have added an extra special entry block B0, for making further analysis simpler. The boxes represent the basic blocks and the arrows indicate the flow of control among the blocks; indeed, all control flow information is captured in the graph. Notice that the boxes and arrows are not really necessary to follow the flow of control among blocks; we put them there in the figure only for emphasis. You will notice that the first line of text within each box identifies the block, a list of any *successor blocks* (labeled by `succ:`) and a list of any *predecessor blocks* (labeled by `pred:`). For example, block B3 has a predecessor B2 indicating that control flows from the end of block B2 into the start of B3; B3 also has a successor B2 indicating that control flows from the end of B3 to the start of block B3. We add an extra beginning block B0 for the method's entry point.

The denotation [LH] on the first line of B2 indicates that B2 is a *loop header*—the first block in a loop. The denotation [LT] on the first line of B3 indicates that B3 is a *loop tail*—the last block in a loop (and a predecessor of the loop header). This information is used later in identifying loops for performing optimizations and ordering blocks for optimal register allocation.

The pairs of numbers within square brackets, for example [7, 14] on the first line of B3, denote the ranges of JVM instructions captured in the block.

The keyword `dom:` labels the basic block's *immediate dominator*. In our graph, a node d is said to *dominate* a node n if every path from the entry node (B0) to n must go through d. A node d *strictly dominates* n if it dominates n but is not the same as n. Finally, node d is an *immediate dominator* of node n if d strictly dominates n but does not dominate any other node that strictly dominates n. That is, it is the node on the path from the entry node to n that is the "closest" to n. Dominators are useful for certain optimizations, including the lifting of loop-invariant code (see Section 6.3.3).

The State Vector

Local variables are tracked in a *state vector* called `locals` and are indexed in this vector by their location in the JVM stack frame. The current state of this vector *at the end of a block's instruction sequence* is printed on the block's second line and labeled with `Locals:`. The values are listed in positional order and each value is represented by the instruction id for the instruction that computes it.

For example, in B0 this vector has just one element, corresponding to the method's

computeIter I(I)

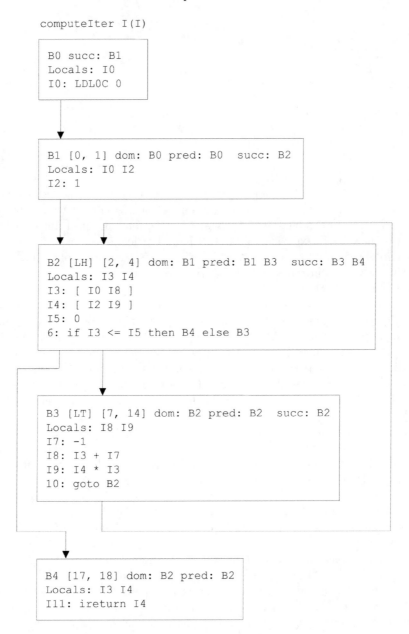

```
B0 succ: B1
Locals: I0
I0: LDLOC 0
```

```
B1 [0, 1] dom: B0 pred: B0  succ: B2
Locals: I0 I2
I2: 1
```

```
B2 [LH] [2, 4] dom: B1 pred: B1 B3  succ: B3 B4
Locals: I3 I4
I3: [ I0 I8 ]
I4: [ I2 I9 ]
I5: 0
6: if I3 <= I5 then B4 else B3
```

```
B3 [LT] [7, 14] dom: B2 pred: B2  succ: B2
Locals: I8 I9
I7: -1
I8: I3 + I7
I9: I4 * I3
10: goto B2
```

```
B4 [17, 18] dom: B2 pred: B2
Locals: I3 I4
I11: ireturn I4
```

FIGURE 6.7 HIR flow graph for `Factorial.computeIter()`.

formal argument n. I0 identifies the instruction `LDLOC 0` that loads the value (for n) from position 0 on the stack[2]; notice the instruction is typed to be integer by the I in I0. In B1, the vector has two elements: the first is I0 for n and the second is I2 for `result`.

HIR Instructions

The instruction sequence within each basic block is of a higher level than is JVM code. These sequences capture the expression trees from the original source. We follow [Wimmer,

---

[2]The argument actually will be in register $a0 in the SPIM code, following SPIM convention.

2004] in calling this a high-level intermediate representation (HIR). For example, the Java statement

```
w = x + y + z;
```

might be represented in HIR by

```
I8:  I0 + I1
I9:  I8 + I2
```

The I0, I1, and I2 refer to the instruction ids labeling instructions that compute values for x, y and z, respectively. So the instruction labeled I8 computes the sum of the values for x and y, and then the instruction labeled I9 sums that with the value for z to produce a value for w. Think of this as an abstract syntax tree such as that illustrated in Figure 6.8. The I is a type label, indicating the type is (in this case) an integer.

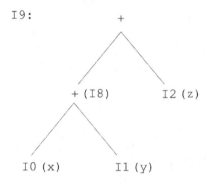

FIGURE 6.8 (HIR) AST for w = x + y + z.

The instruction for loading a constant is simply the constant itself. For example, consider block B1 in Figure 6.7. It has just the single instruction

```
I2: 1
```

which is to say, the temporary I2 (later, I2 will be allocated a register) is loaded with the constant 1. 1 is that instruction that loads the constant 1.

Not all instructions generate values. For example, the instruction

```
6: if I3 <= I5 then B4 else B3
```

in block B2 produces no value but transfers control to either B4 or B3. Of course, the 6: has no type associated with it.

As we shall see in Section 6.3.3, HIR lends itself to various optimizations.

Static Single Assignment (SSA) Form

Our HIR employs static single assignment (SSA) form, where for every variable, there is just one place in the method where that variable is assigned a value. This means that when a variable is re-assigned in the method, one must create a new version for it.

For example, given the simple sequence

```
x = 3;
x = x + y;
```

the second assignment to x requires that we create a new version. For example, we might subscript our variables to distinguish different versions.

```
x₁ = 3;
x₁ = x₁ + y₁;
```

In the HIR, we represent a variable's value by the instruction that computed it and we track these values in the state vector. The value in a state vector's element may change as we sequence through the block's instructions. If the next block has just one predecessor, it can copy the predecessor's state vector at its start; if there are more than two predecessors, the states must be merged.

For example, consider the following *j--* method, where the variables are in SSA form.

```
static int ssa(int w₁) {
    if (w₁ > 0) {
        w₂ = 1;
    }
    else {
        w₃ = 2;
    }
    return w₍;
}
```

In the statement,

```
return w₍;
```

which w do we return, w_2 or w_3? Well, it depends on which of the two paths is taken through the if-then-else statement. In terms of basic blocks, it is as illustrated in Figure 6.9.

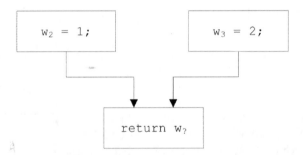

FIGURE 6.9 The SSA merge problem.

We solve this problem using what is called a *Phi function*, a special HIR instruction that captures the possibility of a variable having one of several values. In our example, the final block would contain the following code.

```
w₄ = [w₂  w₃];
return w₄;
```

The $[w_2 \ w_3]$ represents a Phi function with two operands: the operand w_2 captures the possibility that one branch of the if-then-else was taken and the w_3 captures the possibility that the other branch was taken. Of course, the target SPIM has no representation for this but we shall remove these special Phi instructions before attempting to generate SPIM instructions. We can still do analysis on it. The data flow is illustrated in Figure 6.10.

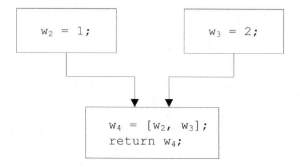

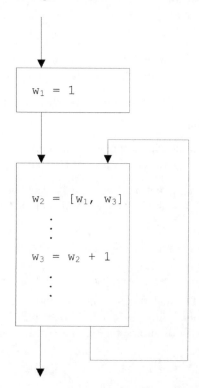

FIGURE 6.10 Phi functions solve the SSA merge problem.

Another place where Phi functions are needed are in loop headers. Recall that, a loop header is a basic block having at least one incoming backward branch and at least two predecessors, as is illustrated in Figure 6.11.

FIGURE 6.11 Phi functions in loop headers.

Unfortunately, a changed variable flowing in from a backward branch is known only after the block has been fully processed. For this reason, at loop headers, we conservatively define Phi functions for all variables and then remove redundant Phi functions later. In the first instance, w_2 can be defined as a Phi function with operands w_1 and w_2:

$$w_2 = [w_1 \ w_2]$$

Then, when w is later incremented, the second operand may be overwritten by the new w_3.

$$w_2 = [w_1 \ w_3]$$

Of course, a redundant Phi function will not be changed. If the w is never modified in the loop body, the Phi function instruction takes the form

$$w_2 = [w_1 \ w_2]$$

and can be removed as it is apparent that w has not been changed.

Phi functions are tightly bound to state vectors. When a block is processed:

- If the block has just a single predecessor, then it may inherit the state vector of that predecessor; the states are simply copied.

- If the block has more than one predecessor, then those states in the vectors that differ must be merged using Phi functions.

- For loop headers we conservatively create Phi functions for all variables, and then later remove redundant Phi functions.

In block B2 in the code for Factorial (Figure 6.7), I3 and I4 identify the instructions

```
I3:  [ I0  I8 ]
I4:  [ I2  I9 ]
```

These are Phi function instructions capturing the local variable n and result respectively. I3 is a Phi function with operands I0 and I8; I4 is a Phi function with operands I2 and I9.

Control-Flow Graph Construction

The best place to look at the translation process is the constructor NEmitter() in class NEmitter. For each method, the control-flow graph of HIR is constructed in several steps:

1. The NControlFlowGraph constructor is invoked on the method. This produces the control-flow graph cfg. In this first step, the JVM code is translated to sequences of tuples:

   (a) Objects of type NBasicBlock represent the basic blocks in the control-flow graph. The control flow is captured by the links successors in each block. There are also the links predecessors for analysis.

   (b) The JVM code is first translated to a list of tuples, corresponding to the JVM instructions. Each block stores its sequence of tuples in an ArrayList, tuples. For example, the blocks of tuples for our Factorial example, computeIter() are as follows:

```
B0

B1
0: iconst_1
1: istore_1

B2
2: iload_0
3: iconst_0
4: if_icmple    0    13

B3
7: iload_1
8: iload_0
9: iinc    0    255
12: imul
```

```
13: istore_1
14: goto    255    244

B4
17: iload_1
18: ireturn
```

This use of tuples is not strictly necessary but it makes the translation to HIR easier.

2. The message expression

```
cfg.detectLoops(cfg.basicBlocks.get(0), null);
```

detects loop headers and loop tails. This information may be used for lifting invariant code from loops during the optimization phase and for ordering blocks for register allocation.

3. The message expression

```
cfg.removeUnreachableBlocks();
```

removes unreachable blocks, for example, blocks resulting from jumps that come after a return instruction.

4. The message expression,

```
cfg.computeDominators(cfg.basicBlocks.get(0), null);
```

computes an *immediate dominator* for each basic block, that closest predecessor through which all paths must pass to reach the target block. It is a useful place to insert invariant code that is lifted out of a loop in optimization.

5. The message expression

```
cfg.tuplesToHir();
```

converts the tuples representation to HIR, stored as a sequence of HIR instructions in the array list, hir for each block. As the tuples are scanned, their execution is simulated, using a stack to keep track of newly created values and instruction ids. The HIR is in SSA form, with (sometimes redundant) Phi function instructions.

6. The message expression

```
cfg.eliminateRedundantPhiFunctions();
```

eliminates unnecessary Phi functions and replaces them with their simpler values as discussed above.

The HIR is now ready for further analysis.

6.3.3 Simple Optimizations on the HIR

That the HIR is in SSA form makes it amenable to several simple optimizations—improvements to the HIR code; these improvements make for fewer instructions and/or faster programs.

Local optimizations are improvements made based on analysis of the linear sequence of instructions within a basic block. *Global optimizations* require an analysis of the whole control graph and involve what is called data-flow analysis[3].

One can make important improvements to the code even with analysis within single basic blocks. Moreover, the SSA form of the HIR allows us to deal with sequences of code that cross basic block boundaries, as long as we do not deal with those values produced by Phi functions.

Inlining

The cost of calling a routine (for invoking a method) can be considerable. In addition the control-flow management and stack frame management, register values must be saved on the stack. But in certain cases, the code of a callee's body can replace the call sequence in the caller's code, saving the overhead of a routine call. We call this *inlining*.

Inlining usually makes sense for methods whose only purpose is to access a field as in getters and setters.

For example, consider the following two methods, defined within a class `Getter`.

```
static int getA() {
    return Getter.a;
}

static void foo() {
    int i;
    i = getA();
}
```

The `a` in `Getter.a` refers to a static field declared in `Getter`. We can replace the `getA()` with the field selection directly:

```
static void foo() {
    int i;
    i = Getter.a;
}
```

Inlining makes sense only when the callee's code is relatively short. Also, when modeling a virtual method call, we must be able to determine the routine that should be called at compile-time. One also should be on the lookout for nested recursive calls; repeated inlining could go on indefinitely.

A simple strategy for enabling inlining is to keep track of the number of instructions within the HIR for a method and whether or not that method makes nested calls; this information can be computed as the HIR for a method is constructed. Inlining method invocations can create further opportunities for inlining. Inlining can also create opportunities for other optimizations.

Constant Folding and Constant Propagation

Expressions having operands that are both constants, for example,

```
3 + 4
```

or even variables whose values are known to be constants, for example,

```
int i = 3;
int j = 4;
... i + j ...
```

---

[3]We describe a detailed data-flow analysis algorithm for computing liveness intervals for variables as part of register allocation in Section 7.4.1

can be *folded*, that is, replaced by their constant value.

For example, consider the Java method

```
static void foo1() {
    int i = 1;
    int j = 2;
    int k = i + j + 3;
}
```

and corresponding HIR code

```
B0 succ: B1
Locals:
   0: 0
   1: 1
   2: 2

B1 [0, 10] dom: B0 pred: B0
Locals:
   0: I3
   1: I4
   2: I7
I3: 1
I4: 2
I5: I3 + I4
I6: 3
I7: I5 + I6
8: return
```

The instruction I3 + I4 at I5: can be replaced by the constant 3, and the I5 + I6 at I7: can replaced by the constant 6.

A strategy for implementing is to associate a hash table with each basic block for storing the constant values associated with instruction ids. As we scan a block's instructions, an instruction defining a constant is added to the hash table, keyed by its id. If we come across an instruction whose operands are already stored in this hash table of constants, we can perform the operation immediately and store the result in that same table, keyed by the instruction's id. This analysis may be carried across basic block boundaries, so long as we do not store instruction ids for instructions representing Phi functions.

Common Sub-Expression Elimination (CSE) within Basic Blocks

Another optimization one may make is *common sub-expression elimination*, where we identify expressions that are re-evaluated even if their operands are unchanged.

For example, consider the following method, unlikely as it may be.

```
void foo(int i) {
    int j = i * i * i;
    int k = i * i * i;
}
```

We can replace

```
int k = i * i * i;
```

in foo() with the more efficient

```
int k = j;
```

To see how we might recognize common sub-expressions, consider the HIR code for the original version of foo():.

```
B0 succ: B1
Locals: I0
I0: LDLOC 0

B1 [0, 12] dom: B0 pred: B0
Locals: I0 I4 I6
I3: I0 * I0
I4: I3 * I0
I5: I0 * I0
I6: I5 * I0
7: return
```

We sequence through the instructions, registering each in the block's hash table, indexed by the instruction and its operand(s); the value stored is the instruction id. As for constant propagation, we cannot trust values defined by Phi functions without further data-flow analysis, so we do not register such instructions.

At I6:, the reference to I5 can replaced by I3. More importantly, any subsequent reference to I6 could be replaced by I4.

Of course, we are not likely to see such a method with such obvious common sub-expressions. But common sub-expressions do arise in places one might not expect them. For example, consider the following C Language fragment:

```
for (i = 0; i < 1000; i++) {
    for (j = 0; j < 1000; j++) {
        c[i][j] = a[i][j] + b[i][j];
    }
}
```

The C compiler represents the matrices as linear byte sequences, one row of elements after another; this is called *row major form*[4]. If a, b, and c are integer matrices, and the base address of c is c', then the byte memory address of c[i][j] is

```
c' + i * 4 * 1000 + j * 4
```

The factor i is there because it is the (counting from zero) i^{th} row; the factor 1000 is there because there are 1,000 elements in a row; and the factor 4 is there (twice) because there are 4 bytes in an integer word. Likewise, the addresses of a[i][j] and b[i][j] are

```
a' + i * 4 * 1000 + j * 4
```

and

```
b' + i * 4 * 1000 + j * 4
```

respectively. So, eliminating the common offsets, i * 4 * 1000 + j * 4 can save us a lot of computation, particularly because the inner loop is executed a million times!

But in Java, matrices are not laid out this way. Rather, in the expression

```
a[i][j]
```

the sub-expression a[i] yields an integer array object, call it ai, from which we can index the jth element,

```
ai[j]
```

So, we cannot expect the same savings. On the other hand, the expression a[i] never changes in the inner loop; only the j is being incremented. We say that, within the inner

---

[4]In FORTRAN, matrices are stored as a sequence of columns, which is in *column major form.*

loop, the expression `a[i]` is invariant; that is, it is a *loop invariant expression*. We deal with these in the next section.

And as we shall see below, there is still much to be gained from common sub-expression elimination when it comes to dealing with Java arrays.

Lifting Loop Invariant Code

Loop invariant expressions can be lifted out of the loop and computed in the predecessor block to the loop header.

For example, consider the *j--* code for summing the two matrices from the previous section[5].

```
int i = 0;
while (i <= 999) {
    int j = 0;
    while (j <= 999) {
        c[i][j] = a[i][j] + b[i][j];
        j = j + 1;;
    }
    i = i + 1;
}
```

This can be rewritten as

```
int i = 0;
while (i <= 999) {
    int[] ai = a[i];
    int[] bi = b[i];
    int[] ci = c[i];
    int j = 0;
    while (j <= 999)
    {
        ci[j] = ai[j] + bi[j];
        j = j + 1;;
    }
    i = i + 1;
}
```

Thus, `a[i]`, `b[i]`, and `c[i]` need be evaluated just 1,000 times instead of 1,000,000 times.

This optimization is more difficult than the others. Basically, the expression `a[i]` is invariant if either

1. `i` is constant, or

2. All the definitions of `i` that reach the expression are outside the loop, or

3. Only one definition of `i` reaches the expression, and the definition is loop invariant.

But that the HIR conforms to SSA makes it possible. Any code that is lifted must be lifted to an additional block, which is inserted before the loop header.

Invariant expressions can be distinguished from expressions defined within the loop by the fact that they are registered in the hash tables for dominators of the loop header. Both a loop header and its immediate dominator are stored in the blocks. To find previous dominators, one just links back through the block list via the immediate dominator links. Computations involving invariant operands may be moved to the end of a loop header's immediate dominator block.

---

[5]The for-loop is not part of the *j--* language so we use the equivalent while-loop notation. If you have implemented for-loops in your compiler by rewriting them as while-loops, then you should produce similar HIR.

Array Bounds Check Elimination

When indexing an array, we must check that the index is within bounds. For example, in our example of matrix multiplication, in the assignment

```
c[i][j] = a[i][j] + b[i][j];
```

The i and j must be tested to make sure each is greater than or equal than zero, and less than 1,000. And this must be done for each of c, a, and b. But if we know that a, b, and c are all of like dimensions, then once the check is done for a, it need not be repeated for b and c. Given that the inner loop in our earlier matrix multiplication example is executed one million times, we have saved two million checks!

Again, a clever compiler can analyze the context in which the statement

```
c[i][j] = a[i][j] + b[i][j];
```

is executed (for example, within a nested for-loop) and show that the indices are never out of bounds. When the arrays for a, b and c are created in a loop header's dominator, this is straightforward; other more complicated control graphs may warrant more extensive data-flow analysis (below). Moreover, an array index operation such as a[i] can be registered in the hash table, ensuring that subsequent occurrences, in the same basic block or in subsequent blocks where the operation is not involved in a Phi function, need not be re-checked.

Null Check Elimination

Every time we send a message to an object, or access a field of an object, we must insure that the object is not the special null object. For example, in

```
...a.f...
```

we will want to make sure that a is non-null before computing the offset to the field f. But we may know that a is non-null; for example, we may have already checked it as in the following.

```
...a.f...
...a.g...
...a.h...
```

Once we have done the null-check for

```
...a.f...
```

there is no reason to do it again for either a.g or a.h.

Again, the ids of checked operands can be stored in the hash tables attached to blocks. If they are found subsequently, they need not be checked again.

A Need for Data-Flow Analysis

The HIR does not expose every opportunity for optimization, particularly those involving back branches (in loops); for this to happen we would need full data-flow analysis where we compute where in the code computed values remain valid. We use data-flow analysis in Section 7.4.1 to compute such liveness intervals for virtual registers (variables) in the LIR as a prerequisite to register allocation; look there for a detailed data-flow analysis algorithm.

On the other hand, the HIR does expose many opportunities for optimization, particularly for the effort involved. The amount of code improvement achieved in proportion to the time spent is important in a just-in-time (or hot spot) compiler such as we use for processing JVM code.

A full discussion of compiler optimization is outside the scope of this text; see the Further Readings section (Section 6.4) at the end of this chapter for detailed discussions of code optimization techniques.

Summary

At a panel discussion on compiler optimization, Kathleen Knobe observed that, rather than to put all of one's efforts into just one or two optimizations, it is best to attempt all of them, even if they cannot be done perfectly.

We have not implemented any of these optimizations, but leave them as exercises.

Of course, one of the best optimizations one can make is to replace references to local variables with references to registers by allocating registers for holding local variable values. We discuss register allocation in Chapter 7.

6.3.4 Low-Level Intermediate Representation (LIR)

The HIR lends itself to optimization but is not necessarily suitable for register allocation. For this reason we translate it into a low-level intermediate representation (LIR) where

- Phi functions are removed from the code and replaced by explicit moves, and

- Instruction operands are expressed as explicit virtual registers.

For example, the LIR for `Factorial.computeIter()` is given below; notice that we retain the control-flow graph from the HIR. Notice also that we have allocated virtual registers to the computation. One may have an arbitrary number of virtual registers. In register allocation, we shall map these virtual registers to actual physical registers on the MIPS architecture. Because there might be many more virtual registers than physical registers, the mapping may require that some physical registers be spilled to stack locations and reloaded when the spilled values are needed again. Notice that each instruction is addressed at a location that is a multiple of five; this leaves room for spills and loads (at intervening locations).

```
computeIter (I)I

  B0

  B1
  0: LDC [1] [V32|I]
  5: MOVE $a0 [V33|I]
  10: MOVE [V32|I] [V34|I]

  B2
  15: LDC [0] [V35|I]
  20: BRANCH [LE] [V33|I] [V35|I] B4

  B3
  25: LDC [-1] [V36|I]
  30: ADD [V33|I] [V36|I] [V37|I]
  35: MUL [V34|I] [V33|I] [V38|I]
  40: MOVE [V38|I] [V34|I]
  45: MOVE [V37|I] [V33|I]
  50: BRANCH B2

  B4
  55: MOVE [V34|I] $v0
  60: RETURN $v0
```

In this case, seven virtual registers V32–V38 are allocated to the LIR computation. Two physical registers, $a0 for the argument n and $v0 for the return value, are referred to by their symbolic names; that we use these physical registers indicates that SPIM expects values to be stored in them. Notice that LIR instructions are read from left to right. For example, the load constant instruction

```
0: LDC [1] [V32|I]
```

in block B1, loads the constant integer 1 into the integer virtual register V32; the |I notation types the virtual register as an integer. Notice that we start enumerating virtual registers beginning at 32; the numbers 0 through 31 are reserved for enumerating physical registers. The next instruction,

```
5: MOVE $a0 [V33|I]
```

copies the value from the integer physical register $a0 into the integer virtual register V33. The add instruction in block B3,

```
30: ADD [V33|I] [V36|I] [V37|I]
```

adds the integer values from registers V33 and V36 and puts the sum in V37.

Notice that we enumerate the LIR instructions by multiples of five. This eases the insertion of spill (and restore) instructions, which may be required for register allocation (and are discussed in Chapter 7).

The process of translating HIR to LIR is relatively straightforward and is a two-step process.

1. First, the NEmitter constructor invokes the NControlFlowGraph method hirToLir() on the control-flow graph:

   ```
   cfg.hirToLir();
   ```

   The method hitToLir() iterates through the array of HIR instructions for the control-flow graph translating each to an LIR instruction, relying on a method toLir(), which is defined for each HIR instruction.

2. NEmitter invokes the NControlFlowGraph method resolvePhiFunctions() on the control-flow graph:

   ```
   cfg.resolvePhiFunctions();
   ```

   This method resolves Phi function instructions, replacing them by move instructions near the end of the predecessor blocks. For example, the Phi function from Figure 6.10 and now repeated in Figure 6.12(a) resolves to the moves in Figure 6.12(b). One must make sure to place any move instructions before any branches in the predecessor blocks.

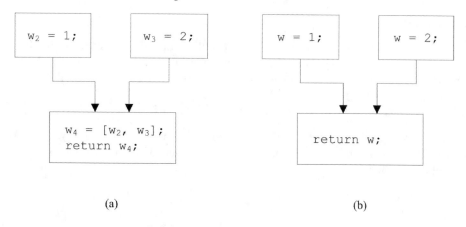

(a)   (b)

FIGURE 6.12 Resolving Phi functions.

6.3.5   Simple Run-Time Environment

Introduction

An actual run-time environment supporting code produced for Java would require:

1. A naming convention.

2. A run-time stack.

3. A representation for arrays and objects.

4. A heap, that is, an area of memory from which arrays and objects may be dynamically allocated. The heap should offer some sort of garbage collection, making objects that are no longer in use available for (re-)allocation.

5. A run-time library of code that supports the Java API.

To do all of this here would be beyond the scope of this text. Given that our goal is to have a taste of native code generation, including register allocation, we do much less. We implement just enough of the run-time environment for supporting our running the `Factorial` example (and a few more examples, which are included in our code tree).

Naming Convention

We use a simple naming mechanism that takes account of just classes and their members. Methods are represented in SPIM by routines with assembly language names of the form `<class>.<name>`, where `<name>` names a method in a class `<class>`. For example, the `computeIter()` method in class `Factorial` would have the entry point

```
Factorial.computeIter:
```

Static fields are similarly named; for example, a static integer field `staticIntField` in class `Factorial` would translate to SPIM assembly language as

```
Factorial.staticIntField:    .word   0
```

String literals in the data segment have labels that suggest what they label, for example, `Constant..String1:`.

Run-Time Stack

Our run-time stack conforms to the run-time convention described for SPIM in [Larus, 2009] and our Section 6.2.4. Each time a method is invoked, a new stack frame of the type illustrated in Figure 6.5 is pushed onto the stack. Upon return from the method, the same frame is popped off the stack.

When the caller wants to invoke a callee, it does the following:

1. Argument passing. The first four arguments are passed in registers $a0 to $a3. Any additional arguments are pushed onto the stack and thus will appear at the beginning of the callee's stack frame.

2. Caller-saved registers. Registers $a0 to $a3 and $t0 to $t9 can be used by the callee without its having to save their values, so it is up to the caller to save any registers whose values it expects to use after the call, within its own stack frame before executing the call.

3. Executing the call. After arguments have passed and caller registers have been saved, we execute the `jal` instruction, which jumps to the callee's first instruction and saves the return address in register $ra.

Once a routine (the callee) has been invoked, it must first do the following:

1. Push stack frame. Allocate memory for its new stack frame by subtracting the frame's size from the stack pointer $sp. (Recall, the stack grows downward in memory.)

2. Callee-saved registers. The callee must save any of $s0 to $s7, $fp, and $ra before altering any of them because the caller expects their values will be preserved. We must always save $fp. $ra must be saved only if the callee invokes another routine.

3. The frame pointer. The new $fp = $sp + stack frame size − 4.

The stack frame for an invoked (callee) routine is illustrated in Figure 6.13.

Once it has done its work, the callee returns control to the caller by doing the following:

1. Return value. If the callee returns a value, it must put that value in register $0.

2. Callee-saved registers. All callee-saved registers that were saved at the start of the routine are restored.

3. Stack frame. The stack frame is popped from the stack, restoring register $fp from its saved location in the frame, and adding the frame size to register $sp.

4. Return. Execute the return instruction, which causes a jump to the address in register $ra.

A Layout for Objects

Although we mean only to support static methods, objects do leak into our implementation. For example, method `main()` takes as its sole argument an array of strings.

Although, we do not really support objects, doing so need not be difficult. [Corliss and Lewis, 2007]) propose layouts, which we use here. For example, an arbitrary object might be organized as in Figure 6.14.

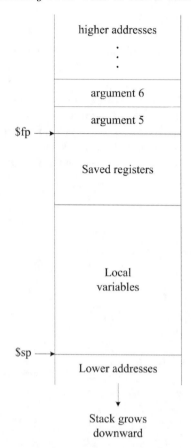

FIGURE 6.13 A stack frame.

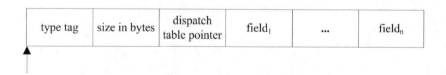

FIGURE 6.14 Layout for an object.

The type tag identifies the type (that is, the class) of the object. The size in bytes is the number of bytes that must be allocated to represent the object; this number is rounded up to a multiple of 4 so that objects begin on word boundaries. The dispatch table pointer is the address of a dispatch table, which contains the address of each method for the object's class.

The dispatch table may be allocated in the data segment and keeps track of both inherited and defined methods. When code is generated for the class, entries in the table are first copied from the superclass. Then entries for newly defined method are added; overriding method entries replace the overridden entries in the table and other method entries are simply added at the end. Entries are added in the order they are declared in the class declaration. In this way, the dispatch table for a class keeps track of the proper (inherited, overriding, or newly defined) methods that are applicable to each class. A dispatch table for class <class> may be labeled (and addressed) in the data segment using the name

```
<class>..Dispatch:
```

The double period distinguishes this name from a method named `Dispatch()`.

The order of fields in an object is also important because a class inherits fields from a superclass. In constructing the format of an object, first any inherited fields are allocated. Then any newly defined fields are allocated. Java semantics require that space is allocated to all newly defined fields, even those with the same names as inherited fields; the "overridden" fields can are always available through casting an object to its superclass.

For each class, we also maintain a class template in the data segment, which among other things may include a typical copy of an object of that class that can be copied to the heap during allocation. We name this

```
<class>..Template:
```

For example, consider the following *j--* code, which defines two classes: `Foo` and `Bar`.

```java
public class Foo {
    int field1 = 1;
    int field2 = 2;

    int f() {
        return field1 + field2;
    }

    int foo() {
        return field1;
    }
}

class Bar extends Foo {
    int field3 = 3;
    int field1 = 4;

    int f() {
        return field1 + field2 + field3;
    }

    int bar() {
        return field1;
    }
}
```

Consider an object of class `Foo`, illustrated in Figure 6.15.

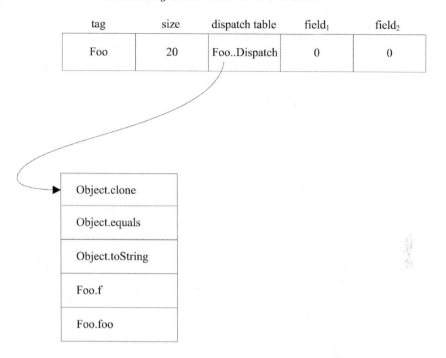

FIGURE 6.15 Layout and dispatch table for Foo.

Assuming the type tag for Foo is 4^6,

```
Foo..Tag:    .word    4
```

The SPIM code for a template for Foo might look like

```
Foo..Template:
    .word    Foo..Tag
    .word    20
    .word    Foo..Dispatch
    .word    0
    .word    0
```

The fields have initial values of zero. It is the responsibility of the `<init>` routine for Foo to initialize them to 1 and 2, respectively.

The construction of the dispatch table for Foo follows these steps:

1. First the methods for Foo's superclass (`Object`) are copied into the table. (Here we assume we have given `Object` just the three methods `clone()`, `equals()` and `toString ()`).

2. Then we add the methods defined in Foo to the dispatch table: `f()` and `foo()`.

The SPIM code for this dispatch table might look something like the following:

```
Foo..Dispatch:
    .word    5 # the number of entries in our table
    .word    Object.clone
    .word    Object.equals
    .word    Object.toString
    .word    Foo.f
    .word    Foo.foo
```

---

[6]This assumes that, 1 represents a string, 2 represents an array, and 3 represents an `Object`.

The layout and dispatch table for class `Bar`, which extends class `Foo`, is illustrated in Figure 6.16.

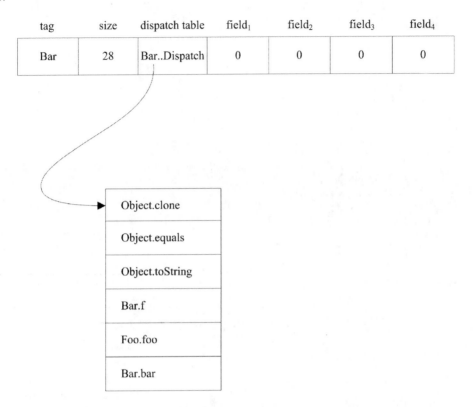

tag	size	dispatch table	field$_1$	field$_2$	field$_3$	field$_4$
Bar	28	Bar..Dispatch	0	0	0	0

FIGURE 6.16 Layout and dispatch table for `Bar`.

Assuming the type tag for `Bar` is 5,

```
Bar..Tag:    .word    5
```

The SPIM code for a template for `Bar` might look like

```
Bar..Template:
    .word    Bar..Tag
    .word    28
    .word    Bar..Dispatch
    .word    0
    .word    0
    .word    0
    .word    0
```

Again, the fields have initial values of zero. It is the responsibility of the `<init>` routine for `Foo` and `Bar` to initialize them to 1,2, 3, and 4, respectively.

The construction of the dispatch table for `Bar` follows these steps:

1. First the methods for `Bar`'s superclass (`Foo`) are copied into the table.

2. Then we add the methods defined in `Bar` to the dispatch table. The method `Bar.f()` overrides, and so replaces `Foo.f()` in the table; the new `Bar.bar()` is simply added to the table.

The SPIM code for this dispatch table might look something like the following:

```
Bar..Dispatch:
    .word    6 # the number of entries in our table
    .word    Object.clone
    .word    Object.equals
    .word    Object.toString
    .word    Bar.f
    .word    Foo.foo
    .word    Bar.bar
```

Arrays are a special kind of object; they are dynamically allocated on the heap but lie outside the `Object` hierarchy. A possible layout for an array is illustrated in Figure 6.17.

FIGURE 6.17 Layout for an array.

Arrays have a type tag of 1. The *size in bytes* is the size of the array object in bytes, that is, $12 + n * bytes$, where n is the number of elements in the array and *bytes* is the number of bytes in each element. The array length is the number of elements in the array; any implementation of the JVM `arraylength` instruction must compute this value. The array elements follow. The size of the array object should be rounded up to a multiple of 4 to make sure the next object is word aligned. Notice that an array indexing expression `a[i]` would translate to the heap address $a + 12 + i * bytes$, where a is the address of the array object on the heap and *bytes* is the number of bytes in a single element.

Strings are also special objects; while they are part of the `Object` hierarchy, they are final and so they may not be sub-classed. This means we can do without a dispatch table pointer in string objects; we can maintain one dispatch table for all string operations:

```
String..Dispatch:
    .word m # m is the number of entries in the table
    .word String.clone
    .word String.equals
    .word String.toString
    ...addresses of methods for Strings...
```

Which `String` methods we implement is arbitrary and is left as an exercise.

A possible layout for a string is illustrated in Figure 6.18.

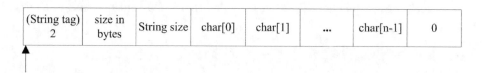

FIGURE 6.18 Layout for a string.

The type tag for a string is 2. Its size in bytes is the number of bytes that must be allocated to represent the string object; it is $12 + size + 1$ rounded up to a multiple of 4, where *size* is the number of characters in the string. String size is the number of characters in the string. The sequence of characters is terminated with an extra null character, with Unicode representation 0; the null character is necessary so that our strings are compatible with SPIM strings.

Constant strings may be allocated in the data segment, which is to say, we may generate constant strings in the data segment; strings in the data segment will then look exactly like those on the heap (the dynamic segment). For example, the constant "Hello World!" might be represented at the label `Constant..String1:`.

```
Constant..String1:
    .word    2   # tag 2 indicates a string.
    .word    28  # size of object in bytes.
    .word    12  # string length (not including null terminator).
    .asciiz  "Hello World!" # string terminated by null character 0.
    .align 2     # next object is on a word boundary.
```

The reader will have noticed that all objects on the heap share three words (12 bytes) of information at the start of each object. This complicates the addressing of components in the object. For example, the first element of an array a is at address $a + 12$; the first character of a string s is at $s + 12$; and the first field of an object o is at address $o + 12$. That 12 must be added to an object's address each time we want an array's element, a string's character or an object's field can be expensive. We can do away with this expense by having our pointer, which we use to reference the object, always point into the object 12 bytes as illustrated in Figure 6.19.

FIGURE 6.19 An alternative addressing scheme for objects on the heap.

Dynamic Allocation

The logic for allocating free space on the heap is pretty simple in our (oversimplified) model.

```
obj = heapPointer;
heapPointer += <object size>;
if (heapPointer >= stackPointer) goto freakOut;
<copy object template to obj>;
```

When allocating a new object, we simply increment a free pointer by the appropriate size and if we have run out of space (that is, when the heap pointer meets the stack pointer), we freak out. A more robust heap management system might perform garbage collection.

Once we have allocated the space for the object, we copy its template onto the heap at the object's location. Proper initialization is left to the SPIM routines that model the constructors.

SPIM Class for IO

SPIM provides a set of built-in system calls for performing simple IO tasks. Our run-time environment includes a class `SPIM`, which is a wrapper that gives us access to these calls as a set of static methods. The wrapper is defined as follows:

```
package spim;

/**
 * This is a Java wrapper class for the SPIM run-time object SPIM.s.
 * Any j-- program that's compiled for the SPIM target must import
 * this class for (file and console) IO operations. Note that the
 * functions have no implementations here which means that if the
```

```
 * programs using this class are compiled using j--, they will
 * compile fine but won't function as desired when run against
 * the JVM. Such programs must be compiled using the j-- compiler
 * for the SPIM target and must be run against the SPIM simulator.
 */

public class SPIM {
    /** Wrapper for SPIM.printInt(). */

    public static void printInt(int value) { }

    /** Wrapper for SPIM.printFloat(). */

    public static void printFloat(float value) { }

    /** Wrapper for SPIM.printDouble(). */

    public static void printDouble(double value) { }

    /** Wrapper for SPIM.printString(). */

    public static void printString(String value) { }

    /** Wrapper for SPIM.printChar(). */

    public static void printChar(char value) { }

    /** Wrapper for SPIM.readInt(). */

    public static int readInt() { return 0; }

    /** Wrapper for SPIM.readFloat(). */

    public static float readFloat() { return 0; }

    /** Wrapper for SPIM.readDouble(). */

    public static double readDouble() { return 0; }

    /** Wrapper for SPIM.readString(). */

    public static String readString(int length) { return null; }

    /** Wrapper for SPIM.readChar(). */

    public static char readChar() { return ' '; }

    /** Wrapper for SPIM.open(). */

    public static int open(String filename, int flags, int mode)
    { return 0; }

    /** Wrapper for SPIM.read(). */

    public static String read(int fd, int length) { return null; }

    /** Wrapper for SPIM.write(). */

    public static int write(int fd, String buffer, int length)
    { return 0; }

    /** Wrapper for SPIM.close(). */

    public static void close(int fd) { }
```

```
     /** Wrapper for SPIM.exit(). */

     public static void exit() { }

     /** Wrapper for SPIM.exit2(). */

     public static void exit2(int status) { }
}
```

Because the SPIM class is defined in the package spim, that package name is part of the label for the entry point to each SPIM method, for example,

```
spim.SPIM.printInt:
```

6.3.6   Generating SPIM Code

After LIR with virtual registers has been generated, we invoke register allocation for mapping the virtual registers to physical registers. In this chapter we use a *naïve* register allocation scheme where physical registers are arbitrarily assigned to virtual registers (see Section 7.2). Register allocation is discussed in greater detail in Chapter 7.

Once virtual registers have been mapped to physical registers, translating LIR to SPIM code is pretty straightforward.

1. We iterate through the list of methods for each class; for each method, we do the following:

   (a) We generate a label for the method's entry point, for example,

   ```
   Factorial.computeIter:
   ```

   (b) We generate code to push a new frame onto the run-time stack (remember this stack grows downward in memory) and then code to save all our registers. In this compiler, we treat all of SPIM's general-purpose registers $t0 to $t9 and $s0 to $s7 as callee-saved registers; that is, we make it the responsibility of the invoked method to save any registers it uses. As we shall see in Chapter 7, some register allocation schemes do otherwise; that is, the caller saves just those registers that contain meaningful values when call is encountered.

   For computeIter(), the LIR uses seven general purpose registers, so we generate the following:

   ```
        subu    $sp,$sp,36    # Stack frame is 36 bytes long
        sw      $ra,32($sp)   # Save return address
        sw      $fp,28($sp)   # Save frame pointer
        sw      $t0,24($sp)   # Save register $t0
        sw      $t1,20($sp)   # Save register $t1
        sw      $t2,16($sp)   # Save register $t2
        sw      $t3,12($sp)   # Save register $t3
        sw      $t4,8($sp)    # Save register $t4
        sw      $t5,4($sp)    # Save register $t5
        sw      $t6,0($sp)    # Save register $t6
        addiu   $fp,$sp,32    # Save frame pointer
   ```

   (c) Because all branches in the code are expressed as branches to basic blocks, a unique label for each basic block is generated into the code; for example,

   ```
   Factorial.computeIter.2:
   ```

(d) We then iterate through the LIR instructions for the block, invoking a method `toSpim()`, which is defined for each LIR instruction; there is a one-to-one translation from each LIR instruction to its SPIM equivalent. For example, the (labeled) code for block B2 is

```
Factorial.computeIter.2:
    li $t3,0
    ble $t1,$t3,Factorial.computeIter.4
    j Factorial.computeIter.3
```

(e) Any string literals that are encountered in the instructions are put into a list, together with appropriate labels. These will be emitted into a data segment at the end of the method (see step 2 below).

(f) At the end of the method, we generate code to restore those registers that had been saved at the start. This code also does a jump to that instruction following the call in the calling code, which had been stored in the $ra register (ra is a mnemonic for return address) at the call. This code is labeled so that, once a return value has been placed in $v0, execution may branch to it to affect the return. For example, the register-restoring code for `computeIter()` is

```
Factorial.computeIter.restore:
    lw      $ra,32($sp)   # Restore return address
    lw      $fp,28($sp)   # Restore frame pointer
    lw      $t0,24($sp)   # Restore register $t0
    lw      $t1,20($sp)   # Restore register $t1
    lw      $t2,16($sp)   # Restore register $t2
    lw      $t3,12($sp)   # Restore register $t3
    lw      $t4,8($sp)    # Restore register $t4
    lw      $t5,4($sp)    # Restore register $t5
    lw      $t6,0($sp)    # Restore register $t6
    addiu   $sp,$sp,36    # Pop stack
    jr      $ra           # Return to caller
```

2. After we have generated the text portion (the program instructions) for the method (notice the `.text` directive at the start of code for each method), we then populate a data area, beginning with the `.data` directive, from the list of string literals constructed in step 1(e). Any other literals that you may wish to implement would be handled in the same way. Notice that `computerIter()` has no string literals; its data segment is empty.

3. Once all of the program code has been generated, we then copy out the SPIM code for implementing the SPIM class. This is where any further system code, which you may wish to implement, should go.

For example, the code for `Factorial.computeIter()` is as follows:

```
.text

Factorial.computeIter:
    subu    $sp,$sp,36    # Stack frame is 36 bytes long
    sw      $ra,32($sp)   # Save return address
    sw      $fp,28($sp)   # Save frame pointer
    sw      $t0,24($sp)   # Save register $t0
    sw      $t1,20($sp)   # Save register $t1
    sw      $t2,16($sp)   # Save register $t2
    sw      $t3,12($sp)   # Save register $t3
    sw      $t4,8($sp)    # Save register $t4
    sw      $t5,4($sp)    # Save register $t5
```

```
    sw      $t6,0($sp)    # Save register $t6
    addiu   $fp,$sp,32    # Save frame pointer

Factorial.computeIter.0:

Factorial.computeIter.1:
    li $t0,1
    move $t1,$a0
    move $t2,$t0

Factorial.computeIter.2:
    li $t3,0
    ble $t1,$t3,Factorial.computeIter.4
    j Factorial.computeIter.3

Factorial.computeIter.3:
    li $t4,-1
    add $t5,$t1,$t4
    mul $t6,$t2,$t1
    move $t2,$t6
    move $t1,$t5
    j Factorial.computeIter.2

Factorial.computeIter.4:
    move $v0,$t2
    j Factorial.computeIter.restore

Factorial.computeIter.restore:
    lw      $ra,32($sp)   # Restore return address
    lw      $fp,28($sp)   # Restore frame pointer
    lw      $t0,24($sp)   # Restore register $t0
    lw      $t1,20($sp)   # Restore register $t1
    lw      $t2,16($sp)   # Restore register $t2
    lw      $t3,12($sp)   # Restore register $t3
    lw      $t4,8($sp)    # Restore register $t4
    lw      $t5,4($sp)    # Restore register $t5
    lw      $t6,0($sp)    # Restore register $t6
    addiu   $sp,$sp,36    # Pop stack
    jr      $ra           # Return to caller
```

This code is emitted, together with the code for other `Factorial` methods and the SPIM class to a file ending in `.s`; in this case, the file would be named `Factorial.s`. This file can be loaded into any SPIM interpreter and executed.

You will notice, there are lots of moving values around among registers. In Chapter 7, we address how that might be minimized.

6.3.7   Peephole Optimization of the SPIM Code

You might have noticed in the SPIM code generated in the previous section, that some jumps were superfluous. For example, at the end of the SPIM code above, the jump at the end of block B4 simply jumps to the very next instruction! There might also be jumps to jumps. Such code can often be simplified. Jumps to the next instruction can be removed, and jumps to jumps can be simplified, (sometimes) removing the intervening jump.

We call such simplification *peephole optimization* because we need to consider just a few instructions at a time as we pass over the program; we need not do any global analysis.

To do peephole optimization on the SPIM code, it would be easier if we were to keep a list (perhaps an array list) of SPIM labels and instructions that we emit, representing the instructions in some easily analyzable way. Once the simplifications are done, we can spit out the SPIM instructions to a file.

6.4 Further Readings

James Larus' SPIM simulator [Larus, 2010] may be freely obtained on the WWW at `http://sourceforge.net/projects/spimsimulator/files/`. QtSpim is the interpreter of choice these days and is easily installed on Windows®, Linux®, and Mac OS X. Larus maintains a resource page at `http://pages.cs.wisc.edu/~larus/spim.html`. Finally, an excellent introduction to SPIM and the MIPS may be found at [Larus, 2009].

Our JVM-to-SPIM translator is based on that described in Christian Wimmer's Master's thesis [Wimmer, 2004]; this is a good overview of the Oracle HotSpot compiler. Another report on a register allocator for the HotSpot compiler is [Mössenböck, 2000].

Our proposed layout of objects is based on that of [Corliss and Lewis, 2007].

For comprehensive coverage of code optimization techniques, see [Morgan, 1998], [Muchnick, 1997], [Allen and Kennedy, 2002], and [Cooper and Torczon, 2011].

[Wilson, 1994] is a good introduction to classical garbage collection techniques. [Jones and Lins, 1996] also presents many memory management strategies.

6.5 Exercises

As we have stated in the narrative, we implement enough of the JVM-to-SPIM translator to implement a small subset of the JVM. The following exercises ask one to expand on this.

Exercise 6.1. Implement all of the relational and equality JVM instructions in the HIR, the LIR, and SPIM. (Notice that some of this has been done.) Test these.

Exercise 6.2. Assuming you have implemented / and % in your j-- compiler, implement them in the HIR, LIR, and SPIM. Test these.

Exercise 6.3. Implement the bitwise JVM instructions in the HIR, LIR and SPIM. Test these.

Exercise 6.4. In many places in the HIR, we refer to the instruction 0, which means 0 is an operand to some computation. In generating the LIR, that constant 0 is loaded into some virtual register. But SPIM has the register $zero, which always contains the constant 0 and so may replace the virtual register in the LIR code. Modify the translation of HIR to LIR to take advantage of this special case.

Exercise 6.5. The JVM instructions `getstatic` and `putstatic` are implemented in the HIR and LIR but not in SPIM. Implement these instructions in SPIM and test them.

Exercise 6.6. Implement the `Object` type, supporting the methods `clone()`, `equals()`, and `toString()`.

Exercise 6.7. Implement the `String` type, which is a subclass of `Object` and implements the methods `charAt()`, `concat()`, `length()`, and `substring()` with two arguments.

Exercise 6.8. In Section 6.3.5, we came up with a naming convention for the SPIM routines for methods that does not deal with overloaded method names, that is, methods having the same name but different signatures. Propose an extension to our convention that deals with overloaded methods.

Exercise 6.9. In Figure 6.19, we suggest an alternative addressing scheme where a pointer to an object actually points 12 bytes into the object, making the addressing of components (fields, array elements, or string characters) simpler. Implement this scheme, changing both the run-time code and the code generated for addressing components.

Exercise 6.10. Implement the JVM instructions `invokevirtual` and `invokespecial`. Test them.

Exercise 6.11. Implement the JVM `invokeinterface` instruction. See [Alpern et al., 2001] for a discussion of this.

Exercise 6.12. Implement arrays in the HIR (`iaload` and `iastore` have been implemented in the HIR), the LIR, and SPIM. Implement the `arraylength` instruction. Test these.

Exercise 6.13. Implement the instance field operations `getfield` and `putfield` in the HIR, the LIR, and SPIM. Test these.

Exercise 6.14. Implement the remaining JVM load and store instructions and test them.

Exercise 6.15. In the current implementation, all general-purpose registers are treated as callee-saved registers; they are saved by the method being invoked. Modify the compiler so that it instead treats all general-purpose registers as caller-saved, generating code to save registers before a call to another method, and restoring those same registers after the call.

The following exercises implement various optimizations, which are discussed in Section 6.3.3. The code already defines a method, `optimize()`, in the class `NControlFlowGraph`. That is a good place to put calls to any methods that perform optimizations on the HIR.

Exercise 6.16. Implement *inlining* in the optimizations phase. Use an arbitrary instruction count in deciding what methods to inline. Be careful in dealing with virtual methods.

Exercise 6.17. Implement *constant folding* in the optimizations phase.

Exercise 6.18. Implement *common sub-expression elimination* in the optimization phase.

Exercise 6.19. Implement the *lifting of invariant code* in the optimization phase.

Exercise 6.20. Design and implement a strategy for performing various forms of *strength reduction* in the optimization phase.

Exercise 6.21. Implement *redundant array bounds check elimination* in the optimization phase.

Exercise 6.22. Implement *redundant null check elimination* in the optimization phase.

Exercise 6.23. Implement a simple version of *peephole optimizations*, simplifying some of the branches discussed in Section 6.3.7.

Exercise 6.24. (Involved[7]) Implement the JVM instructions `lookupswitch` and `lookuptable` in the HIR, the LIR, and SPIM. Test these. Notice that these flow-of-control instructions introduce multiple predecessors and multiple successors to basic blocks in the flow graph.

Exercise 6.25. (Involved) Introduce the `long` types of constants, variables and operations into the HIR, the LIR, and SPIM. This assumes you have already added them to *j--*. Notice that this will complicate all portions of your JVM to SPIM translation, including register allocation, because longs require two registers and complicate their operations in SPIM.

---

[7]Meaning not necessarily difficult but involving some work.

Exercise 6.26. (Involved) Introduce the `float` and `double` types of constants, variables and operations into the HIR, the LIR and SPIM. This assumes you have already added them to *j--*. Notice that this will complicate all portions of your JVM-to-SPIM translation and will require using the special floating-point processor and registers, which are fully documented in [Larus, 2009].

Exercise 6.27. (Involved) Implement exception handling in the HIR, the LIR, and SPIM. This assumes you have already added exception handling to *j--*. Notice that this will complicate all portions of your JVM-to-SPIM translation and will require using the special exception handling instructions, which are fully documented in [Larus, 2009].

Exercise 6.28. (Involved) Read up on and implement a (relatively) simple copy form of garbage collection. See the section on Further Readings for starting points to learning about garbage collection.

Chapter 7

Register Allocation

7.1 Introduction

Register allocation is the process of assigning as many local variables and temporaries to physical registers as possible. The more values that we can keep in registers instead of in memory, the faster our programs will run. This makes register allocation the most effective optimization technique that we have.

With respect to our LIR discussed in Chapter 6, we wish to assign physical registers to each of the virtual registers that serve as operands to instructions. The problem is that there are often many fewer physical registers than there are virtual registers. Sometimes, as program execution progresses, some values in physical registers will have to be *spilled* to memory while the register is used for another purpose, and then reloaded when those values are needed again. Code must be generated for storing spilled values and then for reloading those values at appropriate places. Of course, we wish to minimize this spilling (and reloading) of values to and from memory.

So, any register allocation strategy must determine how to most effectively allocate physical registers to virtual registers and, when spilling is necessary, *which* physical registers to spill to make room for assignment to other virtual registers. This problem has been shown to be NP-complete in general [Sethi, 1973] but there are several allocation strategies that do a reasonable job in near-linear time.

Register allocation that focuses on just a single basic block, or even just a single statement, is said to be *local*. Register allocation that considers the entire flow graph of a method is said to be *global*.

In this chapter we look briefly at some local register allocation strategies but we focus most of our attention on two global register allocation strategies: linear scan register allocation and graph coloring register allocation.

7.2 Naïve Register Allocation

A *naïve register allocation* strategy simply sequences through the operations in the (LIR) code, assigning global registers to virtual registers. Once all physical registers have been assigned, and if there are additional virtual registers to deal with, we begin spilling physical registers to memory. There is no strategy for determining which registers to spill; for example, one might simply sequence through the physical registers a second time in the same order they were assigned the first time, spilling each to memory as it is re-needed. When a spilled value is used again, it must be reloaded into a (possibly different) register. Such a regimen works just fine when there are as many physical registers as there are virtual registers; in fact, it is as effective as any other register allocation scheme in this case. Of course, when there are many more virtual registers than physical registers, its performance degrades rapidly as physical register values must be repeatedly spilled and reloaded.

7.3   Local Register Allocation

Local register allocation can involve allocating registers for a single statement or a single basic block.

In [Aho et al., 2007], the authors provide an algorithm that allocates the minimal number of registers required for processing the computation represented by an abstract syntax tree (AST); the algorithm does a post-order traversal of the AST, determining the minimum number of registers required, and assigns registers, re-using registers as appropriate, to the computation[1] .

Another strategy is to compute, for each virtual register, a live interval (see Section 7.4.1) local to a block. Registers are allocated in the order of the intervals' start positions. When a register must be spilled, we avoid spilling those registers whose values last the longest in the block.

Yet another strategy mixes local and limited global information. It computes and applies local liveness intervals to the allocation of registers, but in those blocks within the deepest nested loops first, the payoff comes from the fact that instructions more deeply nested within loops are executed much more often.

A more effective register allocation regime considers the entire control-flow graph for a method's computation, that is, global register allocation.

7.4   Global Register Allocation

7.4.1   Computing Liveness Intervals

Liveness Intervals

Global register allocation works with a method's entire control-flow graph to map virtual registers to physical registers. One wants to minimize spills to memory; where spills are necessary, one wants to avoid using them within deeply nested loops. The basic tool in global register allocation is the *liveness interval*, the sequence of instructions for which a virtual register holds a meaningful value.

Liveness intervals are required by both of the global register algorithms that we consider: linear scan register allocation and register allocation by graph coloring.

In its roughest form, a liveness interval for a virtual register extends from the first instruction that assigns it a value to the last instruction that uses its value. A more accurate liveness interval has "holes" in this sequence, where a virtual register does not contain a useful value; for example, a hole occurs from where the previously assigned value was last used (or read) to the next assignment (or write) of a new value.

---

[1] The algorithm is similar to that used by `CLEmitter` for computing the minimum number of locations to allocate to a JVM stack frame for computing expressions in the method.

For example, consider the control-flow graph for `Factorial`'s `computeIter()` in Figure 7.1. Here, it is composed of the LIR instructions that we computed in Chapter 6.

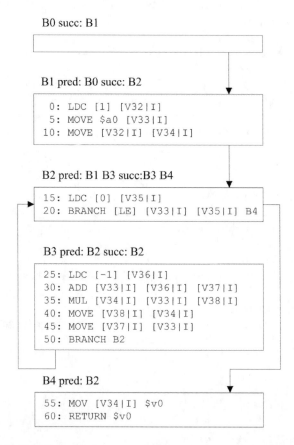

B0 succ: B1

B1 pred: B0 succ: B2

```
 0: LDC [1] [V32|I]
 5: MOVE $a0 [V33|I]
10: MOVE [V32|I] [V34|I]
```

B2 pred: B1 B3 succ:B3 B4

```
15: LDC [0] [V35|I]
20: BRANCH [LE] [V33|I] [V35|I] B4
```

B3 pred: B2 succ: B2

```
25: LDC [-1] [V36|I]
30: ADD [V33|I] [V36|I] [V37|I]
35: MUL [V34|I] [V33|I] [V38|I]
40: MOVE [V38|I] [V34|I]
45: MOVE [V37|I] [V33|I]
50: BRANCH B2
```

B4 pred: B2

```
55: MOV [V34|I] $v0
60: RETURN $v0
```

FIGURE 7.1 Control-flow =graph for `Factorial.computeIter()`.

The liveness intervals for this code are illustrated in Figure 7.2. We shall go through the steps in constructing these intervals in the next sections. The numbers on the horizontal axis represent instruction ids; recall, instruction ids are assigned at increments of 5 to facilitate the insertion of spill code. The vertical axis is labeled with register ids.

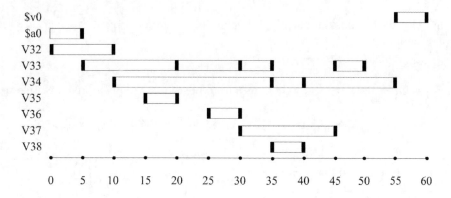

FIGURE 7.2 Liveness intervals for `Factorial.computeIter()`.

There are two physical registers here: $v0 and $a0[2]; $a0 is the first argument to computerIter() and the return value is in $v0. There are seven virtual registers: V32 to V38. Recall that, we begin numbering the virtual registers at 32 because 0 to 31 are reserved for the 32 general-purpose (physical) registers.

An interval for a virtual register extends *from* where it is defined (it could be defined as an argument; for example, $a0) *to* the last time it is used. For example, V37 is defined at instruction position 30 and is last used at instruction position 45.

Some intervals have *holes*. A *hole* might extend from the instruction following the last instruction using the register up until the next time it is defined. For example, consider virtual register V33. It is defined at position 5 and used at positions 20, 30 and 35. But it is not used again until after it is redefined at position 45 (notice the loop again). There is a hole at position 40.

Loops can complicate determining an interval. For example, V34 is first defined at position 10, used at 35, and then redefined at 40. But there is a branch (at position 50) back to basic block B2. At some point in the iteration, the conditional branch at position 20 will bring us to block B4, and V34 will be used a last time at position 55. So we say that the interval for V34 extends from position 10 to position 55 inclusively; its value is always meaningful over that interval.

Finally, the darker vertical segments in the intervals identify use positions: a *use position* is a position in the interval where either the register is defined (that is, written) or the register is being used (that is, read). For example, $v0 in Figure 7.2 is defined at position 55 and used at position 60, so there are two segments. But $a0 is never defined in the interval (actually, we assume it was defined in the calling code) but is used at position 5. From the same figure we can see that V34 is either defined or used at positions 10, 35, 40, and 55.

The compiler prints out these intervals using the following format:

```
v0:   [55, 60]
a0:   [0, 5]
V32:  [0, 10]
V33:  [5, 35]   [45, 50]
V34:  [10, 55]
V35:  [15, 20]
V36:  [25, 30]
V37:  [30, 45]
V38:  [35, 40]
```

Here, just the intervals (and not the use positions) are displayed for each register. The notation

```
[<from> <to>]
```

indicates that the register holds a meaningful value from position <from> to position <to>. For example, the line

```
V37: [30, 45]
```

says the liveness interval for register V37 ranges from position 30 to position 45. Notice that register V33,

```
V33: [5, 35]   [45, 50]
```

has an interval with two ranges: from position 5 to position 35, and from position 45 to position 50; there is a hole in between.

---

[2]The compiler's output leaves out the '$' prefix here but we use it to distinguish physical registers.

Computing Local Liveness Sets

As a first step in building the liveness intervals for LIR operands, we compute, for each block, two local liveness sets: *liveUse* and *liveDef*. The liveUse operands are those operands that are read (or used) before they are written (defined) in the block's instruction sequence; presumably they were defined in a predecessor block (or are arguments to the method). The liveDef operands are those operands that are written to (defined) by some instruction in the block.

Algorithm 7.1 computes liveUse and liveDef for each basic block in a method.

Algorithm 7.1 Computing Local Liveness Information

Input: The control-flow graph g for a method
Output: Two sets for each basic block: liveUse, registers used before they are overwritten (defined) in the block and liveDef, registers that are defined in the block

> **for** block b in g.blocks **do**
>> Set b.liveUse ← {}
>> Set b.liveDef ← {}
>> **for** instruction i in b.instructions **do**
>>> **for** virtual register v in i.readOperands **do**
>>>> **if** $v \notin b$.liveDef **then**
>>>>> b.liveUse.add(v)
>>>> **end if**
>>> **end for**
>>> **for** virtual register v in i.writeOperands **do**
>>>> b.liveDef.add(v)
>>> **end for**
>> **end for**
> **end for**

For example, the control-flow graph for `Factorial`'s `computeIter()` but with its local liveness sets computed is illustrated in Figure 7.3.

In the computation of the local liveness sets, we can sequence through the blocks in any order because all computations are local to each block. Our computation proceeds as follows:

B0 There are no registers in block B0 to be used or defined.

B1 In block B1, there are four registers. V32 is defined in instruction 0. At instruction 5, $a0 is used in the block before it is defined and V33 is defined. At instruction 10, V34 is defined (V32 is used but it is already defined in this block).

B2 In block B2, V35 is defined in instruction 15. V33 is used in instruction 20 without previously being defined in the block.

B3 In block B3, V36 is defined in instruction 25. In instruction 30, V33 is used (before defined) and V37 is defined. In instruction 35, V34 is used (before defined) and V38 is defined. V34 is defined in instruction 40 and V33 is defined in instruction 45. No register is used or defined in the branch at 50.

B4 In block B4, V34 is used and $v0 is defined in instruction 55. $v0 is used in instruction 60 but not before it is defined in the block.

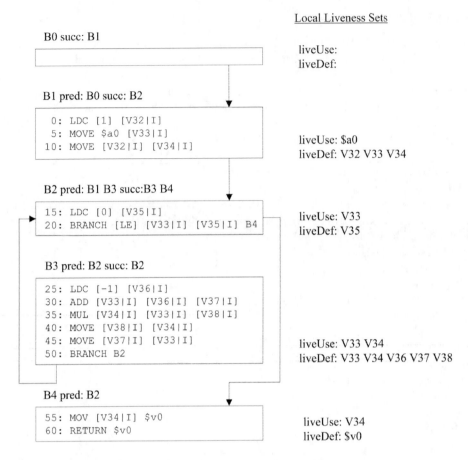

Local Liveness Sets

B0 succ: B1

liveUse:
liveDef:

B1 pred: B0 succ: B2

```
 0: LDC [1] [V32|I]
 5: MOVE $a0 [V33|I]
10: MOVE [V32|I] [V34|I]
```

liveUse: $a0
liveDef: V32 V33 V34

B2 pred: B1 B3 succ:B3 B4

```
15: LDC [0] [V35|I]
20: BRANCH [LE] [V33|I] [V35|I] B4
```

liveUse: V33
liveDef: V35

B3 pred: B2 succ: B2

```
25: LDC [-1] [V36|I]
30: ADD [V33|I] [V36|I] [V37|I]
35: MUL [V34|I] [V33|I] [V38|I]
40: MOVE [V38|I] [V34|I]
45: MOVE [V37|I] [V33|I]
50: BRANCH B2
```

liveUse: V33 V34
liveDef: V33 V34 V36 V37 V38

B4 pred: B2

```
55: MOV [V34|I] $v0
60: RETURN $v0
```

liveUse: V34
liveDef: $v0

FIGURE 7.3 Control-flow graph for `Factorial.computeIter()` with local liveness sets computed.

Our next step is to use this local liveness information to compute global liveness information; for each basic block, we want to know which registers are live, both coming into the block and going out of the block.

Computing Global Liveness Sets

We can compute the set of operands that are live at the beginning and end of a block using a backward data-flow analysis [Aho et al., 2007]. We call the set of operands that are live at the start of a block *liveIn*. We call the set of operands that are live at the end of a block *liveOut*.

Algorithm 7.2 computes this global liveness data for us. Notice we iterate through the blocks, and also the instruction sequence within each block, in a *backwards direction*.

Algorithm 7.2 Computing Global Liveness Information

Input: The control-flow graph g for a method, and the local liveness sets liveUse and liveDef for every basic block

Output: Two sets for each basic block: liveIn, registers live at the beginning of the block, and liveOut, registers that are live at the end of the block

> **repeat**
>> **for** block b in g.blocks in reverse order **do**
>>> b.liveOut $\leftarrow$ {}
>>> **for** block s in b.successors **do**
>>>> b.liveOut $\leftarrow$ b.liveOut $\cup$ s.liveIn
>>> **end for**
>>> b.liveIn $\leftarrow$ (b.liveOut $-$ b.liveDef) $\cup$ b.liveUse
>> **end for**
> **until** no liveOut has changed

We represent the live sets as `BitSets` from the Java API, indexed by the register number of the operands. Recall that registers numbered 0 to 31 are physical registers and registers 32 and higher are virtual registers. By the time we compute liveness intervals, we know how many virtual registers we will need for a method's computation. Later, splitting of intervals may create additional virtual registers but we do not need liveness sets for those.

We cannot compute the live sets for a loop in a single pass; the first time we process a loop end block, we have not yet processed the corresponding loop header block; the loop header's block's liveIn set (and so its predecessor blocks' liveOut sets) is computed correctly only in the second pass. This means our algorithm must make $d + 1$ passes over the blocks, where d is the depth of the deepest loop.

Continuing with our example for `Factorial`'s `computeIter()`, we start with the local liveness sets in Figure 7.3. Our computation requires three iterations:

1. After the first iteration, the sets liveIn and liveOut for each block are

```
B4 liveIn:   V34
B4 liveOut:
B3 liveIn:   V33 V34
B3 liveOut:
B2 liveIn:   V33 V34
B2 liveOut:  V33 V34
B1 liveIn:   $a0
B1 liveOut:  V33 V34
B0 liveIn:   $a0
B0 liveOut:  $a0
```

Recall that at each iteration we process the blocks in reverse order; at the start, liveOut for each block is empty.

(a) B4 has no successors so liveOut remains empty. Because V34 is used in B4 (in B4's liveUse), it is added into B4's liveIn.

(b) B3's only successor B2 has not been processed yet so B3's liveOut is empty at this iteration (the next iteration will add registers). Because B3's liveUse contains V33 and V34 (as they are used before defined in the block), V33 and V34 are added to B3's liveIn.

(c) B2 has successors B3 and B4. Because V33 and V34 are in B3's liveIn, they are added to B2's liveOut; V34 is also in B4's liveIn but it is already in B2's liveOut. Because neither V33 nor V34 are redefined in B2, they are also added to B2's liveIn.

(d) B1 has the single successor B2 and it gets its liveOut (V33 and V34) from B2's liveIn. V33 and V34 are not added to B1's liveIn, even though they are in its liveOut, because they are in its liveDef (defined in B1). Because $a0 is in B1's liveUse, it is added to its liveIn.

(e) B0 has the single successor B1 and gets its liveOut ($a0) from B1's liveIn. Because $a0 is not redefined in B0 (nothing is defined in B0), it is also in B0's liveIn.

Of course, there have been changes to at least one of the sets and we need another iteration.

2. In the second iteration, we again go through the blocks in reverse order to get the following sets:

```
B4 liveIn:   V34
B4 liveOut:
B3 liveIn:   V33 V34
B3 liveOut:  V33 V34
B2 liveIn:   V33 V34
B2 liveOut:  V33 V34
B1 liveIn:   $a0
B1 liveOut:  V33 V34
B0 liveIn:   $a0
B0 liveOut:  $a0
```

(a) Nothing has changed for B4.

(b) But B3's liveOut changes because one of its successor's liveIn changed in the first iteration; V33 and V34 are added to B3's liveOut.

(c) Nothing has changed for B2.

(d) Nothing has changed for B1.

(e) Nothing has changed for B2.

Because there was a change to liveOut in processing B3, we must iterate yet again.

3. But no changes are made to any of the sets in the third iteration, so we can stop. The final global liveness sets are those computed in the second iteration:

```
B4 liveIn:   V34
B4 liveOut:
B3 liveIn:   V33 V34
B3 liveOut:  V33 V34
B2 liveIn:   V33 V34
B2 liveOut:  V33 V34
B1 liveIn:   $a0
B1 liveOut:  V33 V34
B0 liveIn:   $a0
B0 liveOut:  $a0
```

We can now use this global live in and live out information to compute accurate liveness intervals.

Building the Intervals

To build the intervals, we make a single pass over the blocks and instructions, again in reverse order. Algorithm 7.3 computes these intervals with both the ranges and the use positions.

Algorithm 7.3 Building Liveness Intervals

Input: The control-flow graph g for a method with LIR, and the liveIn and liveOut sets for each basic block

Output: A liveness interval for each register, with ranges and use positions

  for block b in g.blocks in reverse order **do**

    int $blockFrom \leftarrow b$.firstInstruction.id

    Set $blockTo \leftarrow b$.lastInstruction.id

    for register r in b.liveOut **do**

      intervals[r].addOrExtendRange($blockFrom$, $blockRange$)

    end for

    for instruction i in b.instructions in reverse order **do**

      if i.isAMethodCall **then**

        for physical register r in the set of physical registers **do**

          intervals[r].addRange(i.id, i.id)

        end for

      end if

      for virtual register r in i.writeOperands **do**

        intervals[r].firstRange.from $\leftarrow i$.id

        intervals[r].addUsePos(i.id)

      end for

      for virtual register r in i.readOperands **do**

        intervals[r].addOrExtendRange($blockFrom$, i.id)

        intervals[r].addUsePos(i.id)

      end for

    end for

  end for

Before we even look at the LIR instructions, we add ranges for all registers that are live out. These ranges must extend to the end of the block because they are live out. Initially, we define the ranges to extend from the start of the block because they may have been defined in a predecessor; if we later find they are defined (or redefined) in the block, then we shorten this range by overwriting the from position with the defining position.

As we iterate through the instructions of each block, in reverse order, we add or modify ranges:

- When we encounter a subroutine call, we add ranges of length 1 at the call's position to the intervals of all physical registers. The reason for this is that we must assume the subroutine itself will use these registers and so we would want to force spills.

  In our current implementation, we do not do this step. Rather, we treat all registers as callee-saved registers, making it the responsibility of the called subroutine to save any physical registers that it uses. We leave alternative implementations as exercises.

- If the instruction has a register that is written to, then we adjust the first (most recent) range's start position to be the position of the (writing) instruction, and we record the use position.

- For each register that is read (or used) in the instruction, we add a new range extending to this instruction's position. Initially, the new range begins at the start of the block; a write may cause the start position to be re-adjusted.

  Notice the addOrExtendRange() operation merges contiguous ranges into one.

The careful reader will notice that this algorithm assumes that no virtual register is defined but not used; this means we do not have local variables that are given values that are never used. We leave dealing with the more peculiar cases as exercises.

Consider how Algorithm 7.3 would deal with basic block B3 of the LIR for `Factorial.computeIter()`.

```
B3
25: LDC    [-1]     [V36|I]
30: ADD    [V33|I]  [V36|I]  [V37|I]
35: MUL    [V34|I]  [V33|I]  [V38|I]
40: MOVE   [V38|I]  [V34|I]
45: MOVE   [V37|I]  [V33|I]
50: BRANCH B2
```

The progress of building intervals for basic block B3, as we sequence backward through its instructions, is illustrated in Figure 7.4. (LiveOut contains V33 and V34.)

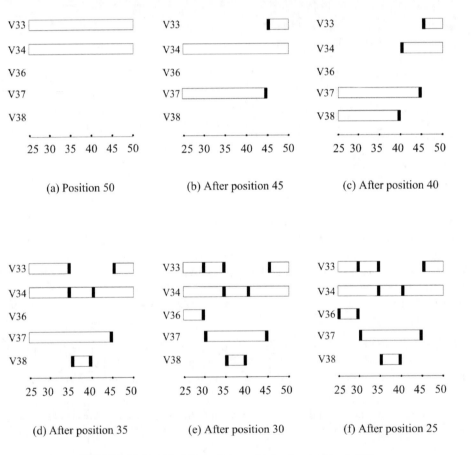

FIGURE 7.4 Building intervals for basic block B3.

At the start of processing B3, virtual registers V33 and V34 are in liveOut and so have intervals with ranges extending from (for now) the start of the block (position 25) to the end of the block (position 50), as illustrated in Figure 7.4(a). We then sequence through the instructions in reverse order.

- At position 50, the branch instruction has no register operands, so no intervals are affected.

- At position 45, the move defines V33 so the range for V33 is shortened to start at 45; V37 is used, so we add a range to V37 extending from (for now) the start of the block to position 45. This is illustrated in Figure 7.4(b).

- At position 40, the move defines V34 so its range is shortened to start at 40; V38 is used so we add a range to V38 extending from (for now) the start of the block to position 40. This is illustrated in Figure 7.4(c).

- At position 35, the multiplication operation defines V38 so its range is shortened to start at 35; V34 is used so we add a range extending from the start of the block to 35, and because it is adjacent to the next segment, we merge the two; V33 is used, so we add a segment for V33 from the start of the block to 35. This is illustrated in Figure 7.4(d).

- At position 30, the add operation defines V37 so we shorten its range to start at 30; V33 is used, so we add a use to its interval at 30; V36 is used so we add a range from the start of the block to 30. This is illustrated in Figure 7.4(e).

- At position 25 the load constant operation defines V36 so we define a use for V36 (definitions are considered uses) at position 25. This is illustrated in Figure 7.4 (f).

Once we have calculated the liveness intervals for a method, we can set about using them to allocate registers.

7.4.2   Linear Scan Register Allocation

Introduction to Linear Scan

Linear scan [Poletto and Sarkar, 1999] is the first register allocation technique we look at, and it is the one we have implemented in the JVM code to SPIM translator. Its principal advantage is that it is fast—it really just does a linear scan through the liveness intervals, in order of their starting positions (earliest first), and maps them to physical registers.

We follow a strategy laid out by Christopher Wimmer in his Master's thesis [Wimmer, 2004], which he used to implement a register allocator for the Oracle HotSpot client JIT compiler. It is fast and it is also intuitive.

Linear Scan Allocation Algorithm

Algorithm 7.4 describes a linear scan register allocation algorithm based on that in [Wimmer, 2004].

First we sort the intervals in increasing order, based on their start positions. At any time, the algorithm is working with one interval called the *current* interval; this interval has a starting *position*, the from field of its first range. This position categorizes the remaining intervals into four lists:

1. A list of *unhandled* intervals sorted on their start positions in increasing order. The unhandled list contains intervals starting after position.

2. A list of *active* intervals, whose intervals cover position and so have a physical register assigned to them.

3. A list of *inactive* intervals, each of which starts before *position* and ends after *position* but do not cover *position*, because *position* lies in a lifetime hole.

4. A list of *handled* intervals, each of which ends before *position* or was spilled to memory. These intervals are no longer needed.

Algorithm 7.4 Linear Scan Register Allocation

Input: The control-flow graph g for a method with its associated liveness intervals
Output: A version of the LIR where all virtual registers are mapped to physical registers,
　　and code for any necessary spills, has been inserted into the LIR

　　List *unhandled* ← a list of all intervals, sorted in increasing start-position order
　　List *active* ← {}
　　List *inactive* ← {}
　　List *handled* ← {}
　　while *unhandled* ≠ {} **do**
　　　　current ← first interval removed from *unhandled*
　　　　position ← *current*.firstRange.from
　　　　for interval *i* in *active* **do**
　　　　　　if *i*.lastRange.to < *position* **then**
　　　　　　　　move *i* from *active* to *handled*
　　　　　　else if *i*.covers(*position*) **then**
　　　　　　　　move *i* from *active* to *inactive*
　　　　　　end if
　　　　end for
　　　　for interval *i* in *inactive* **do**
　　　　　　if *i*.lastRange.to < *position* **then**
　　　　　　　　move *i* from *inactive* to *handled*
　　　　　　else if *i*.covers(*position*) **then**
　　　　　　　　move *i* from *inactive* to *inactive*
　　　　　　end if
　　　　end for
　　　　if foundFreeRegisterFor(*current*) **then**
　　　　　　allocateBlockedRegisterFor(*current*)
　　　　end if
　　end while

The algorithm scans through the unhandled intervals one at a time, finding a physical register for each; in the first instance, it looks for a free physical register and, if it fails at that, it looks for a register to spill. In either case, a split may occur, creating an additional interval and sorting it back into unhandled. Before allocating a register, the algorithm does a little bookkeeping: based on the new position (the start of the current interval), it moves intervals among the lists.

The method foundFreeRegisterFor() takes an interval as its argument and attempts to map it to a freely available register; if it is successful, it returns `true`; if unsuccessful in finding a free register, it returns `false`. In the latter case, the linear scan algorithm calls upon the method allocateBlockedRegisterFor(), which determines which register should be spilled so that it can be allocated to the interval.

Algorithm 7.5 describes the behavior of the method `foundFreeRegisterFor()`.

Algorithm 7.5 Attempting to Allocate a Free Register

Input: The current interval *current*, from Algorithm 7.4

Output: `true` if a free register was assigned to the current interval; `false` if no free register was found

    List *unhandled* ← a list of all intervals, sorted in increasing start-position order

    for physical register *r* in the set of physical registers **do**

        freePos[*r*] ← maxInt // default

    end for

    for interval *i* in *active* **do**

        freePos[*i*.reg] ← 0 // unavailable

    end for

    for interval *i* in *inactive* **do**

        if *i* intersects with *current* **then**

            freePos[*i*.reg] ← min(freePos[*i*.reg], next intersection of *i* with *current*)

        end if

    end for

    register *reg* ← register with highest freePos

    if freePos[*reg*] = 0 **then**

        return `false` // failure

    else if freePos[*reg*] > *current*.lastRange.to **then**

        // available for all of current interval

        current.r ← *reg* // assign it to current interval

    else

        // register available for first part of current

        current.r ←*reg* // assign it to current interval

        // split current at optimal position before freePos[reg]

        put *current* with the split-off part back on unhandled

    end if

    return `true` // success

The array freePos[] is used to compute the next position that each register is used; the register is free up until that position. The element freePos[*r*] records this position for the physical register (numbered) *r*.

By default, all physical registers are free for the lifetime of the method; hence, the freePos is maxInt. But none of the physical registers assigned to intervals that are on the active list are free; hence the freePos is 0. But registers assigned to intervals that are inactive (and so in a hole) are assigned a freePos equal to the next position that the current interval intersects with the interval in question, that is, the position where both intervals would need the same register; the register is free to be assigned to the current interval up until that position. Because the same physical register may (because of previous splits) have several next use positions, we choose the closest (the minimum) for freePos.

So, when we choose the register with the highest freePos, we are choosing that register that will be free the longest.

If the chosen register has a freePos of 0, then neither it nor any other register is free and so we return `false` to signal we will have to spill some register. But if its freePos is greater than the to position of current's last range, then it is available for the duration of current and so is assigned to current. Otherwise, the interval we are considering is in a hole; indeed, it is in a hole that extends the furthest of any other interval in a hole. Its register is available to the current interval up until that position. So we *split* the current interval at *some optimal position* between the current position and that freePos, we assign the interval's register (temporarily) to the first split off part of current, and we put current with the remaining split off part back on unhandled for further processing. So we make the best use of registers that are assigned to intervals that are currently in holes.

If the current and the candidate interval do not intersect at all, then both intervals may make use of the same physical register with no conflict; one interval is in a hole while the other is making use of the shared register.

When we say we split the current interval at *some optimal position,* we are choosing how much of the hole to take up for *current*'s purposes. One could say we want as much as possible, that is, all the way up to freePos. But, in the next section we shall see that this is not always wise.

Consider the intervals for `Factorial.computeIter()` from Figure 7.2 and repeated in Figure 7.5.

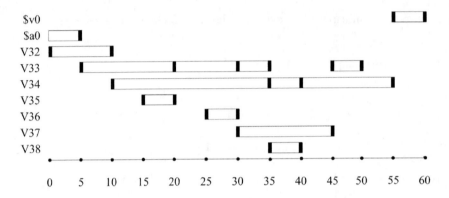

FIGURE 7.5 Liveness intervals for `Factorial.computeIter()`, again.

We first assume we have four free physical registers ($t0–$t3) that we can allocate to our seven virtual registers (V32–V38); that is, we invoke *j--* with

```
> j-- -s linear -r 4 Factorial.java
```

The argument -s specifies that we are using linear scan; the -r 4 says we are working with four physical registers. Algorithm 7.4 does a linear scan through the seven virtual registers, in order of their starting positions V32, V33,..., V38, as follows.

Algorithm 7.4 starts off with the list of seven unhandled intervals, an empty active list and an empty inactive list. A trace of the linear scan process follows:[3].

1. unhandled: [V32, V33, V34, V35, V36, V37, V38]

   The first interval in the list is removed and becomes *current*. The *active* and *inactive* lists remain empty. All four physical registers are available for allocation.

   current = V32
   active:
   inactive:
   free registers: [$t0, $t1, $t2, $t3]

   The current interval, V32, is assigned the first free physical register $t0; V32 is put on the active list for the next iteration.

---

[3]In the trace, we do not italicize variables.

Interval V32 is allocated physical register $t0

2. The next iteration deals with V33.

   unhandled: [V33, V34, V35, V36, V37, V38]
   current = V33
   active: [V32]
   inactive: []
   free registers: [$t1, $t2, $t3]

   Interval V33 is allocated physical register $t1

3. unhandled: [V34, V35, V36, V37, V38]
   current = V34
   active: [V32, V33]
   inactive: []
   free registers: [$t2, $t3]

   Interval V34 is allocated physical register $t2

4. unhandled: [V35, V36, V37, V38]
   current = V35
   active: [V33, V34]
   inactive: []
   free registers: [$t0, $t3]

   Interval V35 is allocated physical register $t0

5. unhandled: [V36, V37, V38]
   current = V36
   active: [V33, V34]
   inactive: []
   free registers: [$t0, $t3]

   Interval V36 is allocated physical register $t0

6. unhandled: [V37, V38]
   current = V37
   active: [V33, V34, V36]
   inactive: []
   free registers: [$t3]

   Interval V37 is allocated physical register $t3

7. unhandled: [V38]
   current = V38
   active: [V33, V34, V37]
   inactive: []
   free registers: [$t0]

   Interval V38 is allocated physical register $t0

At the end of the linear scan allocation, all intervals have been assigned physical registers and the LIR now looks as follows:

```
B0


B1
0:   LDC [1] $t0
5:   MOVE $a0 $t1
10:  MOVE $t0 $t2


B2
15:  LDC [0] $t0
20:  BRANCH [LE] $t1 $t0 B4


B3
25:  LDC [-1] $t0
30:  ADD $t1 $t0 $t3
35:  MUL $t2 $t1 $t0
40:  MOVE $t0 $t2
45:  MOVE $t3 $t1
50:  BRANCH B2


B4
55:  MOVE $t2 $v0
60:  RETURN $v0
```

Now, if no register is free to be assigned to current even for a portion of its lifetime, then we must spill some interval to a memory location on the stack frame. We use Algorithm 7.6 [Wimmer, 2004] to select an interval for spilling and to assign the freed register to *current*. The algorithm chooses to spill that register that is not used for the longest time—that register next used at the highest position in the code.

Algorithm 7.6 describes the behavior of the method foundFreeRegisterFor(). This algorithm collects two positions for each physical register r by iterating through all active and inactive intervals:

1. usePos[r] records the position where an interval with register r assigned is used next. If more than one use position is available, the minimum is used. It is used in selecting a spill candidate.

2. blockPos[r] records a (minimum) hard limit where register r cannot be freed by spilling. Because this is a candidate value for usePos, usePos can never be higher than this hard blockPos.

The register with the highest usePos is selected as a candidate for spilling. Based on the use positions, the algorithm identifies three possibilities:

1. It is better to spill *current* if its first use position is found after the candidate with the highest usePos. We split current before its first use position where it must be reloaded; the active and inactive intervals are left untouched. This case also applies if all registers are blocked at a method-call site.

2. Otherwise, *current* is assigned the selected register. All inactive and active intervals for this register intersecting with *current* are split before the start of *current* and spilled. We need not consider these split children further because they do not have a register assigned. But if they have use positions requiring a register, they will have to be reloaded to some register so they are split a second time before the use positions and the second split children are sorted back into *unhandled*. They will get a register assigned when the allocator has advanced to their position.

3. If the selected register has a blockPos in the middle of current, then it is not available for *current*'s entire lifetime; *current* is split before blockPos and the sorted child is sorted into *unhandled*. (That the current interval starts before a use is probably the result of a previous split.)

4. Notice that neither loads nor stores are actually generated in the LIR. This will be done later based on the split intervals. This differs slightly from [Wimmer, 2004], which (apparently) generates the loads and stores local to a block immediately, and leaves those needed in other blocks to a (global) data-flow resolver.

Either *current* gets spilled or *current* is assigned a register. Many new split children may be sorted into *unhandled*, but the allocator always advances position and so will eventually terminate.

Algorithm 7.6 Spill and Allocate a Blocked Register

Input: The current interval *current*, from Algorithm 7.4
Output: A register is spilled and assigned to the *current* interval
  for physical register r in the set of physical registers **do**
    usePos[r] ← blockPos[r] ← maxInt
  end for
  for non-fixed interval i in *active* **do**
    usePos[i.reg] ← min(usePos[i.reg], next usage of i after *current*.firstRange.from)
  end for
  for non-fixed interval i in *inactive* **do**
    if i intersects with *current* **then**
      usePos[i.reg] ← min(usePos[i.reg], next usage of i after *current*.firstRange.from)
    end if
  end for
  for fixed interval i in *active* **do**
    usePos[i] ← blockPos[i] ← 0
  end for
  for fixed interval i in *inactive* **do**
    blockPos[i.reg] ← next intersection of i with *current*
    usePos[i.reg] ← min(usePos[i.reg], blockPos[i.reg])
  end for
  register *reg* ← register with highest usePos
  if usePos[*reg*] < first usage of current **then**
    // all intervals are used before current ⇒ spill current itself
    assign a spill slot on stack frame to *current*
    split *current* at an optimal position before first use position requiring a register
  else if blockPos[*reg*] > *current*.lastRange.to **then**
    // spilling frees reg for all of current
    assign a spill slot on stack frame to child interval for *reg*
    assign register *reg* to interval *current*
    split, assign a stack slot, and spill intersecting active and inactive intervals for *reg*
  else
    // spilling frees reg for first part of current
    assign a spill slot on stack frame to child interval for *reg*
    assign register *reg* to interval *current*
    split current at optimal position before blockPos[*reg*]
    split, assign a stack slot, and spill intersecting active and inactive intervals for *reg*
  end if

Splitting Intervals

In some cases, where an interval cannot be assigned a physical register for its entire lifetime, we will want to split intervals.

Consider the interval for virtual register V33:

V33: [5, 35] [45, 50]

Originally, this interval is modeled by the `NInterval` object in Figure 7.6(a).

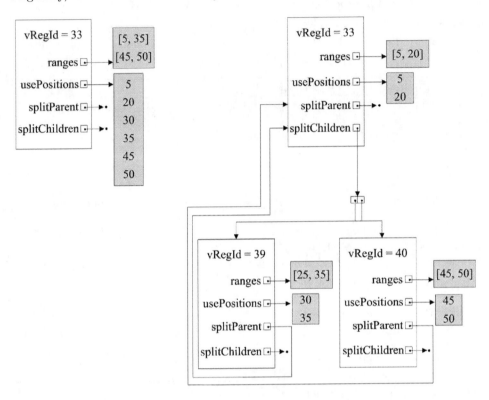

(a) Before splitting   (b) After splitting

FIGURE 7.6 The splitting of interval V33.

Then, we split the interval twice:

1. First, the interval is split at position 20. This creates a new interval V39 (the next available virtual register number) with all ranges and use positions above 20. The original range [5, 35] is split into the range [5, 20] for interval V33 and the range [25, 35] for interval V39.

2. Next, the interval V39 is split at position 40, creating the new interval V40. Although the split occurred at position 40, interval V39 now ends at position 35 and interval V40 starts at position 45 because neither interval is live between 35 and 45.

The result of the two splits is illustrated in Figure 7.6(b). Each interval maintains an array of split children.

Choosing an Optimal Split Position

Once we have burned through all of our registers, we will be forced to split intervals. As we have seen above, we may split current to temporarily assign it a physical register that is attached to an inactive interval. We may split and spill intervals that are active. Usually, we are offered a range of positions where we may split an interval; the position where an interval is spilled or reloaded can be moved to a lower position. Where exactly we choose to split an interval can affect the quality of the code.

Wimmer [Wimmer, 2004] proposes the following three heuristics to apply when choosing a split position:

1. Move the split position out of loops. Code in loops is executed much more often than code outside of loops, so it is best to move register spills and register reloads outside of these loops.

2. Move the split position to block boundaries. Data-flow resolution (in the next section) takes care of spills and reloads at block boundaries automatically.

3. Move split positions to positions that are not multiples of five. These positions are not occupied by normal LIR operations and so are available for spills and reloads.

These rules do not guarantee optimal code but they go a long way toward improving the code quality.

Resolving the Data Flow, Again

The splitting of intervals that occurs as a result of register allocation may cause inconsistencies in the resulting code.

For example, consider the follow of data in an if-then-else statement:

```
if (condition) {
    then part
} else {
    else part
}
subsequent code
```

Consider the case where the register allocator must spill a register (say, for example, $s5) in allocating registers to the LIR code for the *else part*; somewhere in the LIR code for the *else part* is a move instruction (for example, a store) to spill the value of $s5 to a slot on the run-time stack. If the *subsequent code* makes use of the value in $s5, then it will know that the value has been spilled to the stack because the interval will extend at least to this next use; it will use a move to load that value back into a (perhaps different) register.

But what if the flow of control passes through the *then part* code? Again, *subsequent code* will expect the value of $s5 to have been spilled to the stack slot; it will load a value that has not been stored there.

For this reason, an extra store of $s5 must be inserted at the end of the *then part* code so that the (spilled) state of $s5 is consistent in the subsequent code.

Algorithm 7.7 ([Wimmer, 2004] and [Traub et al., 1998]) inserts all moves for spilling, including these necessary extra moves.

Algorithm 7.7 Resolve Data Flow

Input: The version of the control-flow graph g after register allocation (Algorithm 7.4)

Output: The same LIR but with all necessary loads and stores inserted, including those
　required to resolve changes in the data flow

// Resolve local (to a block) data flow

for interval i in g.intervals **do**

　if the interval i is marked for spilling **then**

　　for child interval c in the interval i **do**

　　　if c is the first child **then**

　　　　add a store instruction at the end of i's last range

　　　　add a load instruction at the start of c's first range

　　　else

　　　　add a store instruction at the end of the previous child's last range

　　　　add a load instruction at the start of c's first range

　　　end if

　　end for

　end if

end for

// Resolve global data flow

for block b in g.blocks **do**

　for block s in b.succesors **do**

　　// collect all necessary resolving moves between b and s

　　for operand register r in s.liveIn **do**

　　　interval $parentInterval \leftarrow g$.intervals[$r$]

　　　interval $fromInterval \leftarrow parentInterval$.childAt($b$.lastOp.id)

　　　interval $fromInterval \leftarrow parentInterval$.childAt($s$.firstOp.id)

　　　if $fromInterval \neq toInterval$ **then**

　　　　add a store instruction after the last use position in the from block

　　　　if there is a use in the from block before a define **then**

　　　　　add a load instruction before it in the to block

　　　　end if

　　　end if

　　end for

　end for

end for

The algorithm has two parts. The first part deals with the local basic block, that is, the basic block in which the split occurred. The second part deals with other blocks, adding stores and loads to preserve data flow. Notice the similarity to the algorithm for Phi-resolution when translating HIR to LIR in Chapter 6.

An interval starts off whole, and may consist of multiple ranges with holes between them. Once that interval is split, it is either split at a hole or at a range; in either case, a child is created. The child consists of the remaining range(s). The child's first range's start is guaranteed to be at an index after its parent interval's last range's end. The child is itself an interval with ranges, holes, etc., so it can be split too. The resulting interval however, is considered the second child to the original parent interval and not the child of the child. This keeps interval families only one level deep, allowing easy tracking of which children belong to which parent. Note also that the list of children, if more than one, do not overlap index-wise and are consecutive according to their position in the children array of the parent.

Now let us re-run Algorithm 7.4, but with *three* registers. The first five iterations go exactly as for four registers in our previous example.

1. unhandled: [V32, V33, V34, V35, V36, V37, V38]
   current = V32
   active: []
   inactive: []
   free registers: [$t0, $t1, $t2]

   Interval V32 is allocated physical register $t0

2. unhandled: [V33, V34, V35, V36, V37, V38]
   current = V33
   active: [V32]
   inactive: []
   free registers: [$t1, $t2]

   Interval V33 is allocated physical register $t1

3. unhandled: [V34, V35, V36, V37, V38]
   current = V34
   active: [V32, V33]
   inactive: []
   free registers: [$t2]

   Interval V34 is allocated physical register $t2

4. unhandled: [V35, V36, V37, V38]
   current = V35
   active: [V33, V34]
   inactive: []
   free registers: [$t0]

   Interval V35 is allocated physical register $t0

5. unhandled: [V36, V37, V38]
   current = V36
   active: [V33, V34]
   inactive: []
   free registers: [$t0]

   Interval V36 is allocated physical register $t0

6. But in the sixth iteration, we have used up our three registers ($t1 for V33, $t2 for V34, and $t0 for V36). We decide to spill $t2 because its next use position is the furthest away.[4]

   unhandled: [V37, V38]

---

[4]Actually, a close reading of the instruction at position 30 in the LIR suggests that we might re-use $t0 for V37 because it can be used as the target in the same **ADD** instruction that it is used in as the input V36. We leave this improvement to Algorithm 7.4 as an exercise.

current = V37
active: [V33, V34, V36]
inactive: []
free registers: []

Allocation failed.
Spill $t2 for all of current.

7. unhandled: [V38, V39]
   current = V38
   active: [V33, V37]
   inactive: []
   free registers: [$t0]

   Interval V38 is allocated physical register $t0

8. In the next iteration we spill $t2 again[5].

   unhandled: [V39]
   current = V39
   active: [V33, V37, V38]
   inactive: []
   free registers: []

   Allocation failed.
   Spill $t2 for all of current.

9. unhandled: [V40]
   current = V40
   active: [V33, V39]
   inactive: []
   free registers: [$t0]

   Interval V40 is allocated physical register $t0

Global data-flow resolution requires the insertion of several loads and stores to keep the register values consistent[6].

Local Data-flow Resolution:
Store inserted for V34 at 26 in block [B3]
Load inserted for V39 at 34 in block [B3]
Store inserted for V37 at 31 in block [B3]
Load inserted for V40 at 44 in block [B3]

Global Data-flow Resolution:
Store inserted for V34 at 11 in block [B1]
Load inserted for V39 at 54 in block [B4]

---

[5]The previous footnote applies here also.

[6]As the two previous footnotes observe, we avoid all of these loads and stores if we implement the improvement that distinguishes between reads and writes in an LIR instruction.

Store inserted for V39 at 41 in block [B3]

This gives us the following LIR.

```
B0


B1
0:   LDC [1] $t0
5:   MOVE $a0 $t1
10:  MOVE $t0 $t2
11:  STORE $t2 [stack:0]


B2
15:  LDC [0] $t0
20:  BRANCH [LE] $t1 $t0 B4


B3
25:  LDC [-1] $t0
26:  STORE $t2 [stack:0]
30:  ADD $t1 $t0 $t2
31:  STORE $t2 [stack:1]
34:  LOAD [stack:0] $t2
35:  MUL $t2 $t1 $t0
40:  MOVE $t0 $t2
41:  STORE $t2 [stack:0]
44:  LOAD [stack:1] $t0
45:  MOVE $t0 $t1
50:  BRANCH B2


B4
54:  LOAD [stack:0] $t2
55:  MOVE $t2 $v0
60:  RETURN $v0
```

Physical Register Assignment in the LIR

The final step in Wimmer's strategy replaces the virtual registers in the LIR with assigned physical registers. Algorithm 7.8 [Wimmer, 2004] does this register assignment.

Algorithm 7.8 Assign Physical Register Numbers

Input: The version of the control-flow graph g after resolution (Algorithm 7.7)
Output: The same LIR but with virtual registers replaced by physical registers
  for block b in g.blocks **do**
    for instruction i in b.instructions **do**
      for virtual register r in i.operands **do**
        // determine new operand
        physical register $p \leftarrow$ intervals$[r]$.childAt(i.id).assignedReg
        replace r with p in i
      end for
      if i is a move where the target is the same as the source **then**
        b.instructions.remove(i)
      end if
    end for
  end for

Now the LIR refers only to physical registers and is ready for translation to SPIM.

Some Improvements to the LIR

An examination of the LIR after register allocation (using any global allocation strategy) makes it clear that many (often unnecessary) moves are generated.

Because we have not made fixed registers, for example, $a0 to $a3 and $v0 (and $v1), available for allocation, there is much movement among these and virtual registers in the original LIR; this leads to many meaningless moves from one register to another once physical registers are allocated. Bringing these physical registers (modeled as fixed intervals) into play should eliminate our overreliance on other general-purpose registers and eliminate many moves among them.

There is much room for spill optimization. When a variable is defined once and then used several times, it needs to be spilled just once even if the interval modeling the virtual register holding the value is split several times. A dirty bit may be used to keep track of definitions versus uses. Certain arguments to methods, for example, $a0 when it contains the address of an object for `this`, needs to be spilled just once also.

If all of a block's predecessors end with the same sequence of move instructions, they may be moved to the start of the block itself. Likewise, if all of a block's successors start with the same sequence of move instructions, they can be moved to the end of the (predecessor) block. Such moves are most likely to be introduced either by Phi-resolution (Chapter 6) or by data-flow resolution. Wimmer makes the point that this is best done after virtual registers have been replaced by physical registers because this allows us to combine moves involving like physical registers even if they originated from differing virtual registers.

By examining the output of the register allocation process, one will find many places for improvement. It is best to formulate algorithms that are as general as possible in making these improvements.

What is in the Code Tree; What is Left Undone

Much of Wimmer's strategy, recounted above, is implemented in our code. The following issues are not addressed and thus are left as exercises.

- We do not deal with physical registers in the register allocation process. Although we make use of particular fixed registers, for example, $a0 to $a3 for holding arguments passed to methods and $v0 for holding return values, we do not otherwise make these available for allocation to other computations. Moreover, when we allocate $t0 to $t9 and $s0 to $s7, we assume they are callee-saved. The Wimmer algorithms treat them as caller-saved and depend on the (short) intervals at calls to insure they are spilled before the calls, and reloaded after the calls; otherwise, they are treated like any other general-purpose registers and are available to the allocator.

- Our code does not use heuristics to find the optimal split position.

- Our code does not optimize spills.

- Our code does not move sequences of like moves between successors and predecessors.

7.4.3　Register Allocation by Graph Coloring

Introduction to Graph Coloring Register Allocation

In graph coloring register allocation, we start with an interference graph built from the liveness intervals. An *interference graph* consists of a set of nodes, one for each virtual register or liveness interval, and a set of edges. There is an edge between two nodes if the corresponding intervals interfere, that is, if they are live at the same time.

For example, reconsider the liveness intervals for the virtual registers V32 to V38 of `Factorial.computeIter()`, originally in Figure 7.2 but repeated here as Figure 7.7.

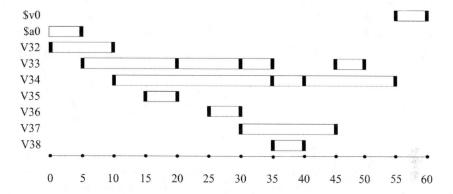

FIGURE 7.7 Liveness intervals for `Factorial.computeIter()`, yet again.

Figure 7.8 shows the corresponding interference graph.

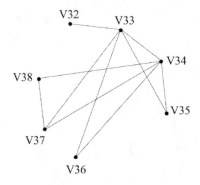

FIGURE 7.8 Interference graph for intervals for `Factorial.computeIter()`.

We need not draw an arc from the node for a virtual register we are using to the node for the virtual register we are defining. For example, at position 10, the instruction

```
10: MOVE [V32|I] [V34|I]
```

uses V32 to define virtual register V34; so the interval for V32 ends where the interval for V34 begins. So we draw no edge from V32 to V34.

Register allocation then becomes coloring the graph but using physical registers as the colors. The question becomes: can the graph be "colored" such that no two adjacent nodes (that is no two intervals that are live at the same time) share the same physical register? John Cocke was the first person to see that register allocation could be modeled as graph coloring [Allen and Kennedy, 2002]. Gregory Chaitin and colleagues ([Chaitin et al., 1981] and [Chaitin, 1982]) implemented such an allocator at IBM in 1981.

We say that a graph has an R-coloring if it can be colored using R distinct colors, or in our case R distinct physical registers. To exhaustively find such an R-coloring for R $\geq$ 2 has long been known to be NP-complete. But there are two heuristics available to us for simplifying the graph.

1. The first, called the *degree* < *R* rule, derives from the fact that a graph with a node of degree <*R* (that is a node with <*R* adjacent nodes) is *R*-colorable if and only if the graph with that node removed is *R*-colorable. We may use this rule to prune the graph, removing one node of degree <*R* at a time and pushing it onto a stack; removing nodes from the graph removes corresponding edges, potentially creating more nodes with degree <*R*. We continue removing nodes until either all nodes have been pruned or until we reach a state where all remaining nodes have degrees ≥ *R*.

2. The second is called the *optimistic heuristic* and allows us to continue pruning nodes even with degree ≥ *R*. We use a heuristic function spillCost() to find a node having the smallest cost of spilling its associated virtual register; we mark that register for possible spilling and remove the node (and its edges) and push it onto the stack in the hope that we will not really have to spill it later. (Just because a node has degree ≥ *R* does not mean the nodes need different colors.)

Let us attempt to prune the interference graph in Figure 7.8 where $R = 3$, meaning we have three physical registers to work with. In this case, we need to invoke just the first degree < *R* rule. The steps to pruning the graph are illustrated in Figure 7.9. Figure 7.9 pretty much speaks for itself. We start with V32, which (in Figure 7.9(a)) has just one adjacent interval, remove it, and push it onto the stack to give us the graph in Figure 7.9(b). We continue in this way until all nodes have been removed and pushed onto the stack. Removing nodes with degree < 3 causes edges to be removed and so more nodes end up with degree < 3.

We may then pop the virtual registers off the list, one at a time, and try to assign physical register numbers (r1, r2, or r3) to each in such a way that adjacent virtual registers (in the graph) are never assigned the same physical register. A possible assignment is

V38	r1
V37	r2
V34	r3
V33	r1
V36	r2
V35	r2
V32	r2

Imposing this mapping onto our LIR for `Factorial.computeIter()` gives us

```
B0


B1
 0: LDC [1] r2
 5: MOVE $a0 r1
10: MOVE r2 r3


B2
15: LDC [0] r2
20: BRANCH [LE] r1 r2 B4


B3
25: LDC [-1] r2
30: ADD r1 r2 r2
35: MUL r3 r1 r1
40: MOVE r1 r3
45: MOVE r2 r1
50: BRANCH B2


B4
55: MOVE r3 $v0
60: RETURN $v0
```

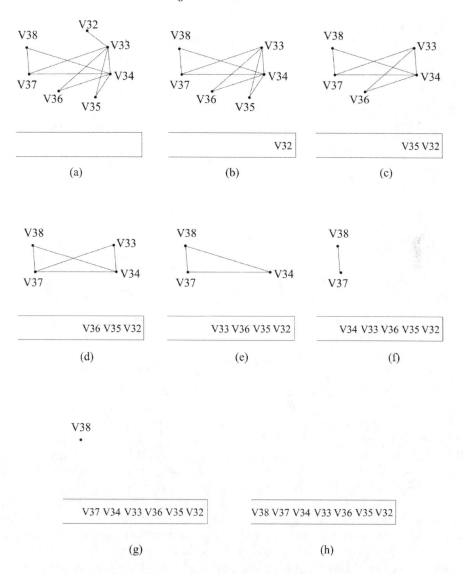

FIGURE 7.9 Pruning an interference graph.

Graph Coloring Register Allocation Algorithm

Algorithm 7.9 performs graph coloring register allocation based on that of ([Muchnick, 1997]). The algorithm attempts to coalesce virtual registers where there is a move from one to the next and the two registers' intervals do not conflict. Also, once spill code is generated, it repeats the register allocation process. It does this repeatedly until it succeeds in assigning physical registers to virtual registers without introducing additional spills.

Algorithm 7.9 Graph Coloring Register Allocation

Input: The control-flow graph g for a method with LIR that makes use of virtual registers
Output: The same g but with virtual registers replaced by physical registers

> $registersAssignedSuccessfully \leftarrow$ `false`
> **repeat**
> > **repeat**
> > > buildIntervals()
> > > buildInterferenceGraph()
> > **until** coalesceRegistersSuccessful()
> > buildAdjacencyLists()
> > computeSpillCosts()
> > pruneGraph()
> > $registersAssignedSuccessfully \leftarrow$ assignRegistersSuccessful()
> > **if** $registersAssignedSuccessfully$ **then**
> > > generateSpillCode()
> > **end if**
> **until** $registersAssignedSuccessfully$

This algorithm looks like it could go on for a long time, but it usually terminates after at most two or three iterations.

Register Coalescing

Coalescing registers reduces both the number of virtual registers and the number of moves. The method coalesceRegistersSuccessful() returns `true` if it is able to coalesce two registers and `false` otherwise; this Boolean result insures that any register coalescing is followed by a rebuilding of the intervals and the interference graph.

Register coalescing also makes for longer intervals and could render the graph uncolorable. [Briggs et al., 1994] and [George and Appel, 1996] propose more conservative conditions under which we may safely coalesce registers, without making the graph uncolorable. Briggs proposes that two nodes in the graph may be coalesced if the resulting node will have fewer than R neighbors of *significant degree* (having R or more edges). George and Appel propose that two nodes may be coalesced if for every neighbor t of one of those nodes, either t already interferes with the other node (so no additional edges are added) or t is of insignificant degree. Appel [Appel, 2002] points out that these rules may prevent the safe removal of certain moves but that extra moves are less costly than spills.

Representing the Interference Graph

The representation of the interference graph is driven by the kind of queries one wants to make of it. Our algorithm wants to ask two things:

1. Whether or not one node is adjacent to another. We want to know this when we coalesce registers and when we want to assign registers.

2. How many nodes, and which nodes, are adjacent to a given node.

The first is best answered by a Boolean adjacency matrix: a matrix with a row and column for each register; adj$[i, j]$ is true if nodes i and j are adjacent and false otherwise. A lower-triangular matrix is sufficient because there is an ordering on the nodes. For example, our interference graph in Figure 7.8 might be captured using the following lower-triangular matrix, where T is true and F is false.

	V32	V33	V34	V35	V36	V37
V33	T					
V34	F	T				
V35	F	T	T			
V36	F	T	F	F		
V37	F	T	T	F	F	
V38	F	F	T	F	F	T

Method buildIntervals() might build such an adjacency matrix.

The second is best answered by a couple of vectors, each indexed by the register number: a vector of integers recording the numbers of neighbors, and a vector of lists of neighboring nodes. These vectors are easily computed from the adjacency matrix. Method buildAdjacencyLists() might build these vectors.

Determining Spill Cost

During the pruning process, we may reach a state where only nodes with degree $\geq R$ remain in the graph. In this state, our algorithm must choose a node with the smallest spill cost.

The spill cost of a register certainly depends on the loop depths of the positions where the register must be stored to or loaded from memory; [Muchnick, 1997] suggests summing up the uses and definitions, using a factor of 10^{depth} for taking loop depth into account. Muchnick also suggests recomputing the value for a register when it is cheaper than spilling and reloading that same value.

Pre-Coloring Nodes

Our example does not address fixed intervals. [Muchnick, 1997] suggests *pre-coloring* nodes in the interference graph that correspond to these pre-chosen physical registers, for example, $a0–$a3, $v1, and $v2, and bring them into the allocation process so that they may be used for other purposes when they are not needed for their specialized functions. Register allocation designers take an aggressive approach to register allocation, where all registers are in play.

Comparison to Linear Scan Register Allocation

The complexity of linear scan register allocation is roughly linear with respect to n—the number of virtual registers, where the complexity of graph coloring register allocation is roughly n^2. At the same time, graph coloring register allocation is claimed to do a much better job at allocating physical registers, producing faster code. For this reason, linear scan is often sold as a solution to many just-in-time compiling strategies where compilation occurs at run-time, and graph coloring is sold as a strategy where code quality is more important than compile time. For example, in the Oracle HotSpot compiler, a client version uses linear scan and a server version uses graph coloring [Wimmer, 2004].

Even so, there has been a kind of duel between advocates for the two strategies, where linear scan register allocation developers are improving the quality of register assignment, for example, [Traub et al., 1998], and graph coloring register allocation developers are improving the speed of register allocation, for example, [Cooper and Dasgupta, 2006].

7.5   Further Readings

The linear scan algorithm discussed in this chapter relies extensively on [Wimmer, 2004] and [Traub et al., 1998]. [Wimmer and Franz, 2010] describe a version of the linear scan that operates on code adhering to SSA.

Graph coloring register allocation was first introduced in [Chaitin et al., 1981] and [Chaitin, 1982]. Two texts that describe graph coloring register allocation in great detail are [Appel, 2002] and [Muchnick, 1997]. The performance of graph coloring register allocation is addressed in [Cooper and Dasgupta, 2006].

7.6   Exercises

Exercise 7.1. Implement a local register allocation strategy that works with liveness intervals computed only for local individual basic blocks.

Exercise 7.2. Implement a local register allocation strategy that works with individual basic blocks but that gives preference to blocks that are most nested within loops.

Exercise 7.3. Modify the linear scan algorithm described in Section 7.3 so that it deals effectively with virtual registers that are defined but are never used.

Exercise 7.4. In a footnote to the 6th iteration in the second example of running Algorithm 7.4, we observe that we need not spill \$t2. Rather, in the instruction

```
30: ADD $t1 $t0 $t2
```

we can re-use \$t0 in place of \$t2, as in

```
30: ADD $t1 $t0 $t0
```

Implement this modification using Algorithm 7.4. Note that this requires that intervals keep track of the instruction that uses them, and distinguishes between reads and writes (two reads cannot share the same physical register); this can be done in Algorithm 7.3, which builds the intervals.

Exercise 7.5. Modify the code to deal with physical registers in the register allocation process. Treat all registers as caller-saved registers and bring registers such as \$a0–\$a3 for holding arguments passed to methods and \$v0 for holding return values into the greater allocation process.

Exercise 7.6. Modify the code to use the heuristics in Section 7.4.2 to find the optimal split position.

Exercise 7.7. Modify the code to optimize spills as discussed in Section 7.4.2.

Exercise 7.8. Modify the code to move sequences of like moves between successors and predecessors, as described in Section 7.4.2.

Exercise 7.9. Implement the linear scan register allocation algorithm that operates on LIR in SSA form and is described in [Wimmer and Franz, 2010].

Exercise 7.10. Repeat the example register allocation performed by hand in Section 7.4.3 but where $R = 2$; that is, where we have two physical registers to work with.

Exercise 7.11. Implement Algorithm 7.9 for performing graph coloring register allocation on our LIR.

Exercise 7.12. Modify the coalesceRegistersSuccessful() method to make use of the more conservative [Briggs et al., 1994] condition for identifying registers that may be coalesced.

Exercise 7.13. Modify the coalesceRegistersSuccessful() method to make use of the more conservative [George and Appel, 1996] condition for identifying registers that may be coalesced.

Exercise 7.14. Compare the run-time speed of your linear scan register allocator and your graph coloring register allocator. How does the code produced compare?

Exercise 7.15. Try pre-coloring fixed intervals, bringing more physical registers into your graph coloring register allocation program.

Chapter 8

Celebrity Compilers

8.1 Introduction

Here we survey some of the popular Java (or Java-like) compilers. For each of these, we discuss issues or features that are peculiar to the compiler in question. The compilers we discuss are the following:

- Oracle's Java HotSpot™ compiler

- IBM's Eclipse compiler for Java

- The GNU Java compiler

- Microsoft's C# compiler

This chapter will only give a taste of these compilers. The reader may wish to consult some of the recommended readings in the Further Readings section (Section 8.6) at the end of this chapter for a deeper understanding of them.

8.2 Java HotSpot Compiler

The original Java Virtual Machine (JVM) was developed by Sun Microsystems. The Oracle Corporation, which acquired Sun in 2010, now maintains it. What makes Oracle's JVM compiler [Oracle, 2010] special, aside from the fact that it is *the* original Java compiler, is its implementation of its just-in-time (JIT) compiler.

A typical JIT compiler will translate a method's (JVM) byte code into native code the first time the method is invoked and then cache the native code. In this manner, only native code is executed. When a method is invoked repeatedly, its native code may be found in the cache. In programs with large loops or recursive methods, JIT compilation drastically reduces the execution time [Kazi et al., 2000].

An obvious drawback of JIT compilation is the initial run-time delay in the execution of a program, which is caused by both the compilation of methods the first time they are encountered and the application of platform-dependent optimizations. JIT compilers are able to perform optimizations that static compilers[1] cannot, because they have access to run-time information, such as input parameters, control flow, and target machine specifics (for example, the compiler knows what processor the program is running on and can tune the generated code accordingly). Other than that, JIT compilers do not spend as much

---

[1]Static compilation is also known as ahead-of-time compilation, which is discussed in Section 8.4.2.

time on optimizing as the static compilers do because the time spent on optimizations adds to the execution time. Also, because classes can be dynamically invoked or each method is compiled on demand, it is difficult for JIT compilers to perform global optimizations.

When compared to interpretation, JIT compilation has another downside: methods are entirely compiled before they are executed. If only a small part of a large method is executed, JIT compilation may be time consuming [Kazi et al., 2000]. The memory usage is increased as well because native code requires more space than the more compact JVM byte code.

The Oracle JVM uses a JIT compilation regimen called HotSpot. Knuth's finding, which states that most programs spend the majority of time executing a small fraction of the code [Knuth, 1971a], lays the foundation for a technique called *adaptive optimization*. Adaptive optimization was pioneered by the Jikes Research Virtual Machine[2] at IBM Watson Research Center. Animorphic, a small start-up company, which was acquired by Sun Microsystems in 1997, adopted the technique and developed the Java HotSpot VM.

The Java HotSpot VM has both an interpreter and a byte code-to-native machine code compiler. Instead of compiling every method it encounters, the VM first runs a profiling interpreter to quickly gather run-time information, detect the critical "hot spots" in the program that are executed most often, and collect information about the program behavior so it may use it to optimize the generated native code in later stages of program execution. The identified hot spots are then compiled into optimized native code, while infrequently executed parts of the program continue to be interpreted. As a result, more time can be spent on optimizing those performance-critical hot spots, and the optimizations are smarter than static compiler optimizations because of all the information gathered during interpretation. Hot spot monitoring continues during run-time to adapt the performance according to program behavior.

There are two versions of the Java HotSpot VM: the Client VM [Kotzmann et al., 2008] and the Server VM. These two VMs are identical in their run-time, interpreter, memory model, and garbage collector components; they differ only in their byte code-to-native machine code compilers.

The client compiler, is simpler and focuses on extracting as much information as possible from the byte code, like locality information and an initial control flow graph, to reduce the compilation time. It aims to reduce the initial start-up time and memory footprint on users' computers.

The server compiler on the other hand, is an advanced dynamic optimizing compiler focusing on the run-time execution speed. It uses an advanced SSA-based IR[3] for optimizations. The optimizer performs classic optimizations like dead code elimination, loop-invariant hoisting, common sub-expression elimination and constant propagation, as well as Java-specific optimizations such as null-check and range check elimination[4]. The register allocator is a global graph coloring allocator[5] [Oracle, 2010]. However, the most important optimizations performed by the Java HotSpot server compiler are *method inlining* and *dynamic de-optimization*.

Static compilers cannot take full advantage of method inlining because they cannot know if a method is overridden in a sub-class. A static compiler can conservatively inline static, final, and private methods because such methods cannot be overridden, but there is no such guarantee for public and protected methods. Moreover, classes can be loaded dynamically during run-time and a static compiler cannot deal with such run-time changes in the program structure [Wilson and Kesselman, 2000].

While method inlining can reduce both compilation time and execution time, it will

---

[2]Initially called the Jalapeño project.

[3]SSA stands for *static single assignment* and IR stands for *Intermediate Representation*. See Chapter 6.

[4]See Chapter 6 for various optimizations techniques.

[5]See Chapter 7 for register allocation techniques, including graph-coloring allocation.

increase the total size of the produced native machine code. That is why it is important to apply this optimization selectively rather than blindly inlining every method call. The HotSpot compiler applies method inlining in the detected program hot spots. HotSpot has the freedom of inlining any method because it can always undo an inlining if the method's inheritance structure changes during run time.

Undoing the inlining optimizations is called dynamic de-optimization. Suppose the compiler comes across to a virtual function call as

```
MyProgram.foo();
```

There are two possibilities here: the compiler can run the `foo()` function implemented in `MyProgram`, or a child class's implementation of `foo()`. If `MyProgram` is defined as a final class, or `foo()` in `MyProgram` is defined as a final method, the compiler knows that `MyProgram`'s `foo()` should be executed. What if that is not the case? Well, then the compiler has to make a guess, and pick whichever seems right to it at the moment. If it is lucky, there is no need to go back; but that decision may turn out to be wrong or invalidated by a class loaded dynamically that extends `MyProgram` during run time [Goetz, 2004]. Every time a class is dynamically loaded, the HotSpot VM checks whether the interclass dependencies of inlined methods have been altered. If there are any altered dependencies, the HotSpot VM can dynamically de-optimize the affected inlined code, switch to interpretation mode and maybe re-optimize later during run time as a result of new class dependencies. Dynamic de-optimization allows HotSpot VM to perform inlining aggressively with the confidence that possible wrong decisions can always be backed out.

If we look at the evolution of Oracle's JVM since the beginning, we see that it has matured in stages. Early VMs always interpreted byte code, which could result in 10 to 20 times slower performance compared to C. Then, a JIT compiler was introduced for the first time with JDK 1.2 as an add-on, which was originally developed by Animorphic Systems. With JDK 1.3, HotSpot became the default compiler. Compared to the initial release, JDK 1.3 showed a 40% improvement in start-up time and a 25% smaller RAM footprint [Shudo, 2004]. Later releases further improved HotSpot even more [Kotzmann et al., 2008]. Thus, the performance of a Java program written in Java code today approaches that of an equivalent C program.

As an example of improvements made to HotSpot, consider the history of the JVM's handling of the transfer of control between byte code and native code. In HotSpot's initial version, two counters were associated with each method: a method-entry counter and a backward-branch counter. The method entry counter was incremented every time the method was called while the backward-branch counter was incremented every time a backward branch was taken (for example, imagine a for-loop; the closing brace of the for-loop's body would correspond to a backward branch); and these counters were combined into a single counter value with some additional heuristics for the method. Once the combined counters hit a threshold (10,000 in HotSpot version 1.0) during interpretation, the method would be considered "hot" and it then would be compiled [Paleczny et al., 2001]. However, the HotSpot compiler could only execute the compiled code for this method when the method is called the next time.

In other words, if the HotSpot compiler detected a big, computationally intensive method as a hot spot and compiled it into native code, say, because of a loop at the beginning of the method, it would not be able to use the compiled version of the method until the next time it was invoked. It was possible that heavy-weight methods were compiled during run-time but the compiled native code would never be used [Goetz, 2004]. Programmers used to "warm up" their programs for HotSpot by adding redundant and very long running loops in order to provoke the HotSpot to produce native code at specific points in their program and obtain better performance at the end.

On-stack replacement (OSR) is used to overcome this problem. It is the opposite of dynamic de-optimization, where the JVM can switch from interpretation mode to compilation mode or swap a better version of compiled code in the middle of a loop, without waiting for the enclosing method to be exited and re-entered [Fink and Qian, 2003]. When the interpreter sees a method looping, it will invoke HotSpot to compile this method. While the interpreter is still running the method, HotSpot will compile the method aside, with an entry point at the target of the backward branch. Then the run-time will replace the interpreted stack activation frame with the compiled frame, so that execution continues with the new frame from that point [Paleczny et al., 2001]. HotSpot can apply aggressive specializations based on the current conditions at any time and re-compile the code that is running if those conditions change during run-time by means of OSR.

Clever techniques like aggressive inlining, dynamic de-optimization, OSR, and many others [Paleczny et al., 2001, Oracle, 2010] allow HotSpot to produce better code when compared to traditional JIT compilation.

8.3   Eclipse Compiler for Java (ECJ)

Eclipse is an open-source development platform comprised of extensible frameworks, tools, and run-times, originally developed by IBM. The Eclipse platform is structured as a set of sub-systems, which are implemented as plug-ins. The Eclipse Java Development Tools (JDT) provide the plug-ins for a Java integrated development environment with its own Java compiler called the *Eclipse Compiler for Java* (ECJ), which is often compared with Oracle's Java compiler.

The ECJ is an incremental compiler, meaning that after the initial compilation of a project, it compiles only the modified (or newly added) file and its dependent files next time, instead of re-compiling the whole project again.

Compilation in Eclipse can be invoked in three different ways [Arthorne, 2004]:

- **Full Compilation** requires that all source files in the project be compiled. This is either performed as the initial compilation, or as a result of a *clean* operation that is performed on the project, which deletes all the .class files and problem markers, thus forcing a re-compilation of an entire project.

- **Incremental Compilation** compiles only the changed files (by visiting the complete resource delta tree) and the files that are affected by the change, for example, classes that implement an interface, the classes calling methods of the changed classes, etc.

- **Auto Compilation** is essentially the same as incremental compilation. Here, incremental compilation is triggered automatically because a change in source code has occurred.

Eclipse's incremental compiler for Java uses a *last build state* to do an optimized build based on the changes in the project since the last compilation. The changes since the last compilation are captured as a *resource delta tree*, which is a hierarchical description of what has changed between two discrete points in the lifetime of the Eclipse workspace. The next time ECJ is called, it uses this resource delta tree to determine the source files that need to be re-compiled because they were changed, removed, or added.

In addition to the resource delta tree, the compiler keeps a *dependency graph* in memory as part of its *last built state*. The dependency graph includes all the references from each type to other types. It is created from scratch in the initial compilation, and updated incrementally with new information on each subsequent compilation. Using this graph, the compiler can decide if any structural changes[6] occur as a consequence of changing, adding, or deleting a file. The computation of these structural changes involves the set of source files that might compile differently as a consequence. The compiler deletes obsolete class files and associated Java problem markers that were added to the source files due to previous compilation errors, and compiles only the computed subset of source files. The dependency graph is saved between sessions with workspace saves. The dependency graph does not have to be re-generated, and the compiler avoids full compilation every time the project is opened. Of course, the last built state is updated with the new reference information for the compiled type, and new problem markers are generated for each compiled type if it has any compilation problems, as the final steps in compilation [Rivieres and Beaton, 2006]. Figure 8.1 shows these steps.

Incremental compilation is very effective, especially in big projects with hundreds of source files, as most of the source files will remain unchanged between two consecutive compilations. Frequent compilations on hundreds or thousands of source files can be performed without delay.

Many times when a Java file is changed, it does not result in any structural changes; there is only a single file to be compiled. Even when structural changes occur and all referencing types need to be re-compiled, those secondary types will almost never have structural changes themselves, so the compilation will be completed in, at most, a couple of iterations. Of course, one cannot claim that there will never be significant structural changes that may cause many files to be re-compiled. ECJ considers this trade-off worth the risk, assuming that the compiler runs very fast for the most common cases; rare occasions of longer delays are acceptable to users.

Incremental compilers are usually accused of not optimizing enough because of the locality and relatively small amount of the code that has been changed since the last build. However, this is not a real problem, as all the heavy-weight optimizations are performed during run-time. Java compilers are not expected to optimize heavily when compiling from source code to bytecode. ECJ performs light optimizations like discarding unused local variables from the generated bytecode and inlining to load the bytecode faster into the VM (since the verification process is then much simpler even though the code size to load increases), but the real run-time performance difference comes from the VM that is used (for example, IBM's JVM, Oracle's JVM, etc.). And if one decides to use, say, Oracle's JVM, choosing the latest version would be a good practice because HotSpot is being improved in each new version.

Perhaps one of the most interesting features of ECJ is its ability to run and debug code that contains errors. Consider the following naïve test:

```java
public class Foo {
    public void foo() {
        System.println("I feel like I forgot something...");
    }

    public static void main(String[] args) {
        System.out.println("It really works!");
    }
}
```

The ECJ will flag the erroneous use of `println()` within `foo()`, marking the line in the editor and underlining the `println`; however, it will still generate the bytecode. And when

---

[6]Structural changes are the changes that can affect the compilation of a referencing type, for example, added or removed methods, fields or types, or changed method signatures.

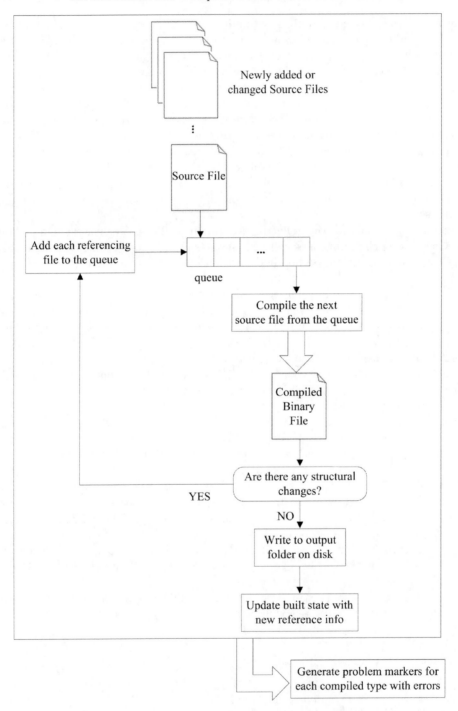

FIGURE 8.1 Steps to ECJ incremental compilation.

the generated bytecode is executed, it will run properly because foo() is never called in the program. If the control reaches the error, the run-time will then throw an exception. This feature is useful when testing individual complete pieces of a project that contains incomplete pieces.

A lesser-known feature of ECJ is its support for scrapbook pages. A scrapbook page allows programmers to test small pieces of Java expressions without creating a surrounding Java class or method for the specific code snippet.

This can be created by either simply creating a file with .jpage extension, or using the *scrapbook page* wizard in Eclipse [Eclipse, 2011]. One can write a single line of code as:

```
new java.util.Date()
```

inside the new scrapbook page, and then evaluate it. Notice that, a semicolon at the end of line is omitted, which would normally cause a compile-time error in a java file; yet ECJ understands what is intended here and evaluates the expression. The scrapbook editor also supports code assist, showing the possible completions when writing a qualified name.

Another convenience is the ability to *Inspect* a highlighted expression, which will open a pop-up window inside the editor and show all the debug information on the highlighted object without having to switch to the Debug perspective.

There is also the ability to *Execute* scrapbook expressions on the JVM. Suppose a scrapbook page has the following lines:

```
java.util.Date now = new java.util.Date();
System.out.println("Current date and time: " + now);
```

Executing these two lines in scrapbook produce the requisite output:

```
Current date and time: Mon Sep 26 23:33:03 EDT 2011
```

The scrapbook provides that immediate feedback that LISP users find so useful.

The ECJ is also used in Apache Tomcat for compiling JSP pages and the Java IDE called IntelliJ IDEA. In addition, GCJ, discussed in the next section, uses ECJ as its front-end compiler (as of GCJ 4.3).

8.4   GNU Java Compiler (GCJ)

8.4.1   Overview

The GNU Java Compiler (GCJ) [GNU, 2012] is one of the compilers provided in the GNU Compiler Collection (GCC). Although GCC started as a C compiler (aka the GNU C Compiler), compilers for other languages including Java were soon added to the collection. Today, GCC is the standard compiler for many Unix-like operating systems.

GCJ may directly produce either native machine code or JVM bytecode class files. GCJ can deal with both Java source files or zip/jar archives, and can package Java applications into both .jar files and Windows executables (.exe files), as well as producing class files.

GCJ uses the Eclipse Java Compiler (ECJ) as its front end. Programs that are compiled into bytecode with ECJ are linked with the GCJ run-time called libgcj. Libgcj provides typical run-time system components such as a class loader, core class libraries, a garbage collector[7], and a bytecode interpreter. Libgcj can also interpret source code to machine code directly. Figure 8.2 shows possible routes a Java source code can take when fed to GCJ.

---

[7]Hans Boehm's conservative garbage collector, which is known as Boehm GC, is used. Detailed information can be found at `http://www.hpl.hp.com/personal/Hans_Boehm/gc/`.

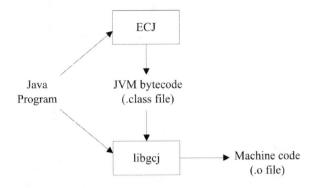

FIGURE 8.2 Possible paths a Java program takes in GCJ.

Compiled Java programs are next fed into the GCJ back end to be optimized and executed. The GCJ back end and the philosophy behind it are discussed in the next section.

8.4.2 GCJ in Detail

In addition to providing a Java implementation under the GPL license, a principal motivation for GCJ was to compete with JIT performance. As we saw in Section 8.2, JIT compilation suffers from issues with start-up overhead and the overhead in detecting hot spots.

GCJ took a "radically traditional" approach to overcome these problems and followed the classical ahead-of-time (AOT) compilation, where bytecode programs are compiled into a system-dependent binary *before* executing the programs.

GCJ developers treated Java like another form of C++. Java programs are translated to the same abstract syntax tree used for other GCC source languages. Thus, all existing optimizations, like common sub-expression elimination, strength reduction, loop optimization and register allocation, that were available for GNU tools could be used in GCJ [Bothner, 2003].

As a result, programs compiled directly to native code using the GCC back ends run at full native speed with low start-up overhead and modest memory usage. This makes GCJ a suitable compiler for embedded systems.

On the other hand, after start-up, Java programs compiled by GCJ do not necessarily always run faster than a modern JIT compiler like HotSpot. There are various reasons for this:

- First of all, some optimizations possible in JIT compilation, such as run-time profile-guided optimizations and virtual function inlining, are not possible in GCJ.

- Programs that allocate many small, short-lived objects can cause the heap to fill quickly and are garbage collected often, which in turn slows down the execution.

- GCJ has sub-optimal run-time checking code, and the compiler is not so smart about automatically removing array checks. A (dangerous) hack for this can be compiling with GCJ's --no-bounds-check option.

- On many platforms, dynamic (PIC[8]) function calls are more expensive than static ones. In particular, the interaction with Boehm BC seems to incur extra overhead

---

[8]Position-independent code (PIC) is machine instruction code that can be copied to an arbitrary location in memory and work without requiring any address relocation. PIC is commonly used for shared libraries.

when shared libraries are used. In such cases, static linking[9] can be used to overcome this problem.

- GCJ does not behave well with threads. Multi-threaded processes are less efficient in GCJ.

A useful feature of GCJ is the flexibility it offers in choosing what optimizations to perform. GCJ categorizes optimizations by level and permits users to choose the level of optimizations they want to apply on the programs. There are five optimization levels, each specified by the -Ox (x representing the optimization level) option [GNU, 2011]. Overall, GCJ is a good choice for embedded systems, and is being improved continually.

8.5 Microsoft C# Compiler for .NET Framework

8.5.1 Introduction to .NET Framework

The .NET Framework [Microsoft, 2012] is an integral Microsoft Windows component for building and running applications. It is a multi-language development platform that comprises a virtual machine called the *Common Language Runtime* (CLR) [Box and Sells, 2002], and a library called the *Framework Class Library* (FCL).

The CLR is the heart of the .NET framework, and it consists of several compnents, including a class loader, metadata engine, garbage collector, debugging services, and security services. The CLR is Microsoft's implementation for the Common Language Infrastructure (CLI) Standard, which defines an execution environment that allows multiple high-level languages to be used on different computer platforms[10] without being rewritten for specific architectures [Miller and Ragsdale, 2004]. It takes care of compiling intermediate language programs to native machine code, memory management (for example, allocation of objects and buffers), thread execution, code execution and exception handling, code safety[11] verification (for example, array bounds and index checking), cross-language integration (that is, following certain rules to ensure the interoperability between languages), garbage collection, and controls the interaction with the OS.

FCL is the other core component of the .NET Framework; it is a library of classes, interfaces, and value types. It is structured as a hierarchical tree and divided into logical groupings of types called *namespaces* (as in the Java API) according to functionality. System is the root for all types in the .NET Framework namespace hierarchy. The class libraries provided with the .NET Framework can be accessed using any language that targets the CLR. One can combine the object-oriented collection of reusable types and common functions from the FCL with her own code, which can be written in any of the .NET-compatible languages. The .NET-compatible languages can altogether be called *.NET languages*. C#, C++, Perl, Python, Ruby, Scheme, Visual Basic, Visual J++, Phalanger[12], and FORTRAN.NET[13] are a few examples of .NET languages with CLR-targeting compilers [Hamilton, 2003].

---

[9]Static linking refers to resolving the calls to libraries in a caller and copying them into the target application program at compile time, rather than loading them in at run-time.

[10]Such platforms would be the combination of any architecture with any version of the Microsoft Windows operating system running atop.

[11]Comprising type, memory, and control-flow safety, meaning that the intermediate code is correctly generated and only accesses the memory locations it is authorized to access, and isolating objects from each other to protect them from accidental or malicious corruption.

[12]An implementation of PHP with extensions for ASP.NET.

[13]Fortran compiling to .NET.

Programs written in .NET languages are first compiled into a stack-based bytecode format named as *Common Intermediate Language* (CIL)[14]. Then this CIL code is fed into the CLR, with the associated *metadata*. Metadata in .NET is binary information that describes the classes and methods (with their declarations and implementations), data types, references to other types and members, and attributes (for example, versioning information, security permissions, etc.) of the program. Metadata is stored in a file called the *Manifest*. CIL code and the Manifest are wrapped in a Portable Executable[15] file called an *Assembly*. An Assembly can consist of one or more program files and a manifest, possibly along with reference files, such as bitmap files that are called within the program. Assemblies can be either *Process Assemblies* (.EXE files) or *Library Assemblies* (.DLL files). (See Figure 8.3)

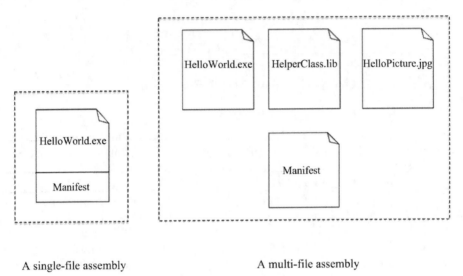

A single-file assembly   A multi-file assembly

FIGURE 8.3 Single-file versus multi-file assembly.

CIL code contained in an assembly with metadata is *managed code*. The program code that executes under the management of a VM is called managed code, whereas the program code that is directly compiled into machine code before execution and executed directly by the CPU is called unmanaged code. Managed code is designed to be more reliable and robust than unmanaged code, with features such as garbage collection and type safety available to it. The managed code running in the CLR cannot be accessed outside the run-time environment and/or cannot call Operating System (OS) services directly from outside the run-time environment. This isolates programs and makes computers more secure. Managed code needs the CLR's JIT compiler to convert it to native executable code, and it relies on CLR to provide services such as security, memory, management, and threading during run-time.

Pre-compiled executables are called *unmanaged code*. One may prefer bypassing CLR and make direct calls to specific OS services through the Win32 API. However, such code would be considered unsafe because it yields security risks, and it will throw an exception if fed to CLR because it is not verifiable. Unmanaged code is directly compiled to architecture-specific machine code. Hence, unmanaged code is not portable.

Figure 8.4 depicts the compilation of programs written in .NET languages for the .NET

---

[14]Also referred to as *Intermediate Language* (IL), formerly *Microsoft Intermediate Language* (MSIL).

[15]The Portable Executable (PE) format is a file format for executables, object code, and DLLs, used in 32-bit and 64-bit versions of Windows operating systems.

Framework. A program is first compiled into an executable file (that is, assembly) by the individual higher-level language compiler, and then the executable file is fed into CLR to be JIT-compiled into native code to be executed [Microsoft, 2012].

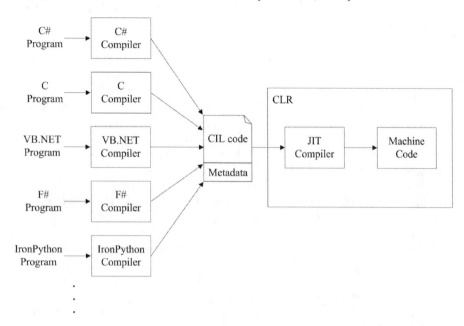

FIGURE 8.4 Language integration in .NET framework.

A significant advantage of the .NET Framework is cross-language interoperability, which is supported on all versions of Windows OS, regardless of the underlying architecture. Language interoperability means that code written in one language can interact with code written in another language. This is made possible in .NET languages and language-specific compilers by following a certain set of rules comprising the *Common Type System* (CTS).

CTS is a unified type system that supports the types found in many programming languages, and is shared by all .NET language-specific compilers and the CLR. It is the model that defines the rules that the CLR follows when declaring, using, and managing runtime types. The CTS also defines casting rules (including boxing and un-boxing operations), scopes, and assemblies. CTS supports object-oriented languages, functional languages, and procedural programming languages. By following the rules of CTS, programmers can mix constructs from different .NET languages, such as passing an object to a method written in a different programming language, or defining a type in one language and deriving from that type in another language. The types defined in CTS can be categorized as the *Value types* and the *Reference Types*.

Value types are stored in the stack rather than the garbage-collected heap. Each value type describes the storage that it occupies, the meanings of the bits in its representation, and the valid operations on that representation. Value types can be either *built-in data types* or *user-defined types* such as **Enum**.

Reference types are passed by reference and stored in the heap. A reference type carries more information than a value type. It has an identity that distinguishes it from all other objects, and it has slots that store other entities, which can be either objects or values. Reference types can be categorized as *self-describing reference types*, *built-in reference types*, *interfaces*, and *pointers*. Self-describing types can be further categorized as arrays and class types. The class types are user-defined classes, boxed value types, and delegates (the managed alternative to unmanaged function pointers). Built-in reference types include `Object` (primary base class of all classes in the .NET Framework, which is the root of the type hierarchy) and `String`.

Of course, a disadvantage of .NET is that it runs only on Windows platforms. One does not get the "write once, run anywhere" flexibility of Java.

8.5.2   Microsoft C# Compiler

When a C# program is fed to Microsoft's C# compiler, the compiler generates metadata, CIL code describing the program. In order to achieve this, the compiler does multiple passes over the code

In a first main pass, it looks for declarations in the code to gather information about the used namespaces, classes, structs, enums, interfaces, delegates, methods, type parameters, formal parameters, constructors, events, and attributes. All this information is extracted over more sub-passes. A lexical analyzer identifies the tokens in the source file, and the parser does a top-level parse, not going inside method bodies. Both the lexical analyzer and parser are hand-written, the parser being a basic recursive-descent parser. Then, a "declaration" pass records the locations of namespaces and type declarations in the program.

After the declaration pass, the compiler does not need to go over the actual program code; it can do the next passes on the symbols it has generated so far to extract further metadata. The next pass verifies that there are no cycles in the base types of the declared types. A similar pass is also done for the generic parameter constraints on generic types, verifying the acyclic hierarchy. After that, another pass checks the members (methods of classes, fields of structs and enum values, etc.) of every type—for example, whether what overriding methods override are actually virtual methods, or enums have no cycles, etc. One last pass takes care of the values of all constant fields.

Then, the compiler generates CIL code. This time the compiler parses the method bodies and determines the type of every expression within the method body. Many more sub-passes follow this. The compiler does not need the actual program code once it has created the annotated parse tree; subsequent passes work on this data structure, rewriting it as necessary. The compiler transforms loops into gotos and labels. Then, more passes are done to look for problems and to do some optimization. The compiler makes a pass for each of the following:

- To search for use of deprecated types and generate warnings if any exist,

- To search for types that have no metadata yet and emit those,

- To check whether *expression trees*[16] are constructed correctly,

- To search for local variables that are defined but not used, and generate warnings,

- To search for illegal patterns inside iterator blocks,

- To search for unreachable code, such as a non-void method with no return statement, and generate warnings,

- To check whether every goto targets a sensible label, and every label is targeted by a reachable goto, and

---

[16]Expression trees are C#-specific tree-like data structures, where each node is an expression in the code, for example, a method call or a binary operation such as `x < y`. For details on expression trees, see [Microsoft, 2012].

- To check whether all local variables have assigned values before their use.

Once the compiler is done with the passes for problem checks, it initiates the set of optimizing passes:

- To transform expression trees into the sequence of factory method calls necessary to create the expression trees at run-time,

- To rewrite all arithmetic that can possibly have null value as code that checks for null values,

- To rewrite references to methods defined in base classes as non-virtual calls,

- To find unreachable code and remove it from the tree, as generating IL for such code is redundant,

- To rewrite switch (constant) expressions as a direct branch to the correct case,

- To optimize arithmetic, and

- To transform iterator blocks into switch-based state machines.

After all these passes[17], the compiler can finally generate IL code by using the latest version of the annotated parse tree. The Microsoft C# compiler is written in unmanaged C++ and generates IL code structures as a sequence of basic blocks. The compiler can apply further optimizations on the basic blocks; for instance, it can rewrite branches between basic blocks for a more efficient call sequence, or remove the basic blocks containing dead code. Generated metadata and IR code are then fed to the CLR as an executable, to be JIT-compiled on any architecture running a Windows OS.

8.5.3 Classic Just-in-Time Compilation in the CLR

Microsoft's .NET run-time has two JIT compilers: the *Standard-JIT compiler* and the *Econo-JIT compiler*. The Econo-JIT compiler works faster than the Standard-JIT compiler. It compiles only Common Intermediate Language (CIL) code for methods that are invoked during run-time, but the native code is not saved for further calls. The Standard-JIT generates more optimized code than does the Econo-JIT and verifies the CIL code; compiled code is cached for subsequent invocations.

When an application is executed, the Windows OS checks whether it is a .NET assembly; if so, the OS starts up the CLR and passes the application to it for execution. The first thing the CLR's JIT compiler does is to subject the code to a verification process to determine whether the code is type safe by examining the CIL code and the associated metadata. In addition to checking that the assembly complies with the CTS, the CLR's JIT uses the metadata to locate and load classes; discover information about the programs classes, members, and inheritance; lay out instances in memory; and resolve method invocations during run-time. The executed assembly may refer to other assemblies; in this case, the referenced assemblies are loaded as needed.

When a type (for example, a structure, interface, enumerated type, or primitive type) is loaded, the loader creates and attaches a stub[18] to each of the types methods. All types have

---

[17]There are some more passes for .NET-specific tasks dealing with COM objects, anonymous types and functions, and dynamic calls using the CLR; however, they are excluded here to avoid introducing extra terminology and unnecessary details.

[18]A *stub* is a routine that only declares itself and the parameters it accepts and returns an expected value for the caller. A stub contains just enough code to allow it to be compiled and linked with the rest of the program; it does not contain the complete implementation.

their own *method tables* to store the addresses of their stubs. All object instances of the same type will refer to the same method table. Each method table contains information about the type, such as whether it is an interface, abstract class, or concrete class; the number of interfaces implemented by the type; the interface map for method dispatch; the number of slots in the method table; and an embedded method slot table, which points to the method implementations called *method descriptors*. Method slot table entries are always ordered as inherited virtuals, introduced virtuals, instance methods, and static methods. As Figure 8.5 shows, `ToString`, `Equals`, `GetHashCode`, and `Finalize` are the first four methods in the method slot table for any type. These are virtual methods inherited from `System.Object`. `.cctor` and `.ctor` are grouped with static methods and instance methods, respectively.

Initially, all entries in the method tables refer to the JIT compiler through method descriptors. Method descriptors are generated during the loading process and they all point to the CIL code (CodeOrIL in Figure 8.5) with a padding of 5 bytes (PreJit Stub in Figure 8.5) containing the instructions to make a call to the JIT Compiler. As discussed in Section 8.2, it is possible that some code might never be called during execution. So instead of converting all of the CIL code into native machine code, the JIT compiler converts the CIL as needed. Once a method is called for the first time, the call instruction contained in the associated stub passes control to the JIT compiler, which converts the CIL code for that method into native code and modifies the stub to direct execution to the location of the native code by overwriting the call instruction with an unconditional jump to the JIT-compiled native code. In other words, while methods are being stubbed out, the JIT compiler overwrites the call instruction, replacing the address in the method table with the compiled codes memory address. The machine code produced for the stubbed out methods is stored in memory. Subsequent calls to the JIT-compiled method directly invoke the machine code that was generated after the first call; thus, time to compile and run the program is reduced [Kommalapati and Christian, 2005].

Conversion of CIL to native code is done in several stages. The first stage is *importing*, in which the CIL code is first converted to JIT's internal representation. The JIT compiler has to make sure that the CIL code is type safe before converting it into machine code—confirming that the code can access memory locations and call methods only through properly defined types.

The second stage is *morphing*, where some transformations are applied to the internal structures to simplify or lightly optimize the code. Examples of such transformations are constant propagation and method inlining.

In the third stage, the JIT compiler performs a traditional *flow-graph analysis* to determine the liveness of variables, loop detection, etc. The information obtained in this stage is used in the subsequent stages.

The fourth stage is about heavy *optimizations* like common sub-expression and range check elimination, loop hoisting, and so on.

Then comes the *register allocation stage*, where the JIT compiler must effectively map variables to registers.

Finally, the *code generation* and *emitting stages* come. While generating the machine code, the JIT compiler must consider what processor it is generating code for and the OS version. For example, it may embed the address of some Win32 APIs in to the native code produced, and the address of these APIs could change between different service packs of a specific Windows OS [Richter, 2005]. After generating the machine code for a specific method, the JIT compiler packs everything together and returns to the CLR, which will then redirect control to the newly generated code. Garbage collection and debugger information is also recorded here. Whenever memory is low, the JIT compiler will free up memory by placing back the stubs of methods that had not been called frequently during program operation up to that point.

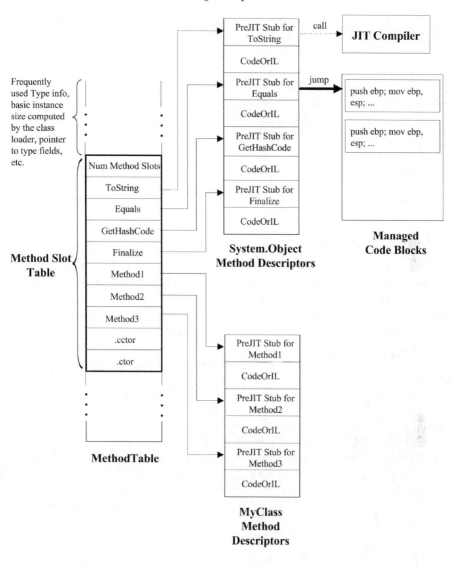

FIGURE 8.5 The method table in .NET.

Knowing that JIT compilation in the CLR works on a method-by-method basis as needed, and gives us an understanding of the performance characteristics and the ability to make better design decisions. For example, if one knows that a piece of code will be needed only in rare or specific cases, one can keep it in a separate assembly. Alternatively, one can keep that rarely needed piece of code in a separate method so that JIT Compiler will not compile it until explicitly invoked.

One important point to emphasize about the CLR's JIT compilation is that the CLR does not provide an interpreter to interpret the CIL Code. A method's CIL code is compiled to machine code at once to run. In Section 8.2 we saw that things work differently in Java's HotSpot JIT compiler. The HotSpot compiler may not JIT compile a method if it expects the overhead of compilation to be lower than the overhead of interpreting the code. It can recompile with heavier optimization than before, based on actual usage. In that sense, CLR's JIT compilation is simpler than HotSpot compilation.

At the end, one may ask how the Java platform compares to .NET Framework. The answer to that, in a nutshell, is that Java platform deals with one language running on multiple operating systems, whereas the .NET framework deals with multiple languages running on a single (Windows) platform.

8.6   Further Readings

[Oracle, 2010] provides comprehensive explanations of the Java HotSpot VM and its compiler. [Liang and Bracha, 1998] explains class loaders in the Java platform in a basic sense, focusing on dynamic class loading and type safety.

[Arthorne, 2004] introduces incremental building in Eclipse. [Clayberg and Rubel, 2009] give a good introduction to plug-in development for Eclipse. [D'Anjou, 2005] is a good source on Java development in Eclipse.

[Bothner, 2003] explains GCJ in depth. Also, [GNU, 2011] gives a full list of optimization options available in GCC.

[Richter and Balena, 2002] and [Richter, 2010] are both excellent and very different introductions to Microsoft's .NET and the CLR. (Anything Jeffrey Richter writes is excellent.) [Hamilton, 2003] presents concepts revolving around language interoperability in the CLR. [Kommalapati and Christian, 2005] examine the internals of the CLR.

[Aycock, 2003] presents the ideas underlying JIT compilation and how it is used in implementing various programming languages.

Appendix A

Setting Up and Running *j--*

A.1   Introduction

This appendix describes where to obtain the *j--* compiler, what is in the distribution, how to set it up for command-line execution, and how to set up, run, and debug the compiler in Eclipse[1].

A.2   Obtaining *j--*

The zip file `j--.zip` containing the *j--* distribution can be downloaded from `http://www.cs.umb.edu/j--`.

A.3   What Is in the Distribution?

The *j--* distribution includes the following files and folders:

---

[1] An open-source IDE; `http://www.eclipse.org`.

File/Folder	Description
`j--/`	*j*-- root directory[2].
`src/`	
`jminusminus/`	
`*.java`	The compiler source files; `J*.java` files define classes representing AST nodes, `CL*.java` files supply back-end code that is used by *j*-- to create JVM byte code, and `N*.java` files implement JVM-to-SPIM translation and register allocation.
`j--.jj`	JavaCC input file for generating a scanner and parser.
`package.html`	Describes the `jminusminus` package for javadoc .
`spim/`	
`*.s, *.java`	SPIM[3] run-time files and their Java wrappers.
`package.html`	Describes the `spim` package for javadoc.
`overview.html`	Describes the *j*-- project for javadoc.
`lib/`	Contains jar files for JavaCC[4], JUnit[5], and Java2HTML[6]. Also contains the jar file generated for the compiler.
`bin/`	Contains UNIX and Windows scripts for running the compiler and the Java class file emitter (`CLEmitter`).
`tests/`	
`clemitter/`	Contains Java programs that programmatically generate class files using the `CLEmitter` interface.
`pass/`	Contains *j*-- conformance tests.
`fail/`	Contains *j*-- deviance tests. None of these tests should compile successfully, that is, the compiler should not produce class files for any of these tests.
`junit/`	Contains JUnit test cases for compiling and running *j*-- test programs.
`spim/`	Contains *j*-- test programs that compile to SPIM.
`lexicalgrammar`	Lexical grammar for *j*--.
`grammar`	Syntactic grammar for *j*--.
`build.xml`	Ant[7] file for building and testing the compiler.
`.externalToolBuilders/`	Contains project settings for Eclipse.
`.classpath`	
`.project`	

---

[2]The directory that contains the *j*-- root directory is referred to as `$j`.

[3]A self-contained simulator that runs MIPS32 programs; `http://spimsimulator.sourceforge.net/`.

[4]A lexer and parser generator for Java; `http://javacc.dev.java.net/`.

[5]A regression testing framework; `http://www.junit.org`.

[6]Converts Java source code into a colorized and browsable HTML representation. `http://www.java2html.com`.

[7]A Java-based build tool; `http://ant.apache.org`.

A.3.1 Scripts

We provide three scripts[8]: The first script, `$j/j--/bin/j--`, is a wrapper for `jminusminus` `.Main`. This is the *j--* compiler. It has the following command-line syntax:

```
Usage: j-- <options> <source file>
where possible options include:
  -t          Only tokenize input and print tokens to STDOUT
  -p          Only parse input and print AST to STDOUT
  -pa         Only parse and pre-analyze input and print AST to STDOUT
  -a          Only parse, pre-analyze, and analyze input and print AST to STDOUT
  -s <naive|linear|graph> Generate SPIM code.
  -r <num> Max. physical registers (1-18) available for allocation; default=8
  -d <dir> Specify where to place generated class files; default=.
```

The second script, `$j/j--/bin/javaccj--`, is a wrapper for `jminusminus.JavaCCMain`. This is also the *j--* compiler, but uses the front end generated by JavaCC. Its command-line syntax is similar to the script discussed above.

The third script, `$j/j--/bin/clemitter`, is separate from the compiler, and is a wrapper for `jminusminus.CLEmitter`. This is the Java class file emitter. It has the following command-line syntax:

```
Usage: clemitter <file>
```

where file is a Java program that uses the `CLEmitter` interface to programmatically generate a JVM class file. The output is a class file for the program, and the class files the program was programmed to generate. For example, running the following command:

```
> $j/j--/bin/clemitter $j/j--/tests/clemitter/GenHelloWorld.java
```

produces `GenHelloWorld.class` and `HelloWorld.class` files. `HelloWorld.class` can then be run as,

```
> java HelloWorld
```

producing as output,

```
> Hello, World!
```

`$j/j--/bin` may be added to the environment variable `PATH` for convenience. This will allow invoking the scripts without specifying their fully qualified path, as follows:

```
> j-- $j/j--/tests/pass/HelloWorld.java
```

A.3.2 Ant Targets

The Ant file `$j/j--/build.xml` advertises the following targets:

---

[8] For the scripts to run, *j--* must be compiled and `j--.jar` file must exist in `$j/j--/lib`.

Target	Description
package	Makes a distributable package for the compiler that includes the source files, binaries, documentation, and JUnit test framework.
runCompilerTests	This is the default target. It first compiles the *j*--test programs under $j/j--/tests/pass and $j/j--/tests/fail using the *j*-- compiler. Second, it runs the *j*-- pass tests.
runCompilerTestsJavaCC	Same as the above target, but using JavaCC scanner and parser.
testScanner	Tokenizes the *j*-- test programs in $j/j--/tests/pass.
testJavaCCScanner	Same as the above target, but using JavaCC scanner.
testParser	Parses the *j*-- test programs in $j/j--/tests/pass.
testJavaCCParser	Same as the above target, but using JavaCC scanner and parser.
testPreAnalysis	Pre-analyzes the *j*-- test programs in $j/j--/tests/pass.
testAnalysis	Analyzes the *j*-- test programs in $j/j--/tests/pass.
help	Lists main targets.

A.4   Setting Up *j*-- for Command-Line Execution

We assume the following:

- J2SE[9] 7 or later (the latest version is preferred) has been installed and the environment variable PATH has been configured to include the path to the Java binaries. For example, on Windows, PATH might include C:\Program Files\Java\jdk1.7.0\bin.

- Ant 1.8.2 or later has been installed, the environment variable ANT_HOME has been set to point to the folder where it is installed, and environment variable PATH is configured to include $ANT_HOME/bin.

Now, the Ant targets listed in A.3.2 can be run as

```
> ant <target>
```

For example, the following command runs the target named testScanner:

```
> ant testScanner
```

If no target is specified, the default (runCompilerTests) target is executed.

A.5   Setting Up *j*-- in Eclipse

In order to be able to set up *j*-- in Eclipse, we assume the following:

---

[9]Java Development Kit (JDK); http://www.oracle.com/technetwork/java/index.html.

- J2SE 7 or later (the latest version is preferred) has been installed and the environment variable JAVA_HOME has been set to point to the installation folder. For example, on Windows, JAVA_HOME might be set to C:\Program Files\Java\jdk1.7.0.

- Eclipse 3.7 or later has been installed.

An Eclipse project for *j--* can be set up as follows:

1. Unzip the *j--* distribution into a temporary folder, say /tmp.

2. In Eclipse, click the *File → New → Project ...* menu [10] to bring up the *New Project* dialog box. Select *Java Project* and click *Next*. In the *New Java Project* dialog box, type "j--" for *Project Name*, make sure JRE (Java Run-time Environment) used is 1.7 or later, and click *Finish*. This creates an empty Java project called "j--" in the current workspace.

3. In the *Project Explorer* pane on the left, select the "j--" project. Click *File → Import ...* menu to bring up the *Import* dialog box. Under *General*, select *File System* as the *Import Source*, and click *Next*. Choose "/tmp/j--" as the *From directory*, select "j--" below, and click *Finish*. Answer *Yes to All* to the question on the *Question* pop-up menu. This imports the *j--* files into the "j--" project. Once the import is complete, Eclipse will automatically build the project. The automatic build feature can be turned off, if necessary, by clicking the *Project → Build Automatically* menu.

To build the *j--* project manually, select the project in the *Project Explorer* window and click the *Project → Build Project* menu. This runs the default (runCompilerTests) target in the Ant file $j/j--/build.xml. To run a different target, edit the following line in the Ant file:

```
<project default="runCompilerTests">
```

and change the value of the default attribute to the desired target.

A.6 Running/Debugging the Compiler

In order to run and debug the compiler within Eclipse, a *Launch Configuration* must be created, which can be done as follows:

1. Click the *Run → Run ...* menu to bring up the *Run* dialog box. Select *Java Application*, and click the *New Launch Configuration* button.

2. Give a suitable name for the configuration. Select "j--" for *Project* and type "jminus-minus.Main" for *Main class*.

3. In the *(x) = Arguments* tab, type appropriate values for *Program arguments*; these are the same as the arguments for the $j/j--/bin/j-- script described in Section A.3.1. For example, type "-t tests/pass/HelloWorld.java" to tokenize the *j--* test program HelloWorld.java using the hand-written scanner.

4. Click *Apply* and *Close*.

---

[10]Note that Eclipse menus have slightly different names under different operating systems.

You can have as many configurations as you like. To run or debug a particular configuration, click *Run → Run ...* or *Run → Debug ...* menu, select the configuration, and click *Run* or *Debug*. The output (STDOUT and STDERR) messages from the compiler are redirected to the *Console* pane.

A.7   Testing Extensions to *j--*

We test extensions to the *j--* language that compile to the JVM target using a JUnit test framework. The framework supports writing conformance and deviance tests in order to test the compiler; it also supports writing JUnit test cases to run the pass tests.

A *pass* test is a *j--* test program that is syntactically and semantically correct. When such a program is compiled using *j--*, the compiler should report no error(s) and should generate class file(s) for the program. A *fail* test is a *j--* test program that has syntactic and/or semantic error(s). When such a program is compiled using *j--*, the compiler should report the error(s) and exit gracefully. It should not produce class files for such programs.

The Ant targets, runCompilerTests and runCompilerTestsUsingJavaCC, attempt at compiling the pass and fail tests using *j--*, and running the test suite for the pass tests. Any pass test that compiles with errors or produces incorrect results would result in failed JUnit assertions. Any fail test that compiles successfully would also result in a failed JUnit assertions.

Chapter 1 describes how to add a simple extension (the division operator) to *j--* and how to test the extension using the JUnit test framework. Appendix E describes the SPIM target and how to compile *j--* programs for that target.

A.8   Further Readings

See the *Java Development User Guide* in [Eclipse, 2011] for more on how to run, debug, and test your programs using Eclipse.

Appendix B

j-- Language

B.1 Introduction

j-- is a subset of Java and is the language that our example compiler translates to JVM code. It has a little less than half the syntax of Java. It has classes; it has `ints`, `booleans`, `chars`, and `Strings`; and it has many Java operators. The 'j' is in the name because *j--* is derived from Java; the '--' is there because *j--* has less functionality than does Java. The exercises in the text involve adding to this language. We add fragments of Java that are not already in *j--*.

B.2 *j--* Program and Its Class Declarations

A *j--* program looks very much like a Java program. It can have an optional package statement, followed by zero or more import declarations, followed by zero or more type declarations. But in *j--*, the only kind of type declaration we have is the class declaration; *j--* has neither interfaces nor enumerations.

We may have only single-type-import declarations in *j--*; it does not support import-on-demand declarations (for example, `java.util.*`). The only Java types that are implicitly imported are `java.lang.Object` and `java.lang.String`. All other external Java types must be explicitly imported.

For example, the following is a legal *j--* program:

```
package pass;

import java.lang.Integer;
import java.lang.System;

public class Series {

    public static int ARITHMETIC = 1;

    public static int GEOMETRIC = 2;

    private int a; // first term

    private int d; // common sum or multiple

    private int n; // number of terms

    public Series() {
        this(1, 1, 10);
    }
```

```
    public Series(int a, int d, int n) {
        this.a = a;
        this.d = d;
        this.n = n;
    }

    public int computeSum(int kind) {
        int sum = a, t = a, i = n;
        while (i-- > 1) {
            if (kind == ARITHMETIC) {
                t += d;
            } else if (kind == GEOMETRIC) {
                t = t * d;
            }
            sum += t;
        }
        return sum;
    }

    public static void main(String[] args) {
        int a = Integer.parseInt(args[0]);
        int d = Integer.parseInt(args[1]);
        int n = Integer.parseInt(args[2]);
        Series s = new Series(a, d, n);
        System.out.println("Arithmetic sum = "
            + s.computeSum(Series.ARITHMETIC));
        System.out.println("Geometric sum = "
            + s.computeSum(Series.GEOMETRIC));
    }
}
```

j-- is quite rich. Although *j*-- is a subset of Java, it can interact with the Java API. Of course, it can only interact to the extent that it has language for talking about things in the Java API. For example, it can send messages to Java objects that take int, boolean, or char arguments, and which return int, boolean, and char values, but it cannot deal with floats, doubles, or even longs.

As for Java, only one of the type declarations in the compilation unit (the file containing the program) can be public, and that class's main() method is the program's entry point[1].

Although *j*-- does not support interface classes, it does support abstract classes. For example,

```
package pass;

import java.lang.System;

abstract class Animal {

    protected String scientificName;

    protected Animal(String scientificName) {
        this.scientificName = scientificName;
    }

    public String scientificName() {
        return scientificName;
    }
}

class FruitFly
```

---

[1] A program's entry point is where the program's execution commences.

```
        extends Animal {

    public FruitFly() {
        super("Drosophila melanogaster");
    }
}

class Tiger
    extends Animal {

    public Tiger() {
        super("Panthera tigris corbetti");
    }
}

public class Animalia {

    public static void main(String[] args) {
        FruitFly fruitFly = new FruitFly();
        Tiger tiger = new Tiger();
        System.out.println(fruitFly.scientificName());
        System.out.println(tiger.scientificName());
    }
}
```

Abstract classes in *j--* conform to the Java rules for abstract classes.

B.3   *j--* Types

j-- has fewer types than does Java. This is an area where *j--* is a much smaller language than is Java.

For example, *j--* primitives include just the `ints`, `chars`, and `booleans`. It excludes many of the Java primitives: `byte`, `short`, `long`, `float`, and `double`.

As indicated above, *j--* has neither interfaces nor enumerations. On the other hand, *j--* has all the reference types that can be defined using classes, including the implicitly imported `String` and `Object` types.

j-- is stricter than is Java when it comes to assignment. The types of actual arguments must exactly match the types of formal parameters, and the type of right-hand side of an assignment statement must exactly match the type of the left-hand side. The only implicit conversion is the Java `String` conversion for the + operator; if any operand of a + is a `String` or if the left-hand side of a += is a `String`, the other operands are converted to `Strings`. *j--* has no other implicit conversions. But *j--* does provide casting; *j--* casting follows the same rules as Java.

That *j--* has fewer types than Java is in fact a rich source of exercises for the student. Many exercises involve the introduction of new types, the introduction of appropriate casts, and implicit type conversion.

B.4　*j--* Expressions and Operators

j-- supports the following Java expressions and operators.

Expression	Operators
Assignment	=, +=
Conditional	&&
Equality	==
Relational	>, <=, instanceof[2]
Additive	+, -
Multiplicative	*
Unary (prefix)	++, -
Simple unary	!
Postfix	--

It also supports casting expressions, field selection, and message expressions. Both `this` and `super` may be the targets of field selection and message expressions.

j-- also provides literals for the types it can talk about, including `Strings`.

B.5　*j--* Statements and Declarations

In addition to statement expressions[3], *j--* provides for the `if` statement, `if-else` statement, `while` statement, `return` statement, and blocks. All of these statements follow the Java rules.

Static and instance field declarations, local variable declarations, and variable initializations are supported, and follow the Java rules.

B.6　Syntax

This section describes the lexical and syntactic grammars for the *j--* programming language; the former specifies how individual tokens in the language are composed, and the latter specifies how language constructs are formed.

We employ the following notation to describe the grammars.

- // indicates comments;

- Non-terminals are written in the form of Java (mixed-case) identifiers;

- Terminals are written in **bold**;

---

[2] Technically, `instanceof` is a keyword.

[3] A statement expression is an expression that can act as a statement. Examples include, `i--;`, `x = y + z;` and `x.doSomething();`.

- Token representations are enclosed in <>;

- [x] indicates x is optional, that is, zero or one occurrence of x;

- {x} indicates zero or more occurrences of x;

- $x|y$ indicates x or y;

- $\tilde{x}$ indicates negation of x;

- Parentheses are used for grouping;

- Level numbers in expressions indicate precedence.

B.6.1 Lexical Grammar

White Space

White space in *j--* is defined as the ASCII space (SP), horizontal tab (HT), and form feed (FF) characters, as well as line terminators: line feed (LF), carriage return (CR), and carriage return (CR) followed by line feed (LF).

Comments

j-- supports single-line comments; all the text from the ASCII characters // to the end of the line is ignored.

Reserved Words

The following tokens are reserved for use as keywords in *j--* and cannot be used as identifiers:

```
abstract    extends     int       protected   this
boolean     false       new       public      true
char        import      null      return      void
class       if          package   static      while
else        instanceof  private   super
```

Operators

The following tokens serve as operators in *j--*:

```
=    ==    >    ++    &&    <=    !    -    --    +    +=    *
```

Separators

The following tokens serve as separators in *j--*:

```
,    .    [    {    (    )    }    ]    ;
```

Identifiers

The following regular expression describes identifiers in *j--*:

```
<identifier> = (a-z|A-Z|_|$){a-z|A-Z|_|0-9|$}
```

int, char and String Literals

An escape (ESC) character in *j--* is a \ followed by n, r, t, b, f, ', ", or \. In addition to the true, false, and null literals, *j--* supports int, char, and String literals as described by the following regular expressions:

```
<int_literal> = 0|(1-9){0-9}
<char_literal> = '(ESC|~('|\|LF|CR))'
<string_literal> = "{ESC|~("|\|LF|CR)}"
```

B.6.2   Syntactic Grammar

compilationUnit ::= [package qualifiedIdentifier ;]
               {import qualifiedIdentifier ;}
               {typeDeclaration} EOF

qualifiedIdentifier ::= <identifier> {. <identifier>}

typeDeclaration ::= modifiers classDeclaration

modifiers ::= {public | protected | private | static | abstract}

classDeclaration ::= class <identifier> [extends qualifiedIdentifier] classBody

classBody ::= { {modifiers memberDecl} }

memberDecl ::= <identifier> // constructor
             formalParameters block
      | (void | type) <identifier> // method
             formalParameters (block | ;)
      | type variableDeclarators ; // field

block ::= { {blockStatement} }

blockStatement ::= localVariableDeclarationStatement
          | statement

statement ::= block
      | <identifier> : statement
      | if parExpression statement [else statement]
      | while parExpression statement
      | return [expression] ;
      | ;
      | statementExpression ;

formalParameters ::= ([formalParameter {, formalParameter}])

formalParameter ::= type <identifier>

parExpression ::= (expression)

localVariableDeclarationStatement ::= type variableDeclarators ;

variableDeclarators ::= variableDeclarator {, variableDeclarator}

variableDeclarator ::= `<identifier>` [= variableInitializer]

variableInitializer ::= arrayInitializer | expression

arrayInitializer ::= { [variableInitializer {, variableInitializer}] }

arguments ::= ([expression {, expression}])

type ::= referenceType | basicType

basicType ::= `boolean` | `char` | `int`

referenceType ::= basicType `[ ]` `{[ ]}`
             | qualifiedIdentifier `{[ ]}`

statementExpression ::= expression // but must have side-effect, eg `i++`

expression ::= assignmentExpression

assignmentExpression ::= conditionalAndExpression // must be a valid lhs
                         [(`=` | `+=`) assignmentExpression]

conditionalAndExpression ::= equalityExpression // level 10
                          {`&&` equalityExpression}

equalityExpression ::= relationalExpression // level 6
                  {`==` relationalExpression}

relationalExpression ::= additiveExpression // level 5
                   [(`>` | `<=`) additiveExpression | `instanceof` referenceType]

additiveExpression ::= multiplicativeExpression // level 3
               {(`+` | `-`) multiplicativeExpression}

multiplicativeExpression ::= unaryExpression // level 2
                     {`*` unaryExpression}

unaryExpression ::= `++` unaryExpression // level 1
            | `-` unaryExpression
            | simpleUnaryExpression

simpleUnaryExpression ::= `!` unaryExpression
                    | (basicType) unaryExpression //cast
                    | (referenceType) simpleUnaryExpression // cast
                    | postfixExpression

postfixExpression ::= primary {selector} {`--`}

selector ::= . qualifiedIdentifier [arguments]
      | [expression]

primary ::= parExpression
      | this [arguments]
      | super (arguments | . \<identifier> [arguments])
      | literal
      | new creator
      | qualifiedIdentifier [arguments]

creator ::= (basicType | qualifiedIdentifier)
        (arguments
        | [ ] {[ ]} [arrayInitializer]
        | newArrayDeclarator)

newArrayDeclarator ::= [expression] {[expression]} {[ ]}

literal ::= \<int_literal> | \<char_literal> | \<string_literal> | true | false | null

B.6.3 Relationship of *j--* to Java

As was said earlier, *j--* is a subset of the Java programming language. Those constructs of Java that are in *j--* roughly conform to their behavior in Java. There are several reasons for defining *j--* in this way:

- Because many students know Java, *j--* will not be totally unfamiliar.

- The exercises involve adding Java features that are not already there to *j--*. Again, because one knows Java, the behavior of these new features should be familiar.

- One learns even more about a programming language by implementing its behavior.

- Because our compiler is written in Java, the student will get more practice in Java programming.

For reasons of history and performance, most compilers are written in C or C++. One might ask: Then why don't we work in one of those languages? Fair question.

Most computer science students study compilers, not because they will write compilers (although some may) but to learn how to program better: to make parts of a program work together, to learn how to apply some of the theory they have learned to their programs, and to learn how to work within an existing framework. We hope that one's working with the *j--* compiler will help in all of these areas.

Appendix C

Java Syntax

C.1 Introduction

This appendix describes the lexical and syntactic grammars[1] for the Java programming language; the former specifies how individual tokens in the language are composed, and the latter specifies how language constructs are formed. The grammars are based on the second edition of the *Java Language Specification*.

C.2 Syntax

We employ the following notation to describe the lexical and syntactic grammars, similar to what we used in Appendix B to specify *j--* syntax.

- // indicates comments;

- Non-terminals are written in the form of Java (mixed-case) identifiers;

- Terminals are written in **bold**;

- Token representations are enclosed in <>;

- [*x*] indicates *x* is optional, that is, zero or one occurrence of *x*;

- {*x*} indicates zero or more occurrences of *x*;

- *x*|*y* indicates *x* or *y*;

- ~*x* indicates negation of *x*;

- Parentheses are used for grouping;

- Level numbers in expressions indicate precedence.

C.2.1 Lexical Grammar

White Space

White space in Java is defined as the ASCII space (SP), horizontal tab (HT), and form feed (FF) characters, as well as line terminators: line feed (LF), carriage return (CR), and carriage return (CR) followed by line feed (LF).

---

[1]Adapted from ANTLR (http://www.antlr.org) Parser Generator examples.

Comments

Java supports two kinds of comments: the traditional comment where all the text from the ASCII characters /* to the ASCII characters */ is ignored, and single-line comments where all the text from the ASCII characters // to the end of the line is ignored.

Reserved Words

The following tokens are reserved for use as keywords in Java and cannot be used as identifiers:

abstract	const	finally	int	public	this
boolean	continue	float	interface	return	throw
break	default	for	long	short	throws
byte	do	goto	native	static	transient
case	double	if	new	strictfp	try
catch	else	implements	package	super	void
char	extends	import	private	switch	volatile
class	final	instanceof	protected	synchronized	while

Operators

The following tokens serve as operators in Java:

?	=	==	!	~	!=	/	/=	+	+=	++	-
-=	--	*	*=	%	%=	>>	>>=	>>>	>>>=	>=	>
<<	<<=	<=	<	^	^=	|	|=	||	&	&=	&&

Separators

The following tokens serve as separators in Java:

,	.	[	{	(	)	}	]	;

Identifiers

The following regular expression describes identifiers in Java:

```
<identifier> = (a-z|A-Z|_|$){a-z|A-Z|_|0-9|$}
```

Literals

An escape (ESC) character in Java is a \ followed by n, r, t, b, f, ', ", or \. An octal escape (OCTAL_ESC)—provided for compatibility with C—is an octal digit (0–7), or an octal digit followed by another octal digit, or one of 0, 1, 2, or 3 followed by two octal digits. In addition to the true, false, and null literals, Java supports int, long, float, double, char, and String literals as described by the following regular expressions:

```
<int_literal> = 0|(1-9) {0-9} // decimal
              | 0 (x|X) ((0-9)|(A-F)|(a-f)) {(0-9)|(A-F)|(a-f)} // hexadecimal
              | 0 (0-7) {0-7} // octal

<long_literal> = <int_literal> (l|L)

<float_literal> = (0   9 ) {0   9 } . {0-9} [(e|E) [+|-] (0-9) {0-9}] [f|F]
                | . {0-9} [(e|E) [+|-] (0-9) {0-9}] [f|F]
                | (0-9) {0-9} [(e|E) [+|-] (0-9) {0-9}] (f|F)
                | (0-9) {0-9} ((e|E) ([+|-] (0-9) {0-9}) [f|F]
```

```
              | (0x|0X) . (0-9)|(a-f)|(A-F) {(0-9)|(a-f)|(A-F)}
                    (p|P) [+|-] (0-9){0-9} [f|F] // hexadecimal
              | (0x|0X) (0-9)|(a-f)|(A-F) {(0-9)|(a-f)|(A-F)}
                [.{(0-9)|(a-f)|(A-F)}]
                    (p|P) [+|-] (0-9){0-9} [f|F] // hexadecimal
<double_literal> = {0-9} [ [ . ] {0-9} [(e|E) [ +|-] (0-9) {0-9} ]] [ d|D ]

<char_literal> = '(ESC|OCTAL_ESC|~('|\|LF|CR))'

<string_literal> = "{ESC|OCTAL_ESC|~("|\|LF|CR)}"
```

C.2.2   Syntactic Grammar

compilationUnit ::= [package qualifiedIdentifier ;]
　　　　　　　　　{import qualifiedIdentifierStar ;}
　　　　　　　　　{typeDeclaration}
　　　　　　　　　EOF

qualifiedIdentifier ::= <identifier> {. <identifier>}

qualifiedIdentifierStar ::= <identifier> {. <identifier>} [. *]

tvpeDeclaration ::= typeDeclarationModifiers (classDeclaration | interfaceDeclaration)
　　　　　　　　　| ;

typeDeclarationModifiers ::= { public | protected | private | static | abstract
　　　　　　　　　　　　　　| final | strictfp }

classDeclaration ::= class <identifier> [extends qualifiedIdentifier]
　　　　　　　　　　[implements qualifiedIdentifier {, qualifiedIdentifier}]
　　　　　　　　　　classBody

interfaceDeclaration ::= interface <identifier> // can't be final
　　　　　　　　　　　[extends qualifiedIdentifier {, qualifiedIdentifier}]
　　　　　　　　　　interfaceBody

modifiers ::= { public | protected | private | static | abstract
　　　　　| transient | final | native | threadsafe | synchronized
　　　　　| const | volatile | strictfp} // const is reserved, but not valid

classBody ::= { { ;
　　　　　| static block
　　　　　| block
　　　　　| modifiers memberDecl
　　　　　}
　　　　　}

interfaceBody ::= { { ;
　　　　　　| modifiers interfaceMemberDecl
　　　　　　}
　　　　　}

memberDecl ::= classDeclaration // inner class
         | interfaceDeclaration // inner interface
         | `<identifier>` // constructor
            formalParameters
                 [`throws` qualifiedIdentifier {, qualifiedIdentifier}] block
         | (`void` | type) `<identiier>` // method
            formalParameters { [ ] }
                 [`throws` qualifiedIdentifier {, qualifiedIdentifier}] (block | ;)
         | type variableDeclarators ; // field

interfaceMemberDecl ::= classDeclaration // inner class
                 | interfaceDeclaration // inner interface
                 | (`void` | type) `<identifier>` // method
                     formalParameters { [ ] }
                         [`throws` qualifiedIdentifier {, qualifiedIdentifier}] ;
                 | type variableDeclarators ; // fields; must have inits

block ::= { {blockStatement} }

blockStatement ::= localVariableDeclarationStatement
            | typeDeclarationModifiers classDeclaration
            | statement

statement ::= block
         | `if` parExpression statement [`else` statement]
         | `for` ([forInit] ; [expression] ; [forUpdate]) statement
         | `while` parExpression statement
         | `do` statement `while` parExpression ;
         | `try` block
            {`catch` (formalParameter) block}
              [`finally` block] // must be present if no catches
         | `switch` parExpression { {switchBlockStatementGroup} }
         | `synchronized` parExpression block
         | `return` [expression] ;
         | `throw` expression ;
         | `break` [`<identifier>`] ;
         | `continue` [`<identifier>`] ;
         | ;
         | `<identifier>` : statement
         | statementExpression ;

formalParameters ::= ([formalParameter {, formalParameter}])

formalParameter ::= [`final`] type `<identifier>` { [ ] }

parExpression ::= (expression)

forInit ::= statementExpression {, statementExpression}
       | [`final`] type variableDeclarators

forUpdate ::= statementExpression {, statementExpression}

switchBlockStatementGroup ::= switchLabel {switchLabel} {blockStatement}

switchLabel ::= `case` expression : // must be constant
          | `default` :

localVariableDeclarationStatement ::= [`final`] type variableDeclarators ;

variableDeclarators ::= variableDeclarator {, variableDeclarator}

variableDeclarator ::= `<identifier>` { [ ] } [= variableInitializer]

variableInitializer ::= arrayInitializer | expression

arrayInitializer ::= { [variableInitializer {, variableInitializer} [ ]]

arguments ::= ([expression {, expression}])

type ::= referenceType | basicType

basicType ::= `boolean` | `byte` | `char` | `short` | `int` | `float` | `long` | `double`

referenceType ::= basicType [ ] {[ ]}
          | qualifiedIdentifier {[ ]}

statementExpression ::= expression // but must have side-effect, eg `i++`

expression ::= assignmentExpression

// level 13
assignmentExpression ::= conditionalExpression // must be a valid lhs
          [
           (`=`
           | `+=`
           | `-=`
           | `*=`
           | `/=`
           | `%=`
           | `>>=`
           | `>>>=`
           | `<<=`
           | `&=`
           | `|=`
           | `^=`
) assignmentExpression]

// level 12
conditionalExpression ::= conditionalOrExpression
              [? assignmentExpression : conditionalExpression]

```
// level 11
conditionalOrExpression ::= conditionalAndExpression { || conditionalAndExpression }

// level 10
conditionalAndExpression ::= inclusiveOrExpression { && inclusiveOrExpression }

// level 9
inclusiveOrExpression ::= exclusiveOrExpression { | exclusiveOrExpression }

// level 8
exclusiveOrExpression ::= andExpression { ^ andExpression }

// level 7
andExpression ::= equalityExpression { & equalityExpression }

// level 6
equalityExpression ::= relationalExpression { ( == | != ) relationalExpression }

// level 5
relationalExpression ::= shiftExpression
                         ( { ( < | > | <= | >= ) shiftExpression }
                         | instanceof referenceType )

// level 4
shiftExpression ::= additiveExpression { ( << | >> | >>> ) additiveExpression }

// level 3
additiveExpression ::= multiplicativeExpression { ( + | - ) multiplicativeExpression }

// level 2
multiplicativeExpression ::= unaryExpression { ( * | / | % ) unaryExpression }

// level 1
unaryExpression ::= ++ unaryExpression
                  | -- unaryExpression
                  | ( + | - ) unaryExpression
                  | simpleUnaryExpression

simpleUnaryExpression ::= ~ unaryExpression
                        | ! unaryExpression
                        | ( basicType ) unaryExpression //cast
                        | ( referenceType ) simpleUnaryExpression // cast
                        | postfixExpression

postfixExpression ::= primary {selector} { -- | ++ }

selector ::= . ( <identifier> [arguments]
               | this
               | super superSuffix
               | new innerCreator
               )
```

| [expression]

superSuffix ::= arguments
        | . `<identifier>` [arguments]

primary ::= (assignmentExpression)
      | `this` [arguments]
      | `super` superSuffix
      | literal
      | `new` creator
      | `<identifier>` . `<identifier>` [identifierSuffix]
      | basicType `{[ ]}` . `class`
      | `void` . `class`

identifierSuffix ::= `[ ] {[ ]}` . `class`
        | [expression]
        | . (`class`
           | `this`
           | `super` arguments
           | `new` innerCreator
)
        | arguments

creator ::= type
        (arguments [classBody]
        | newArrayDeclarator [arrayInitializer]
)

newArrayDeclarator ::= `[` [expression] `] {` `[` [expression] `] }`

innerCreator ::= `<identifier>` arguments [classBody]

literal ::= `<int_literal>` | `<char_literal>` | `<string_literal>` | `<float_literal>`
    | `<long_literal>` | `<double_literal>` | `true` | `false` | `null`

C.3   Further Readings

See Chapter 2 of [Gosling et al., 2005] for a detailed description of the lexical and syntactic grammars for the Java language.

Appendix D

JVM, Class Files, and the `CLEmitter`

D.1 Introduction

In the first instance, our compiler's target is the Java Virtual Machine (JVM), a *virtual* byte-code machine. In this appendix, we provide a brief description of the JVM, and then describe the `CLEmitter`, a high-level interface for producing the byte code.

D.2 Java Virtual Machine (JVM)

The JVM is an abstract architecture that can have any number of implementations. For example, Oracle has implemented a Java Run-time Environment (JRE[1]) that interprets JVM programs, but uses Hotspot technology for further compiling code that is executed repeatedly to native machine code. We say it is a *byte-code* machine, because the programs it executes are sequences of bytes that represent the instructions and the operands.

Although virtual, the JVM has a definite architecture. It has an instruction set and it has an internal organization.

The JVM starts up by creating an initial class, which is specified in an implementation-dependent manner, using the bootstrap class loader. The JVM then links the initial class, initializes it, and invokes its `public` class method `void main(String[] args)`. The invocation of this method drives all further execution. Execution of JVM instructions constituting the `main()` method may cause linking (and consequently creation) of additional classes and interfaces, as well as invocation of additional methods.

A JVM instruction consists of a one-byte opcode (operation code) specifying the operation to be performed, followed by zero or more operands supplying the arguments or data that are used by the operation.

The inner loop of the JVM interpreter is effectively:

```
do {
    fetch an opcode;
    if (operands) fetch operands;
    execute the action for the opcode;
} while (there is more to do);
```

The JVM defines various run-time data areas that are used during execution of a program. Some of these data areas are created at the JVM start-up and are destroyed only when the JVM exits. Other data areas are per thread[2]. Per-thread data areas are created when a thread is created and destroyed when the thread exits.

---

[1] Available at `http://www.oracle.com/technetwork/java/javase/downloads/`.
[2] *j--* does not support implementation of multi-threaded programs.

D.2.1 pc **Register**

The JVM can support many threads of execution at once, and each thread has its own pc (program counter) register. At any point, each JVM thread is executing the code of a single method, the current method for that thread. If the method is not native, the pc register contains the address of the JVM instruction currently being executed.

D.2.2 JVM Stacks and Stack Frames

The JVM is not a register machine, but a *stack machine*. Each JVM thread has a run-time data stack, created at the same time as the thread. The JVM stack is analogous to the stack of a conventional language such as C; it holds local variables and partial results, and plays a role in method invocation and return. There are instructions for loading data values onto the stack, for performing operations on the value(s) that are on top of the stack, and there are instructions for storing the results of computations back in variables.

For example, consider the following simple expression.

```
34 + 6 * 11
```

If the compiler does no constant folding, it might produce the following JVM code for performing the calculation.

```
ldc 34
ldc 6
ldc 11
imul
iadd
```

Executing this sequence of instructions takes a run-time stack through the sequence of states illustrated in Figure D.1

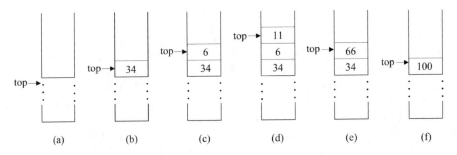

FIGURE D.1 The stack states for computing 34 + 6 * 11.

In (a), the run-time stack is in some initial state, where top points to its top value. Executing the first ldc (load constant) instruction causes the JVM to push the value 34 onto the stack, leaving it in state (b). The second ldc, pushes 6 onto the stack, leaving it in state (c). The third ldc pushes 11 onto the stack, leaving it in state (d). Then, executing the imul (integer multiplication) instruction causes the JVM to pop off the top two values (11 and 6) from the stack, multiply them together, and push the resultant 66 back onto the stack, leaving it in state (e). Finally, the iadd (integer addition) instruction pops off the top two values (66 and 34) the stack, adds them together, and pushes the resultant 100 back onto the stack, leaving it in the state (f).

As we saw in Chapter 1, the stack is organized into *stack frames*. Each time a method is invoked, a new stack frame for the method invocation is pushed onto the stack. All actual arguments that correspond to the methods formal parameters, and all local variables in the method body are allocated space in this stack frame. Also, all the method computations, like that illustrated in Figure D.1, are carried out in the same stack frame. Computations take place in the space above the arguments and local variables.

For example, consider the following instance method, `add()`.

```
int add(int x, int y) {
    int z;
    z = x + y;
    return z;
}
```

Now, say `add()` is a method defined in a class named `Foo`, and further assume that `f` is a variable of type `Foo`. Consider the message expression

```
f.add(2, 3);
```

When `add()` is invoked, a stack frame like that illustrated in Figure D.2 is pushed onto the run-time stack.

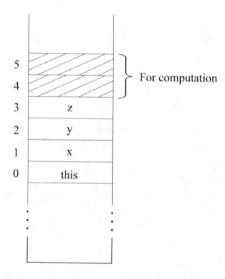

FIGURE D.2 The stack frame for an invocation of `add()`.

Because `add()` is an instance method, the object itself, that is, `this`, must be passed as an implicit argument in the method invocation; so `this` occupies the first location in the stack frame at offset 0. Then the actual parameter values 2 and 3, for formal parameters `x` and `y`, occupy the next two locations at offsets 1 and 2, respectively. The local variable `z` is allocated space for its value at offset 3. Finally, two locations are allocated above the parameters and local variable for the computation.

Here is a symbolic version[3] of the code produced for `add()` by our *j--* compiler.

```
int add(int, int);
  Code:
    Stack=2, Locals=4, Args_size=3
    0: iload_1
    1: iload_2
    2: iadd
    3: istore_3
```

---

[3]Produced using `javap -v Foo`.

```
4:  iload_3
5:  ireturn
```

Here is how the JVM executes this code.

- The `iload_1` instruction loads the integer value at offset 1 (for x) onto the stack at frame offset 4.

- The next `iload_2` instruction loads the integer value at offset 2 (for y) onto the stack at frame offset 5.

- The `iadd` instruction pops the two integers off the top of the stack (from frame offsets 4 and 5), adds them using integer addition, and then pushes the result (x + y) back onto the stack at frame offset 4.

- The `istore_3` instruction pops the top value (at frame offset 4) off the stack and stores it at offset 3 (for z).

- The `iload_3` instruction loads the value at frame offset 3 (for z) onto the stack at frame offset 4.

- Finally, the `ireturn` instruction pops the top integer value from the stack (at frame location 4), pops the stack frame from the stack, and returns control to the invoking method, pushing the returned value onto the invoking method's frame.

Notice that the instruction set takes advantage of common offsets within the stack frame. For example, the `iload_1` instruction is really shorthand for `iload 1`.

The `iload_1` instruction occupies just one byte for the opcode; the opcode for `iload_1` is 27. But the other version requires two bytes: one for the opcode (21) and one byte for the operand's offset (1). The JVM is trading opcode space—a byte may only represent up to 256 different operations—to save code space.

A frame may be extended to contain additional implementation-specific data such as the information required to support a run-time debugger.

D.2.3   Heap

Objects are represented on the stack as pointers into the *heap*, which is shared among all JVM threads. The heap is the run-time data area from which memory for all class instances and arrays is allocated. It is created during the JVM start-up. Heap storage for objects is reclaimed by an automatic storage management system called the *garbage collector*.

D.2.4   Method Area

The JVM has a method area that is shared among all JVM threads. It is created during the JVM start-up. It stores per-class structures such as the run-time constant pool, field and method data, and the code for methods and constructors, including the special methods used in class and instance initialization and interface type initialization.

D.2.5   Run-Time Constant Pool

The *run-time constant pool* is a per-class or per-interface run-time representation of the `constant_pool` table in a class file. It contains several kinds of constants, ranging from numeric literals known at compile time to method and field references that must be resolved at run-time. It is constructed when the class or interface is created by the JVM.

D.2.6 Abrupt Method Invocation Completion

A method invocation completes abruptly if execution of a JVM instruction within the method causes the JVM to throw an exception and that exception is not handled within the method. Execution of an ATHROW instruction also causes an exception to be explicitly thrown and, if the exception is not caught by the current method, results in abrupt method invocation completion. A method invocation that completes abruptly never returns a value to its invoker.

D.3 Class File

D.3.1 Structure of a Class File

The byte code that *j--* generates from a source program is stored in a binary file with a .class extension. We refer to such files as *class files*. Each class file contains the definition of a single class or interface. A class file consists of a stream of 8-bit bytes. All 16-bit, 32-bit, and 64-bit quantities are constructed by reading in two, four, and eight consecutive 8-bit bytes, respectively. Multi-byte data items are always stored in big-endian order, where the high bytes come first.

A class file consists of a single ClassFile structure; in the C language it would be:

```
ClassFile {
    u4 magic;
    u2 minor_version;
    u2 major_version;
    u2 constant_pool_count;
    cp_info constant_pool[constant_pool_count -1];
    u2 access_flags;
    u2 this_class;
    u2 super_class;
    u2 interfaces_count;
    u2 interfaces[interfaces_count];
    u2 fields_count;
    field_info fields[fields_count];
    u2 methods_count;
    method_info methods[methods_count];
    u2 attributes_count;
    attribute_info attributes[attributes_count];
}
```

The types u1, u2, and u4 represent an unsigned one-, two-, or four-byte quantity, respectively. The items of the ClassFile structure are described below.

Attribute	Description
`magic`	A magic number (`0xCAFEBABE`) identifying the class file format.
`minor_version, major_version`	Together, a major and minor version number determine the version of the class file format. A JVM implementation can support a class file format of version v if and only if v lies in some contiguous range of versions. Only Oracle specifies the range of versions a JVM implementation may support.
`constant_pool_count`	Number of entries in the `constant_pool` table plus one.
`constant_pool[]`	A table of structures representing various string constants, class and interface names, field names, and other constants that are referred to within the `ClassFile` structure and its sub-structures.
`access_flags`	Mask of flags used to denote access permissions to and properties of this class or interface.
`this_class`	Must be a valid index into the `constant_pool` table. The entry at that index must be a structure representing the class or interface defined by this class file.
`super_class`	Must be a valid index into the `constant_pool` table. The entry at that index must be the structure representing the direct super class of the class or interface defined by this class file.
`interfaces_count`	The number of direct super interfaces of the class or interface defined by this class file.
`interfaces[]`	Each value in the table must be a valid index into the `constant_pool` table. The entry at each index must be a structure representing an interface that is a direct super interface of the class or interface defined by this class file.
`fields_count`	Number of entries in the `fields` table.
`fields[]`	Each value in the table must be a `field_info` structure giving complete description of a field in the class or interface defined by this class file.
`methods_count`	Number of entries in the `methods` table.
`methods[]`	Each value in the table must be a `method_info` structure giving complete description of a method in the class or interface defined by this class file.
`attributes_count`	Number of entries in the `attributes` table.
`attributes[]`	Must be a table of class attributes.

The internals for all of these are fully described in [Lindholm and Yellin, 1999].

One may certainly create class files by directly working with a binary output stream. However, this approach is rather arcane, and involves a tremendous amount of housekeeping; one has to maintain a representation for the `constant_pool` table, the program counter `pc`, compute branch offsets, compute stack depths, perform various bitwise operations, and do much more.

It would be much easier if there were a high-level interface that would abstract out the gory details of the class file structure. The `CLEmitter` does exactly this.

D.3.2 Names and Descriptors

Class and interface names that appear in the `ClassFile` structure are always represented in a fully qualified form, with identifiers making up the fully qualified name separated by forward slashes ('/')[4]. This is the so-called *internal form* of class or interface names. For example, the name of class `Thread` in internal form is `java/lang/Thread`.

The JVM expects the types of fields and methods to be specified in a certain format called *descriptors*.

A *field descriptor* is a string representing the type of a class, instance, or local variable. It is a series of characters generated by the grammar[5]:

FieldDescriptor ::= FieldType

ComponentType ::= FieldType

FieldType ::= BaseType | ObjectType | ArrayType

BaseType ::= B | C | D | F | I | J | S | Z

ObjectType ::= L <class name> ; // class name is in internal form

ArrayType ::= [ComponentType

The interpretation of the base types is shown in the table below:

Basic Type Character	Type
B	byte
C	char
D	double
F	float
I	int
J	long
S	short
Z	boolean

For example, the table below indicates the field descriptors for various field declarations:

Field	Descriptor
`int i;`	`I`
`Object o;`	`Ljava/lang/Object;`
`double[][][] d;`	`[[[D`
`Long[][] l;`	`[[Ljava/lang/Long;`

A *method descriptor* is a string that represents the types of the parameters that the method takes and the type of the value that it returns. It is a series of characters generated by the grammar:

---

[4]This is different from the familiar syntax of fully qualified names, where the identifiers are separated by periods ('.').

[5]This is the so-called EBNF (Extended Backus Naur Form) notation for describing the syntax of languages.

MethodDescriptor ::= ({ParameterDescriptor}) ReturnDescriptor

ParameterDescriptor ::= FieldType

ReturnDescriptor ::= FieldType | V

For example, the table below indicates the method descriptors for various constructor and method declarations:

Constructor/Method	Descriptor
`public Queue()`	`()V`
`public File[] listFiles()`	`()[Ljava/io/File;`
`public Boolean isPrime(int n)`	`(I)Ljava/lang/Boolean;`
`public static void main(String[] args)`	`([L/java/lang/String;)V`

D.4 `CLEmitter`

D.4.1 `CLEmitter` Operation

The *j--* compiler's purpose is to produce a class file. Given the complexity of class files we supply a tool called the `CLEmitter` to ease the generation of code and the creation of class files. The `CLEmitter`[6] has a relatively small set of methods that support

- The creation of a class or an interface;

- The addition of fields and methods to the class;

- The addition of instructions, exception handlers, and code attributes to methods;

- The addition of inner classes;

- Optional field, method, and class attributes;

- Checking for errors; and

- The writing of the class file to the file system.

While it is much simpler to work with an interface like `CLEmitter`, one still must be aware of certain aspects of the target machine, such as the instruction set.

Figure D.3 outlines the necessary steps for creating an in-memory representation of a class file using the `CLEmitter` interface, and then writing that class file to the file system.

---

[6]This is a class in the `jminusminus` package under `$j/j--/src` folder. The classes that `CLEmitter` depends on are also in that package and have a `CL` prefix.

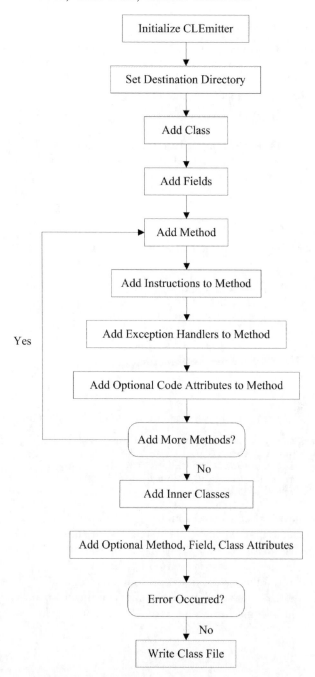

FIGURE D.3 A recipe for creating a class file.

D.4.2 `CLEmitter` **Interface**

The `CLEmitter` interface support: creating a Java class representation in memory; adding inner classes, fields, methods, exception handlers, and attributes to the class; and converting the in-memory representation of the class to a `java.lang.Class` representation, both in memory and on the file system.

Instantiate `CLEmitter`

In order to create a class, one must create an instance of `CLEmitter` as the first step in the process. All subsequent steps involve sending an appropriate message to that instance. Each instance corresponds to a single class.

To instantiate a `CLEmitter`, one simply invokes its constructor:

```
CLEmitter output = new CLEmitter(true);
```

Change the argument to the constructor to false if only an in-memory representation of the class is needed, in which case the class will not be written to the file system.

One then goes about adding classes, method, fields, and instructions. There are methods for adding all of these.

Set Destination Directory for the Class

The destination directory for the class file can be set by sending to the `CLEmitter` instance the following message:

```
public void destinationDir(String destDir)
```

where `destDir` is the directory where the class file will be written. If the class that is being created specifies a package, then the class file will be written to the directory obtained by appending the package name to the destination directory. If a destination directory is not set explicitly, the default is ".", the current directory.

Add Class

A class can be added by sending to the `CLEmitter` instance the following message:

```
public void addClass(ArrayList<String> accessFlags,
                     String thisClass,
                     String superClass,
                     ArrayList<String> superInterfaces,
                     boolean isSynthetic)
```

where `accessFlags`[7] is a list of class access and property flags, `thisClass` is the name of the class in internal form, `superClass` is the name of the super class in internal form, `superInterfaces` is a list of direct super interfaces of the class in internal form, and `isSynthetic` specifies whether the class is created by the compiler.

If the class being added is an interface, `accessFlags` must contain appropriate ("interactive" and "abstract") modifiers, `superInterfaces` must contain the names of the interface's super interfaces (if any) in internal form, and `superClass` must always be "java/lang/Object".

Add Inner Classes

While an inner class C is just another class and can be created using the `CLEmitter` interface, the parent class P that contains the class C must be informed about C, which can be done by sending the `CLEmitter` instance for P the following message:

```
public void addInnerClass(ArrayList<String> accessFlags,
                          String innerClass,
                          String outerClass,
                          String innerName)
```

---

[7]Note that the `CLEmitter` expects the access and property flags as `String`s and internally translates them to a mask of flags. For example, "public" is translated to `ACC_PUBLIC` (0x0001).

where `accessFlags` is a list of inner class access and property flags, `innerClass` is the name of the inner class in internal form, `outerClass` is the name of the outer class in internal form, and `innerName` is the simple name of the inner class.

Add Field

After a class is added, fields can be added to the class by sending to the `CLEmitter` instance the following message:

```
public void addField(ArrayList<String> accessFlags,
                     String name,
                     String type,
                     boolean isSynthetic)
```

where `accessFlags` is a list of field access and property flags, `name` is the name of the field, `type` is the type descriptor for the field, and `isSynthetic` specifies whether the field is created by the compiler.

A `final` field of type `int`, `short`, `byte`, `char`, `long`, `float`, `double`, or `String` with an initialization must be added to the class by sending to the `CLEmitter` instance the respective message from the list of messages below[8]:

```
public void addField(ArrayList<String> accessFlags,
                     String name,
                     String type,
                     boolean isSynthetic,
                     int i)

public void addField(ArrayList<String> accessFlags,
                     String name,
                     String type,
                     boolean isSynthetic,
                     float f)

public void addField(ArrayList<String> accessFlags,
                     String name,
                     String type,
                     boolean isSynthetic,
                     long l)

public void addField(ArrayList<String> accessFlags,
                     String name,
                     String type,
                     boolean isSynthetic,
                     double d)

public void addField(ArrayList<String> accessFlags,
                     String name,
                     String type,
                     boolean isSynthetic,
                     String s)
```

The last parameter in each of the above messages is the value of the field. Note that the JVM treats `short`, `byte`, and `char` types as `int`.

Add Method

A method can be added to the class by sending to the `CLEmitter` instance the following message:

---

[8]The `field_info` structure for such fields must specify a `ConstantValueAttribute` reflecting the value of the constant, and these `addField()` variants take care of that.

```
public void addMethod(ArrayList<String> accessFlags,
                      String name,
                      String descriptor,
                      ArrayList<String> exceptions,
                      boolean isSynthetic)
```

where `accessFlags` is a list of method access and property flags, `name` is the name[9] of the method, `descriptor` is the method descriptor, `exceptions` is a list of exceptions in internal form that this method throws, and `isSythetic` specifies whether the method is created by the compiler.

For example, one may add a method using

```
accessFlags.add("public");
output.addMethod(accessFlags, "factorial", "(I)I", exceptions, false);
```

where `accessFlags` is a list of method access and property flags, "factorial" is the name[10] of the method, "(I)I" is the method descriptor, `exceptions` is a list of exceptions in internal form that this method throws, and the argument `false` indicates that the method appears in source code.

A comment on the method descriptor is warranted. The method descriptor describes the method's signature in a format internal to the JVM. The `I` is the internal type descriptor for the primitive type `int`, so the "(I)I" specifies that the method takes one integer argument and returns a value having integer type.

Add Exception Handlers to Method

An exception handler to code a `try-catch` block can be added to a method by sending to the `CLEmitter` instance the following message:

```
public void addExceptionHandler(String startLabel,
                                String endLabel,
                                String handlerLabel,
                                String catchType)
```

where `startLabel` marks the beginning of the `try` block, `endLabel` marks the end of the `try` block, `handlerLabel` marks the beginning of a `catch` block, and `catchType` specifies the exception that is to be caught in internal form. If `catchType` is `null`, this exception handler is called for all exceptions; this is used to implement `finally`.

`createLabel()` and `addLabel(String label)` can be invoked on the `CLEmitter` instance to create unique labels and for adding them to mark instructions in the code indicating where to jump to.

A method can specify as many exception handlers as there are exceptions that are being caught in the method.

Add Optional Method, Field, Class, Code Attributes

Attributes are used in the ClassFile (CLFile), `field_info`(CLFieldInfo), `method_info` (CLMethodInfo), and `Code_attribute` (CLCodeAttribute) structures of the class file format. While there are many kinds of attributes, only some are mandatory; these include `InnerClasses_attribute` (class attribute), `Synthetic_attribute` (class, field, and method attribute), `Code_attribute` (method attribute), and `Exceptions_attribute` (method attribute).

`CLEmitter` implicitly adds the required attributes to the appropriate structures. The

---

[9]Instance initializers must have the name `<init>` and static initializers must have the name `<clinit>`.
[10]Instance constructors must have the name `<init>`.

optional attributes can be added by sending to the CLEmitter instance one of the following messages:

```
public void addMethodAttribute(CLAttributeInfo attribute)
public void addFieldAttribute(CLAttributeInfo attribute)
public void addClassAttribute(CLAttributeInfo attribute)
public void addCodeAttribute(CLAttributeInfo attribute)
```

Note that for adding optional attributes, you need access to the constant pool table, which the CLEmitter exposes through its constantPool() method, and also the program counter pc, which it exposes through its pc() method. The abstractions for all the attributes (code, method, field, and class) are defined in the CLAttributeInfo class.

Checking for Errors

The caller, at any point during the creation of the class, can check if there was an error, by sending to the CLEmitter the following message:

```
public boolean errorHasOccurred()
```

Write Class File

The in-memory representation of the class can be written to the file system by sending to the CLEmitter instance the following message:

```
public void write()
```

The destination directory for the class is either the default (current) directory or the one specified by invoking destinationDir(String destDir) method. If the class specifies a package, then the class will be written to the directory obtained by appending the package information to the destination directory.

Alternatively, the representation can be converted to java.lang.Class representation in memory by sending to the CLEmitter instance the following message:

```
public Class toClass()
```

D.5 JVM Instruction Set

The instructions supported by the JVM can be categorized into various groups: object, field, method, array, arithmetic, bit, comparison, conversion, flow control, load and store, and stack instructions. In this section, we provide a brief summary of the instructions belonging to each group. The summary includes the mnemonic[11] for the instruction, a one-line description of what the instruction does, and how the instruction affects the operand stack. For each set of instructions, we also specify the CLEmitter method to invoke while generating class files, to add instructions from that set to the code section of methods.

For each instruction, we represent the operand stack as follows:

$$\cdots, value1, value2 \Rightarrow \cdots, result$$

which means that the instruction begins by having $value2$ on top of the operand stack with

---

[11]These mnemonics (symbolic opcodes) are defined in jminusminus.CLConstants.

*value*1 just beneath it. As a result of the execution of the instruction, *value*1 and *value*2 are popped from the operand stack and replaced by *result* value, which has been calculated by the instruction. The remainder of the operand stack, represented by an ellipsis ($\cdots$), is unaffected by the instruction's execution. Values of types `long` and `double` are represented by a single entry on the operand stack.

D.5.1   Object Instructions

Mnemonic	Operation	Operand Stack
new	Create new object	$\cdots \Rightarrow \cdots, objectref$
instanceof	Determine if object is of given type	$\cdots, objectref \Rightarrow \cdots, result$
checkcast	Check whether object is of given type	$\cdots, objectref \Rightarrow \cdots, objectref$

The above instructions can be added to the code section of a method by sending the `CLEmitter` instance the following message:

```
public void addReferenceInstruction(int opcode,
                                    String type)
```

where `opcode` is the mnemonic of the instruction to be added, and `type` is the reference type in internal form or a type descriptor if it is an array type.

D.5.2   Field Instructions

Mnemonic	Operation	Operand Stack
getfield	Get field from object	$\cdots, objectref \Rightarrow \cdots, value$
putfield	Set field in object	$\cdots, objectref, value \Rightarrow \cdots$
getstatic	Get static field from class	$\cdots \Rightarrow \cdots, value$
putstatic	Set static field in class	$\cdots, value \Rightarrow \cdots$

The above instructions can be added to the code section of a method by sending the `CLEmitter` instance the following message:

```
public void addMemberAccessInstruction(int opcode,
                                       String target,
                                       String name,
                                       String type)
```

where `opcode` is the mnemonic of the instruction to be added, `target` is the name (in internal form) of the class to which the field belongs, `name` is the name of the field, and `type` is the type descriptor of the field.

D.5.3 Method Instructions

Mnemonic	Operation	Operand Stack
invokevirtual	Invoke instance method; dispatch based on class	$\cdots, objectref, [arg1, [arg2 \cdots]] \Rightarrow \cdots$
invokeinterface	Invoke interface method	$\cdots, objectref, [arg1, [arg2 \cdots]] \Rightarrow \cdots$
invokespecial	Invoke instance method; special handling for superclass, private, and instance initialization method invocations	$\cdots, objectref, [arg1, [arg2 \cdots]] \Rightarrow \cdots$
invokestatic	Invoke a class (static) method	$\cdots, [arg1, [arg2 \cdots]] \Rightarrow \cdots$
invokedynamic	Invoke instance method; dispatch based on class	$\cdots, objectref, [arg1, [arg2 \cdots]] \Rightarrow \cdots$

The above instructions can be added to the code section of a method by sending the CLEmitter instance the following message:

```
public void addMemberAccessInstruction(int opcode,
                                       String target,
                                       String name,
                                       String type)
```

where opcode is the mnemonic of the instruction to be added, target is the name (in internal form) of the class to which the method belongs, name is the name of the method, and type is the type descriptor of the method.

Mnemonic	Operation	Operand Stack
ireturn	Return int from method	$\cdots, value \Rightarrow [empty]$
lreturn	Return long from method	$\cdots, value \Rightarrow [empty]$
freturn	Return float from method	$\cdots, value \Rightarrow [empty]$
dreturn	Return double from method	$\cdots, value \Rightarrow [empty]$
areturn	Return reference from method	$\cdots, objectref \Rightarrow [empty]$
return	Return void from method	$\cdots \Rightarrow [empty]$

The above instructions can be added to the code section of a method by sending the CLEmitter instance the following message:

```
public void addNoArgInstruction(int opcode)
```

where opcode is the mnemonic of the instruction to be added.

D.5.4   Array Instructions

Mnemonic	Operation	Operand Stack
newarray	Create new array	$\cdots, count \Rightarrow \cdots, arrayref$
anewarray	Create new array of reference type	$\cdots, count \Rightarrow \cdots, arrayref$

The above instructions can be added to the code section of a method by sending the `CLEmitter` instance the following message:

```
public void addArrayInstruction(int opcode,
                                String type)
```

where `opcode` is the mnemonic of the instruction to be added, and `type` is the type descriptor of the array.

Mnemonic	Operation	Operand Stack
multianewarray	Create new multidimensional array	$\cdots, count1, [count2, \cdots]$ $\Rightarrow \cdots, arrayref$

The above instruction can be added to the code section of a method by sending the `CLEmitter` instance the following message:

```
public void addMULTIANEWARRAYInstruction(String type,
                                         int dim)
```

where `type` is the type descriptor of the array, and `dim` is the number of dimensions.

Mnemonic	Operation	Operand Stack
baload	Load byte or boolean from array	$\cdots, arrayref, index \Rightarrow \cdots, value$
caload	Load char from array	$\cdots, arrayref, index \Rightarrow \cdots, value$
saload	Load short from array	$\cdots, arrayref, index \Rightarrow \cdots, value$
iaload	Load int from array	$\cdots, arrayref, index \Rightarrow \cdots, value$
laload	Load long from array	$\cdots, arrayref, index \Rightarrow \cdots, value$
faload	Load float from array	$\cdots, arrayref, index \Rightarrow \cdots, value$
daload	Load double from array	$\cdots, arrayref, index \Rightarrow \cdots, value$
aaload	Load from reference array	$\cdots, arrayref, index \Rightarrow \cdots, value$
bastore	Store into byte or boolean array	$\cdots, arrayref, index, value \Rightarrow \cdots$
castore	Store into char array	$\cdots, arrayref, index, value \Rightarrow \cdots$
sastore	Store into short array	$\cdots, arrayref, index, value \Rightarrow \cdots$
iastore	Store into int array	$\cdots, arrayref, index, value \Rightarrow \cdots$
lastore	Store into long array	$\cdots, arrayref, index, value \Rightarrow \cdots$
fastore	Store into float array	$\cdots, arrayref, index, value \Rightarrow \cdots$
dastore	Store into double array	$\cdots, arrayref, index, value \Rightarrow \cdots$
aastore	Store into reference array	$\cdots, arrayref, index, value \Rightarrow \cdots$
arraylength	Get length of array	$\cdots, arrayref \Rightarrow \cdots, length$

The above instructions can be added to the code section of a method by sending the `CLEmitter` instance the following message:

```
public void addNoArgInstruction(int opcode)
```

where `opcode` is the mnemonic of the instruction to be added.

D.5.5   Arithmetic Instructions

Mnemonic	Operation	Operand Stack
iadd	Add `int`	$\cdots, value1, value2 \Rightarrow \cdots, result$
ladd	Add `long`	$\cdots, value1, value2 \Rightarrow \cdots, result$
fadd	Add `float`	$\cdots, value1, value2 \Rightarrow \cdots, result$
dadd	Add `double`	$\cdots, value1, value2 \Rightarrow \cdots, result$
isub	Subtract `int`	$\cdots, value1, value2 \Rightarrow \cdots, result$
lsub	Subtract `long`	$\cdots, value1, value2 \Rightarrow \cdots, result$
fsub	Subtract `float`	$\cdots, value1, value2 \Rightarrow \cdots, result$
dsub	Subtract `double`	$\cdots, value1, value2 \Rightarrow \cdots, result$
imul	Multiply `int`	$\cdots, value1, value2 \Rightarrow \cdots, result$
lmul	Multiply `long`	$\cdots, value1, value2 \Rightarrow \cdots, result$
fmul	Multiply `float`	$\cdots, value1, value2 \Rightarrow \cdots, result$
dmul	Multiply `double`	$\cdots, value1, value2 \Rightarrow \cdots, result$
idiv	Divide `int`	$\cdots, value1, value2 \Rightarrow \cdots, result$
ldiv	Divide `long`	$\cdots, value1, value2 \Rightarrow \cdots, result$
fdiv	Divide `float`	$\cdots, value1, value2 \Rightarrow \cdots, result$
ddiv	Divide `double`	$\cdots, value1, value2 \Rightarrow \cdots, result$
irem	Remainder `int`	$\cdots, value1, value2 \Rightarrow \cdots, result$
lrem	Remainder `long`	$\cdots, value1, value2 \Rightarrow \cdots, result$
frem	Remainder `float`	$\cdots, value1, value2 \Rightarrow \cdots, result$
drem	Remainder `double`	$\cdots, value1, value2 \Rightarrow \cdots, result$
ineg	Negate `int`	$\cdots, value \Rightarrow \cdots, result$
lneg	Negate `long`	$\cdots, value \Rightarrow \cdots, result$
fneg	Negate `float`	$\cdots, value \Rightarrow \cdots, result$
dneg	Negate `double`	$\cdots, value \Rightarrow \cdots, result$

The above instructions can be added to the code section of a method by sending the `CLEmitter` instance the following message:

```
public void addNoArgInstruction(int opcode)
```

where `opcode` is the mnemonic of the instruction to be added.

D.5.6 Bit Instructions

Mnemonic	Operation	Operand Stack
ishl	Shift left int	$\cdots, value1, value2 \Rightarrow \cdots, result$
ishr	Arithmetic shift right int	$\cdots, value1, value2 \Rightarrow \cdots, result$
iushr	Logical shift right int	$\cdots, value1, value2 \Rightarrow \cdots, result$
lshl	Shift left long	$\cdots, value1, value2 \Rightarrow \cdots, result$
lshr	Arithmetic shift right long	$\cdots, value1, value2 \Rightarrow \cdots, result$
lushr	Logical shift right long	$\cdots, value1, value2 \Rightarrow \cdots, result$
ior	Boolean OR int	$\cdots, value1, value2 \Rightarrow \cdots, result$
lor	Boolean OR long	$\cdots, value1, value2 \Rightarrow \cdots, result$
iand	Boolean AND int	$\cdots, value1, value2 \Rightarrow \cdots, result$
land	Boolean AND long	$\cdots, value1, value2 \Rightarrow \cdots, result$
ixor	Boolean XOR int	$\cdots, value1, value2 \Rightarrow \cdots, result$
lxor	Boolean XOR long	$\cdots, value1, value2 \Rightarrow \cdots, result$

The above instructions can be added to the code section of a method by sending the `CLEmitter` instance the following message:

```
public void addNoArgInstruction(int opcode)
```

where `opcode` is the mnemonic of the instruction to be added.

D.5.7 Comparison Instructions

Mnemonic	Operation	Operand Stack
dcmpg	Compare double; result is 1 if at least one value is NaN	$\cdots, value1, value2 \Rightarrow \cdots, result$
dcmpl	Compare double; result is -1 if at least one value is NaN	$\cdots, value1, value2 \Rightarrow \cdots, result$
fcmpg	Compare float; result is 1 if at least one value is NaN	$\cdots, value1, value2 \Rightarrow \cdots, result$
fcmpl	Compare float; result is -1 if at least one value is NaN	$\cdots, value1, value2 \Rightarrow \cdots, result$
lcmp	Compare long	$\cdots, value1, value2 \Rightarrow \cdots, result$

The above instructions can be added to the code section of a method by sending the `CLEmitter` instance the following message:

```
public void addNoArgInstruction(int opcode)
```

where `opcode` is the mnemonic of the instruction to be added.

D.5.8 Conversion Instructions

Mnemonic	Operation	Operand Stack
i2b	Convert `int` to `byte`	$\cdots, value \Rightarrow \cdots, result$
i2c	Convert `int` to `char`	$\cdots, value \Rightarrow \cdots, result$
i2s	Convert `int` to `short`	$\cdots, value \Rightarrow \cdots, result$
i2l	Convert `int` to `long`	$\cdots, value \Rightarrow \cdots, result$
i2f	Convert `int` to `float`	$\cdots, value \Rightarrow \cdots, result$
i2d	Convert `int` to `double`	$\cdots, value \Rightarrow \cdots, result$
l2f	Convert `long` to `float`	$\cdots, value \Rightarrow \cdots, result$
l2d	Convert `long` to `double`	$\cdots, value \Rightarrow \cdots, result$
l2i	Convert `long` to `int`	$\cdots, value \Rightarrow \cdots, result$
f2d	Convert `float` to `double`	$\cdots, value \Rightarrow \cdots, result$
f2i	Convert `float` to `int`	$\cdots, value \Rightarrow \cdots, result$
f2l	Convert `float` to `long`	$\cdots, value \Rightarrow \cdots, result$
d2i	Convert `double` to `int`	$\cdots, value \Rightarrow \cdots, result$
d2l	Convert `double` to `long`	$\cdots, value \Rightarrow \cdots, result$
d2f	Convert `double` to `float`	$\cdots, value \Rightarrow \cdots, result$

The above instructions can be added to the code section of a method by sending the `CLEmitter` instance the following message:

```
public void addNoArgInstruction(int opcode)
```

where `opcode` is the mnemonic of the instruction to be added.

D.5.9 Flow Control Instructions

Mnemonic	Operation	Operand Stack
ifeq	Branch if `int` comparison value $== 0$ true	$\cdots, value \Rightarrow \cdots$
ifne	Branch if `int` comparison value $!= 0$ true	$\cdots, value \Rightarrow \cdots$
iflt	Branch if `int` comparison value < 0 true	$\cdots, value \Rightarrow \cdots$
ifgt	Branch if `int` comparison value > 0 true	$\cdots, value \Rightarrow \cdots$
ifle	Branch if `int` comparison value ≤ 0 true	$\cdots, value \Rightarrow \cdots$
ifge	Branch if `int` comparison value ≥ 0 true	$\cdots, value \Rightarrow \cdots$
ifnull	Branch if `reference` is null	$\cdots, value \Rightarrow \cdots$
ifnonnull	Branch if `reference` is not null	$\cdots, value \Rightarrow \cdots$
if_icmpeq	Branch if `int` comparison value1 $==$ value2 true	$\cdots, value1, value2 \Rightarrow \cdots$
if_icmpne	Branch if `int` comparison value1 $!=$ value2 true	$\cdots, value1, value2 \Rightarrow \cdots$

Mnemonic	Operation	Operand Stack
if_icmplt	Branch if int comparison value1 < value2 true	$\cdots, value1, value2 \Rightarrow \cdots$
if_icmpgt	Branch if int comparison value1 > value2 true	$\cdots, value1, value2 \Rightarrow \cdots$
if_icmple	Branch if int comparison value1 ≤ value2 true	$\cdots, value1, value2 \Rightarrow \cdots$
if_icmpge	Branch if int comparison value1 ≥ value2 true	$\cdots, value1, value2 \Rightarrow \cdots$
if_acmpeq	Branch if reference comparison value1 == value2 true	$\cdots, value1, value2 \Rightarrow \cdots$
if_acmpne	Branch if reference comparison value1 != value2 true	$\cdots, value1, value2 \Rightarrow \cdots$
goto	Branch always	No change
goto_w	Branch always (wide index)	No change
jsr	Jump subroutine	$\cdots \Rightarrow \cdots, address$
jsr_w	Jump subroutine (wide index)	$\cdots \Rightarrow \cdots, address$

The above instructions can be added to the code section of a method by sending the `CLEmitter` instance the following message:

```
public void addBranchInstruction(int opcode,
                                 String label)
```

where `opcode` is the mnemonic of the instruction to be added, and `label` is the target instruction label.

Mnemonic	Operation	Operand Stack
ret	Return from subroutine	No change

The above instruction can be added to the code section of a method by sending the `CLEmitter` instance the following message:

```
public void addOneArgInstruction(int opcode,
                                 int arg)
```

where `opcode` is the mnemonic of the instruction to be added, and `arg` is the index of the local variable containing the return address.

Mnemonic	Operation	Operand Stack
lookupswitch	Access jump table by key match and jump	$\cdots, key \Rightarrow \cdots$

The above instruction can be added to the code section of a method by sending the `CLEmitter` instance the following message:

```
public void addLOOKUPSWITCHInstruction(String defaultLabel,
                                       int numPairs,
                                       TreeMap<Integer, String>
                                       matchLabelPairs)
```

where `defaultLabel` is the jump label for the default value, `numPairs` is the number of pairs in the match table, and the `matchLabelPairs` is the key match table.

Mnemonic	Operation	Operand Stack
tableswitch	Access jump table by index match and jump	$\cdots, index \Rightarrow \cdots$

The above instruction can be added to the code section of a method by sending the `CLEmitter` instance the following message:

```
public void addTABLESWITCHInstruction(String defaultLabel,
                                      int low,
                                      int high,
                                      ArrayList<String> labels)
```

where `defaultLabel` is the jump label for the default value, `low` is smallest value of index, `high` is the highest value of index, and `labels` is a list of jump labels for each index value from low to high, end values included.

D.5.10 Load Store Instructions

Mnemonic	Operation	Operand Stack
iload_n; $n \in [0 \ldots 3]$	Load `int` from local variable at index n	$\cdots \Rightarrow \cdots, value$
lload_n; $n \in [0 \ldots 3]$	Load `long` from local variable at index n	$\cdots \Rightarrow \cdots, value$
fload_n; $n \in [0 \ldots 3]$	Load `float` from local variable at index n	$\cdots \Rightarrow \cdots, value$
dload_n; $n \in [0 \ldots 3]$	Load `double` from local variable at index n	$\cdots \Rightarrow \cdots, value$
aload_n; $n \in [0 \ldots 3]$	Load reference from local variable at index n	$\cdots \Rightarrow \cdots, objectref$
istore_n; $n \in [0 \ldots 3]$	Store `int` into local variable at index n	$\cdots, value \Rightarrow \cdots$
lstore_n; $n \in [0 \ldots 3]$	Store `long` into local variable at index n	$\cdots, value \Rightarrow \cdots$
fstore_n; $n \in [0 \ldots 3]$	Store `float` into local variable at index n	$\cdots, value \Rightarrow \cdots$
dstore_n; $n \in [0 \ldots 3]$	Store `double` into local variable at index n	$\cdots, value \Rightarrow \cdots$
astore_n; $n \in [0 \ldots 3]$	Store reference into local variable at index n	$\cdots, objectref \Rightarrow \cdots$

Mnemonic	Operation	Operand Stack
iconst_n; $n \in [0 \ldots 5]$	Push int constant n	$\cdots \Rightarrow \cdots, value$
iconst_m1	Push int constant -1	$\cdots \Rightarrow \cdots, value$
lconst_n; $n \in [0 \ldots 1]$	Push long constant n	$\cdots \Rightarrow \cdots, value$
fconst_n; $n \in [0 \ldots 2]$	Push float constant n	$\cdots \Rightarrow \cdots, value$
dconst_n; $n \in [0 \ldots 1]$	Push double constant n	$\cdots \Rightarrow \cdots, value$
aconst_null	Push null	$\cdots \Rightarrow \cdots, null$
wide[12]	Modifies the behavior another instruction[13] by extending local variable index by additional bytes	Same as modified instruction

The above instructions can be added to the code section of a method by sending the `CLEmitter` instance the following message:

```
public void addNoArgInstruction(int opcode)
```

where `opcode` is the mnemonic of the instruction to be added.

Mnemonic	Operation	Operand Stack
iload	Load int from local variable at an index	$\cdots \Rightarrow \cdots, value$
lload	Load long from local variable at an index	$\cdots \Rightarrow \cdots, value$
fload	Load float from local variable at an index	$\cdots \Rightarrow \cdots, value$
dload	Load double from local variable at an index	$\cdots \Rightarrow \cdots, value$
aload	Load reference from local variable at an index	$\cdots \Rightarrow \cdots, objectref$
istore	Store int into local variable at an index	$\cdots, value \Rightarrow \cdots$
lstore	Store long into local variable at an index	$\cdots, value \Rightarrow \cdots$
fstore	Store float into local variable at an index	$\cdots, value \Rightarrow \cdots$
dstore	Store double into local variable at an index	$\cdots, value \Rightarrow \cdots$
astore	Store reference into local variable at an index	$\cdots, objectref \Rightarrow \cdots$

The above instructions can be added to the code section of a method by sending the `CLEmitter` instance the following message:

```
public void addOneArgInstruction(int opcode,
                                 int arg)
```

---

[12]The `CLEmitter` interface implicitly adds this instruction where necessary.

[13]Instructions that can be widened are `iinc`, `iload`, `fload`, `aload`, `lload`, `dload`, `istore`, `fstore`, `astore`, `lstore`, and `ret`.

where `opcode` is the mnemonic of the instruction to be added, and `arg` is the index of the local variable.

Mnemonic	Operation	Operand Stack
iinc	Increment local variable by constant	No change

The above instruction can be added to the code section of a method by sending the `CLEmitter` instance the following message:

```
public void addIINCInstruction(int index,
                               int constVal)
```

where `index` is the local variable index, and `constVal` is the constant by which to increment.

Mnemonic	Operation	Operand Stack
bipush	Push byte	$\cdots \Rightarrow \cdots, value$
sipush	Push short	$\cdots \Rightarrow \cdots, value$

The above instructions can be added to the code section of a method by sending the `CLEmitter` instance the following message:

```
public void addOneArgInstruction(int opcode,
                                 int arg)
```

where `opcode` is the mnemonic of the instruction to be added, and `arg` is `byte` or `short` value to push.

`ldc`[15] instruction can be added to the code section of a method using one of the following `CLEmitter` functions:

```
public void addLDCInstruction(int i)
public void addLDCInstruction(long l)
public void addLDCInstruction(float f)
public void addLDCInstruction(double d)
public void addLDCInstruction(String s)
```

where the argument is the type of the item.

D.5.11 Stack Instructions

Mnemonic	Operation	Operand Stack
pop	Pop the top operand stack value	$\cdots, value \Rightarrow \cdots$
pop2	Pop the top one or two operand stack values	$\cdots, value2, value1 \Rightarrow \cdots$
		$\cdots, value \Rightarrow \cdots$
dup	Duplicate the top operand stack value	$\cdots, value \Rightarrow \cdots, value, value$

---

[15]The `CLEmitter` interface implicitly adds `ldc_w` and `ldc2_w` instructions where necessary.

Mnemonic	Operation	Operand Stack
dup_x1	Duplicate the top operand stack value and insert two values down	$\cdots, value2, value1 \Rightarrow \cdots, value1, value2, value1$
dup_x2	Duplicate the top operand stack value and insert two or three values down	$\cdots, value3, value2, value1 \Rightarrow$ $\cdots, value1, value3, value2, value1$
		$\cdots, value2, value1 \Rightarrow \cdots, value1, value2, value1$
dup2	Duplicate the top one or two operand stack values	$\cdots, value2, value1 \Rightarrow$ $\cdots, value2, value1, value2, value1$
		$\cdots, value \Rightarrow \cdots, value, value$
dup2_x1	Duplicate the top one or two operand stack values and insert two or three values down	$\cdots, value3, value2, value1 \Rightarrow$ $\cdots, value2, value1, value3, value2, value1$
		$\cdots, value2, value1 \Rightarrow \cdots, value1, value2, value1$
dup2_x2	Duplicate the top one or two operand stack values and insert two, three, or four values down	$\cdots, value4, value3, value2, value1 \Rightarrow$ $\cdots, value2, value1, value4, value3, value2, value1$
		$\cdots, value3, value2, value1 \Rightarrow$ $\cdots, value1, value3, value2, value1$
		$\cdots, value3, value2, value1 \Rightarrow$ $\cdots, value2, value1, value3, value2, value1$
		$\cdots, value2, value1 \Rightarrow \cdots, value1, value2, value1$
swap	Swap the top two operand stack values	$\cdots, value2, value1 \Rightarrow \cdots, value1, value2$

The above instructions can be added to the code section of a method by sending the CLEmitter instance the following message:

```
public void addNoArgInstruction(int opcode)
```

where opcode is the mnemonic of the instruction to be added.

D.5.12 Other Instructions

Mnemonic	Operation	Operand Stack
nop	Do nothing	No change
athrow	Throw exception or error	$\cdots, objectref \Rightarrow \cdots, objectref$
monitorenter	Enter monitor for object	$\cdots, objectref \Rightarrow \cdots$
monitorexit	Exit monitor for object	$\cdots, objectref \Rightarrow \cdots$

The above instructions can be added to the code section of a method by sending the CLEmitter instance the following message:

```
public void addNoArgInstruction(int opcode)
```

where `opcode` is the mnemonic of the instruction to be added.

D.6   Further Readings

The *JVM Specification* [Lindholm and Yellin, 1999] describes the JVM in detail. See Chapter 4 of the specification for a detailed description of the class file format. See Chapter 6 for detailed information about the format of each JVM instruction and the operation it performs. See Chapter 7 in that text for hints on compiling various Java language constructs to JVM byte code.

Appendix E

MIPS and the SPIM Simulator

E.1 Introduction

In addition to compiling j-- source programs to JVM byte code, the j-- compiler can also produce native code for the MIPS architecture. The compiler does not actually produce binary code that can be run on such a machine, but it produces human-readable assembly code that can be executed by a program that simulates a MIPS machine. The advantage of using a simulator is that one can easily debug the translated program, and does not need access to an actual MIPS machine in order to run the program. The simulator that we will use is called SPIM (MIPS spelled backward). Chapter 7 describes the MIPS architecture and the SPIM simulator in some detail. In this appendix we describe how one can obtain and run SPIM, compile j-- programs to SPIM code, and extend the JVM to SPIM translator to convert more JVM instructions to SPIM code. The MIPS (and so, SPIM) instruction set is described very nicely in [Larus, 2009].

E.2 Obtaining and Running SPIM

SPIM implements both a command-line and a graphical user interface (GUI). The GUI for SPIM, called QtSpim, is developed using the Qt UI Framework[1]. Qt being cross-platform, QtSpim will run on Windows, Linux, and Mac OS X.

One can download the compiled version of SPIM for Windows, Mac OS X, or Debian-based (including Ubuntu) Linux from http://sourceforge.net/projects /spimsimulator/files/, or build SPIM from source files obtained from http:// spimsimulator.svn.sourceforge.net/viewvc/spimsimulator/. The README file in the source distribution documents the installation steps.

E.3 Compiling j-- Programs to SPIM Code

The j-- compiler supports the following command-line syntax:

```
Usage: j-- <options> <source file>
where possible options include:
  -t Only tokenize input and print tokens to STDOUT
```

---

[1]A cross-platform application and UI framework with APIs for C++ programming and Qt Quick for rapid UI creation.

```
-p Only parse input and print AST to STDOUT
-pa Only parse and pre-analyze input and print AST to STDOUT
-a Only parse, pre-analyze, and analyze input and print AST to STDOUT
-s <naive|linear|graph> Generate SPIM code
-r <num> Max. physical registers (1-18) available for allocation; default=8
-d <dir> Specify where to place output files; default=.
```

In order to compile the *j*-- source program to SPIM code, one must specify the -s and the optional -r switches along with appropriate arguments. The -s switch tells the compiler that we want SPIM code for output and the associated argument specifies the register allocation mechanism that must be used; the options are "naive" for a naïve round robin allocation, "linear" for allocation using the linear scan algorithm, or "graph" for allocation using the graph coloring algorithm. See Chapter 7 for a detailed description of these register allocation algorithms. The optional -r switch forces the compiler to allocate only up to a certain number of physical registers, specified as argument; the argument can take values between 3 and 18, inclusive, for linear register allocation, and between 1 and 18, inclusive, for the other two register allocation methods. If this switch is not specified, then a default value of 8 is used. Registers are allocated starting at $t0[2].

For example, here is the command-line syntax for compiling the Factorial program to SPIM code allocating up to 18 (the maximum) physical registers using the naïve allocation procedure:

```
> $j/j--/bin/j-- -s naive -r 18 $j/j--/tests/spim/Factorial.java
```

The compiler first translates the source program to JVM byte code in memory, which in turn is translated into SPIM code. The compiler prints to STDOUT the details of the translation process for each method, such as the basic block structure with JVM instructions represented as tuples, the high-level (HIR) instructions in each basic block, the low-level (LIR) instructions for each basic block before allocation of physical registers to virtual registers, the intervals and associated ranges for each register, and the LIR instructions for each basic block after the physical registers have been allocated and spills have been handled where needed. In addition, the compiler also produces a target .s file containing the SPIM code (Factorial.s in above example) in the folder in which the command is run. One can specify an alternate folder name for the target .s file using the -d command-line switch.

Once the .s file has been produced, there are several ways to execute it using the SPIM simulator. The simplest way is to run the following command:

```
> spim -file <input file>
```

where spim is the command-line version of the SPIM simulator. The -file argument specifies the name of the file with a .s extension that contains the SPIM code to execute. For example, we can execute the Factorial.s file from above as

```
> spim -file Factorial.s
```

which produces the following output:

```
SPIM Version 8.0 of January 8, 2010
Copyright 1990-2010, James R. Larus.
All Rights Reserved.
See the file README for a full copyright notice.
Loaded: /usr/lib/spim/exceptions.s
5040
5040
```

---

[2]That is a total of up to eighteen registers ($t0, $t1, ..., $t9, $s0, ..., $s7) available for allocation.

Yet another way of running the .s file is to launch the SPIM command-line interpreter, and at the (spim) prompt, use SPIM commands `load` and `run` to load and execute a .s file. The `quit()` command exits the SPIM interpreter. For example, `Factorial.s` can be executed as follows:

```
> spim
SPIM Version 8.0 of January 8, 2010
Copyright 1990-2010, James R. Larus.
All Rights Reserved.
See the file README for a full copyright notice.
Loaded: /usr/lib/spim/exceptions.s
(spim) load "Factorial.s"
(spim) run
5040
5040
(spim) quit()
>
```

For a complete list of all the SPIM commands that are supported by the command-line interpreter, execute the command `help` at the (spim) prompt.

Finally, one can load a .s file into the QtSpim user interface and click the run button to execute the program. QtSpim offers a convenient interface for inspecting values in registers and memory, and also facilitates easy debugging of programs by stepping through the instructions, one at a time.

E.4   Extending the JVM-to-SPIM Translator

The *j--* compiler translates only a limited number of the JVM instructions to SPIM code, just enough to compile and run the four *j--* programs (`Factorial`, `GCD`, `Fibonacci`, and `Formals`) under `$j/j--/tests/spim`. Nevertheless, it provides sufficient machinery with which one could build on to translate more instructions to SPIM code.

Unlike the *j--* programs that are compiled for the JVM, the ones that target SPIM cannot import any Java libraries. Any low-level library support for SPIM must be made available in the form of SPIM run-time files: a .s file and a corresponding Java wrapper for each library. For example, if *j--* programs that target SPIM need to perform input/output (IO) operations, it must import `spim.SPIM`, which is a Java wrapper for the `SPIM.s` file, the run-time library for IO operations in SPIM. These two files can be found under `$j/j--/src/spim`. The Java wrapper only needs to provide the function headers and not the implementation, as it is only needed to make the compiler happy. The actual implementation of the functions resides in the .s file. The compiler copies the run-time .s files verbatim to the end of the compiled .s program. This saves us the trouble of loading the run-time files separately into SPIM before loading the compiled .s program.

If one needs to extend the SPIM run-time to, say, support objects, strings, and arrays, then for each one of them, a .s file and a Java wrapper must be implemented under `$j/j--/src/spim`, and the following code in `NEmitter.write()` must be updated to copy the new run-time libraries to the end of the compiled *j--* program.

```
// Emit SPIM runtime code; just SPIM.s for now.
String[] libs = { "SPIM.s" };
out.printf("# SPIM Runtime\n\n");
for (String lib : libs) {
    ...
```

}

See Chapter 7 for hints on providing run-time representational support for objects, strings, and arrays.

Converting additional JVM instructions, in addition to the ones that already are, into SPIM code, involves the following steps:

- Adding/modifying the HIR representation: This is done in `$j/j--/src/jminusminus /NHIRInstruction.java`. Each class that represents an HIR instruction inherits from the base class `NHIRInstruction`. The additions will be to the constructor of the HIR instruction; the inherited `toLir()` method that converts and returns the low-level (LIR) representation; and the inherited `toString()` method that returns a string representation for the HIR instruction.

- Adding/modifying the LIR representation: This is done in `$j/j--/src/jminusminus /NLIRInstruction.java`. Each class representing an LIR instruction inherits from the base class `NLIRInstruction`. The additions will be to the constructor of the LIR instruction; the inherited `allocatePhysicalRegisters()` method that replaces references to virtual registers in the LIR instruction with references to physical registers and generates LIR spill instructions where needed; the `toSpim()` method that emits SPIM instructions to an output stream (a .s file); and the `toString()` method that returns a string representation for the LIR instruction.

Both `NHIRInstruction` and `NLIRInstruction` have plenty of code in place that can serve as a guideline in implementing the above steps for the new JVM instructions being translated to SPIM code.

E.5   Further Readings

The website [Larus, 2010] for the SPIM simulator provides a wealth of information about the simulator, especially on how to obtain, compile, install, and run the simulator.

For a detailed description of the SPIM simulator, its memory usage, procedure call convention, exceptions and interrupts, input and output, and the MIPS assembly language, see [Larus, 2009].

Bibliography

[Aho et al., 2007] Aho, A., Lam, M., Sethi, R., and Ullman, J. (2007). *Compilers: Principles, Techniques, and Tools*. Prentice-Hall, Inc., Upper Saddle River, NJ.

[Aho et al., 1975] Aho, A. V., Johnson, S. C., and Ullman, J. D. (1975). Deterministic parsing of ambiguous grammars. *Communications of the ACM*, 18(8):441–452.

[Allen and Kennedy, 2002] Allen, R. and Kennedy, K. (2002). *Optimizing Compilers for Modern Architectures: A Dependence-Based Approach*. Morgan Kaufmann Publishers, San Francisco, CA.

[Alpern et al., 2001] Alpern, B., Cocchi, A., Fink, S., and Grove, D. (2001). Efficient implementation of Java interfaces: invokeinterface considered harmless. *ACM SIGPLAN Notices*, 36(11):108–124.

[Appel, 2002] Appel, A. (2002). *Modern Compiler Implementation in Java, Second Edition*. Cambridge University Press, New York, NY.

[Arthorne, 2004] Arthorne, J. (2004). Project builders and natures. *Eclipse Corner Articles*. http://www.eclipse.org/articles/Article-Builders/builders.html.

[Aycock, 2003] Aycock, J. (2003). A brief history of just-in-time. *ACM Computing Surveys*, 35(2):97–113.

[Backus et al., 1963] Backus, J., Bauer, F., Green, J., Katz, C., McCarthy, J., Naur, P., Perlis, A., Rutishauser, H., Samelson, K., Vauquois, B., et al. (1963). Revised report on the algorithmic language Algol 60. *The Computer Journal*, 5(4):349–367.

[Beck and Andres, 2004] Beck, K. and Andres, C. (2004). *Extreme Programming Explained: Embrace Change*. Addison-Wesley Professional, Boston, MA.

[Bothner, 2003] Bothner, P. (2003). Compiling Java with GCJ. *Linux Journal*, 2003(105).

[Box and Sells, 2002] Box, D. and Sells, C. (2002). *Essential .NET: The Common Language Runtime*, volume 1. Addison-Wesley Professional, Boston, MA.

[Bracha, 2004] Bracha, G. (2004). Generics in the Java programming language. *Sun Microsystems Journal*, pages 1–23. http://java.sun.com/j2se/1.5/pdf/generics-tutorial.pdf.

[Briggs et al., 1994] Briggs, P., Cooper, K., and Torczon, L. (1994). Improvements to graph coloring register allocation. *ACM Transactions on Programming Languages and Systems*, 16(3):428–455.

[Burke and Fisher, 1987] Burke, M. and Fisher, G. (1987). A practical method for LR and LL syntactic error diagnosis and recovery. *ACM Transactions on Programming Languages and Systems*, 9(2):164–197.

[Chaitin et al., 1981] Chaitin, G., Auslander, M., Chandra, A., Cocke, J., Hopkins, M., and Markstein, P. (1981). Register allocation via coloring. *Computer Languages*, 6(1):47–57.

[Chaitin, 1982] Chaitin, G. J. (1982). Register allocation & spilling via graph coloring. *ACM SIGPLAN Notices*, 17(6):98–101.

[Clayberg and Rubel, 2009] Clayberg, E. and Rubel, D. (2009). *Eclipse Plug-ins*. Addison-Wesley Professional, Boston, MA.

[Cooper and Dasgupta, 2006] Cooper, K. and Dasgupta, A. (2006). Tailoring graph-coloring register allocation for runtime compilation. In *Proceedings of the International Symposium on Code Generation and Optimization*, pages 39–49. IEEE Computer Society.

[Cooper and Torczon, 2011] Cooper, K. and Torczon, L. (2011). *Engineering a Compiler*. Morgan Kaufmann Publishing, San Francisco, CA.

[Copeland, 2007] Copeland, T. (2007). *Generating Parsers with JavaCC*. Centennial Books, Alexandria, VA.

[Corliss and Lewis, 2007] Corliss, M. and Lewis, E. (2007). *Bantam Java Runtime System Manual*. Bantam Project. http://www.bantamjava.com.

[D'Anjou, 2005] D'Anjou, J. (2005). *The Java Developer's Guide to Eclipse*. Addison-Wesley Professional, Boston, MA.

[Donnelly and Stallman, 2011] Donnelly, C. and Stallman, R. (2011). *Bison, the YACC-compatible Parser Generator*. Free Software Foundation, Boston, MA, 2.5 edition.

[Eclipse, 2011] Eclipse (2011). Eclipse documentation. http://help.eclipse.org/indigo/index.jsp.

[Fink and Qian, 2003] Fink, S. and Qian, F. (2003). Design, implementation and evaluation of adaptive recompilation with on-stack replacement. In *International Symposium on Code Generation and Optimization, 2003*, pages 241–252. IEEE Computer Society.

[Freeman et al., 2004] Freeman, E., Freeman, E., Sierra, K., and Bates, B. (2004). *Head First Design Patterns*. O'Reilly Media, Sebastopol, CA.

[Gamma, 1995] Gamma, E. (1995). *Design Patterns: Elements of Reusable Object-Oriented Software*. Addison-Wesley Professional, Boston, MA.

[George and Appel, 1996] George, L. and Appel, A. W. (1996). Iterated register coalescing. *ACM Transactions on Programming Languages and Systems*, 18(3):300–324.

[GNU, 2011] GNU (2011). *GNU Manual*. Free Software Foundation, Boston, MA. http://gcc.gnu.org/onlinedocs/gcc.

[GNU, 2012] GNU (2012). *The GNU Compiler for the Java Programming Language*. Free Software Foundation, Boston, MA. http://gcc.gnu.org/java/.

[Goetz, 2004] Goetz, B. (2004). Java theory and practice: dynamic compilation and performance measurement - The perils of benchmarking under dynamic compilation. *IBM DeveloperWorks*. http://www.ibm.com/developerworks/library/j-jtp12214/.

[Gosling et al., 2005] Gosling, J., Joy, B., Steele, G., and Bracha, G. (2005). *Java^{TM} Language Specification, Third Edition*. Addison-Wesley Professional, Boston, MA.

[Gough and Gough, 2001] Gough, J. and Gough, K. (2001). *Compiling for the .NET Common Language Runtime*. Prentice Hall PTR, Upper Saddle River, NJ.

[Gries, 1971] Gries, D. (1971). *Compiler Construction for Digital Computers*. Wiley, New York, NY.

[Hamilton, 2003] Hamilton, J. (2003). Language integration in the common language runtime. *ACM SIGPLAN Notices*, 38(2):19–28.

[Hopcroft and Ullman, 1969] Hopcroft, J. and Ullman, J. (1969). *Formal Languages and their Relation to Automata*. Addison-Wesley Longman Publishing, Reading, MA.

[IEEE, 2004] IEEE (2004). IEEE Standard 1003.1, Chapter 9, "Regular Expressions". `http://pubs.opengroup.org/onlinepubs/009695399/basedefs/xbd_chap09.html`.

[Jia, 2003] Jia, X. (2003). *Object-Oriented Software Development Using Java, Second Edition*. Addison Wesley, Reading, MA.

[Johnson, 1975] Johnson, S. (1975). *YACC: Yet Another Compiler-Compiler*. Bell Laboratories, Murray Hill, NJ.

[Jones and Lins, 1996] Jones, R. and Lins, R. (1996). *Garbage Collection: Algorithms for Automatic Dynamic Memory Management*. Wiley, New York, NY.

[Kazi et al., 2000] Kazi, I. H., Chen, H. H., Stanley, B., and Lilja, D. J. (2000). Techniques for obtaining high performance in Java programs. *ACM Computing Surveys*, 32(3):213–240.

[Knuth, 1968] Knuth, D. (1968). Semantics of context-free languages. *Theory of Computing Systems*, 2(2):127–145.

[Knuth, 1971a] Knuth, D. (1971a). An empirical study of FORTRAN programs. *Software: Practice and Experience*, 1(2):105–133.

[Knuth, 1971b] Knuth, D. (1971b). Top-down syntax analysis. *Acta Informatica*, 1(2):79–110.

[Kommalapati and Christian, 2005] Kommalapati, H. and Christian, T. (2005). JIT and Run-Drill into .NET Framework Internals to See How the CLR Creates Runtime Objects. *MSDN Magazine*, pages 52–63.

[Kotzmann et al., 2008] Kotzmann, T., Wimmer, C., Mössenböck, H., Rodriguez, T., Russell, K., and Cox, D. (2008). Design of the Java HotSpot client compiler for Java 6. *ACM Transactions on Architecture and Code Optimization (TACO)*, 5(1).

[Larus, 2009] Larus, J. (2009). Assemblers, linkers, and the SPIM simulator. *Appendix B of Hennessy and Patterson's Computer Organization and Design: The Hardware/Software Interface*. `http://pages.cs.wisc.edu/~larus/HP_AppA.pdf`.

[Larus, 2010] Larus, J. (2010). SPIM: A MIPS32 Simulator. `http://spimsimulator.sourceforge.net/`.

[Lesk and Schmidt, 1975] Lesk, M. and Schmidt, E. (1975). Lex — A Lexical Analyzer Generator. Technical report, Bell Laboratories, Murray Hill, NJ.

[Lewis et al., 1976] Lewis, P., Rosenkrantz, D., and Stearns, R. (1976). *Compiler Design Theory*. Addison-Wesley, Reading, MA.

[Lewis and Stearns, 1968] Lewis, P. and Stearns, R. (1968). Syntax-directed transduction. *Journal of the ACM*, 15(3):465–488.

[Liang and Bracha, 1998] Liang, S. and Bracha, G. (1998). Dynamic class loading in the Java virtual machine. *ACM SIGPLAN Notices*, 33(10):36–44.

[Lindholm and Yellin, 1999] Lindholm, T. and Yellin, F. (1999). *Java Virtual Machine Specification*. Addison-Wesley Longman Publishing, Reading, MA.

[Link and Fröhlich, 2003] Link, J. and Fröhlich, P. (2003). *Unit Testing in Java: How Tests Drive the Code*. Morgan Kaufmann Publishers, San Francisco, CA.

[Linz, 2011] Linz, P. (2011). *An Introduction to Formal Languages and Automata*. Jones & Bartlett Publishers, Burlington, MA.

[Microsoft, 2012] Microsoft (2012). .NET Home Page. http://www.microsoft.com/net.

[Miller and Ragsdale, 2004] Miller, J. and Ragsdale, S. (2004). *The Common Language Infrastructure Annotated Standard*. Addison-Wesley Professional, Boston, MA.

[Morgan, 1998] Morgan, R. (1998). *Building an Optimizing Compiler*. Elsevier Science & Technology Books, Burlington, MA.

[Mössenböck, 2000] Mössenböck, H. (2000). Adding static single assignment form and a graph coloring register allocator to the Java HotSpot client compiler. Technical report, Technical Report 15, Institute for Practical Computer Science, Johannes Kepler University Linz.

[Muchnick, 1997] Muchnick, S. S. (1997). *Advanced Compiler Design and Implementation*. Morgan Kaufmann Publishing, San Francisco, CA.

[Norvell, 2011] Norvell, T. S. (2011). The JavaCC FAQ. http://www.engr.mun.ca/~theo/JavaCC-FAQ/.

[Oracle, 2010] Oracle (2010). The Java HotSpot Performance Engine Architecture. Whitepaper. http://www.oracle.com/technetwork/java/whitepaper-135217.html.

[Paleczny et al., 2001] Paleczny, M., Vick, C., and Click, C. (2001). The Java HotSpot™ server compiler. In *Proceedings of the 2001 Symposium on Java Virtual Machine Research and Technology Symposium – Volume 1*, JVM'01, Berkeley, CA. USENIX Association.

[Paxton, 2008] Paxton, V. (2008). FLex – The Fast Lexical Analyzer. http://flex.sourceforge.net/.

[Poletto and Sarkar, 1999] Poletto, M. and Sarkar, V. (1999). Linear scan register allocation. *ACM Transactions on Programming Language Systems*, 21(5):895–913.

[Rainsberger and Stirling, 2005] Rainsberger, J. and Stirling, S. (2005). *JUnit Recipes*. Manning, Shelter Island, NY.

[Richter, 2005] Richter, J. (2005). .NET Questions regarding JIT compiler/strong-naming security. Blog. http://www.wintellect.com/cs/blogs/jeffreyr/archive/2005/07/25/net-questions-regarding-jit-compiler-strong-naming-security.aspx.

[Richter, 2010] Richter, J. (2010). *CLR Via C#, Third Edition*. Microsoft Press, Portland, OR.

[Richter and Balena, 2002] Richter, J. and Balena, F. (2002). *Applied Microsoft .NET Framework Programming*, volume 1. Microsoft Press, Portland, OR.

[Rivieres and Beaton, 2006] Rivieres, J. and Beaton, W. (2006). Eclipse platform technical overview. `http://www.eclipse.org/`.

[Sethi, 1973] Sethi, R. (1973). Complete register allocation problems. In *Proceedings of the Fifth Annual ACM Symposium on Theory of Computing*, STOC '73, pages 182–195, New York, NY. ACM.

[Shudo, 2004] Shudo, K. (2004). Performance Comparison of Java/.NET Runtimes. `http ://www.shudo.net/jit/perf`.

[Sipser, 2006] Sipser, M. (2006). *Introduction to the Theory of Computation*. Thomson Course Technology, Tampa, FL.

[Slonneger and Kurtz, 1995] Slonneger, K. and Kurtz, B. (1995). *Formal Syntax and Semantics of Programming Languages*, volume 1. Addison-Wesley, Reading, MA.

[Strachey, 2000] Strachey, C. (2000). Fundamental concepts in programming languages. *Higher-Order and Symbolic Computation*, 13(1):11–49.

[Traub et al., 1998] Traub, O., Holloway, G., and Smith, M. D. (1998). Quality and speed in linear-scan register allocation. *ACM SIGPLAN Notices*, 33(5):142–151.

[Turner, 1977] Turner, D. (1977). Error diagnosis and recovery in one pass compilers. *Information Processing Letters*, 6(4):113–115.

[Turner, 1979] Turner, D. (1979). A new implementation technique for applicative languages. *Software: Practice and Experience*, 9(1):31–49.

[van der Spek et al., 2005] van der Spek, P., Plat, N., and Pronk, C. (2005). Syntax error repair for a Java-based parser generator. *ACM SIGPLAN Notices*, 40(4):47–50.

[Waite and Goos, 1984] Waite, W. and Goos, G. (1984). *Compiler Construction*. Springer-Verlag, New York, NY.

[Whittaker, 2003] Whittaker, J. (2003). *How to Break Software*. Addison Wesley Longman Publishing, Reading, MA.

[Wilson and Kesselman, 2000] Wilson, S. and Kesselman, J. (2000). *Java Platform Performance: Strategies and Tactics*. Prentice Hall, Upper Saddle River, NJ.

[Wimmer, 2004] Wimmer, C. (2004). Linear scan register allocation for the Java HotSpot™client compiler. *Master's thesis, Johannes Kepler University Linz*.

[Wimmer and Franz, 2010] Wimmer, C. and Franz, M. (2010). Linear scan register allocation on SSA form. In *Proceedings of the 8th Annual IEEE/ACM International Symposium on Code Generation and Optimization*, pages 170–179, New York, NY. ACM.

Index